Lecture Notes in Mathematics 1877

Editors:
J.-M. Morel, Cachan
F. Takens, Groningen
B. Teissier, Paris

T0202506

Jörn Steuding

Value-Distribution
of *L*-Functions

Springer

Author

Jörn Steuding
Department of Mathematics
Chair of Complex Analysis
University of Würzburg
Am Hubland
97074 Würzburg
Germany
e-mail: steuding@mathematik.uni-wuerzburg.de

Library of Congress Control Number: 2007921875

Mathematics Subject Classification (2000): 11M06, 11M26, 11M41, 30D35, 30E10, 60B05, 60F05

ISSN print edition: 0075-8434
ISSN electronic edition: 1617-9692

ISBN 978-3-540-26526-9 Springer Berlin Heidelberg New York

DOI 10.1007/978-3-540-44822-8

Springer is a part of Springer Science+Business Media
springer.com
© Springer-Verlag Berlin Heidelberg 2007

Typesetting by the author and SPi using a Springer LaTeX macro package
Cover design: WMXDesign GmbH, Heidelberg

Printed on acid-free paper   SPIN: 11504160   VA41/3100/SPi   5 4 3 2 1 0

Contents

Preface

L-functions are important objects in modern number theory. They are generating functions formed out of local data associated with either an arithmetic object or with an automorphic form. They can be attached to smooth projective varieties defined over number fields, to irreducible (complex or p-adic) representations of the Galois group of a number field, to a cusp form or to an irreducible cuspidal automorphic representation. All the L-functions have in common that they can be described by an Euler product, i.e., a product taken over prime numbers. In view of the unique prime factorization of integers L-functions also have a Dirichlet series representation. The famous Riemann zeta-function

$$\zeta(s) = \sum_{n=1}^{\infty} \frac{1}{n^s} = \prod_{p \text{ prime}} \left(1 - \frac{1}{p^s}\right)^{-1}$$

may be regarded as the prototype. L-functions encode in their value-distribution information on the underlying arithmetic or algebraic structure that is often not obtainable by elementary or algebraic methods. For instance, Dirichlet's class number formula gives information on the deviation from unique prime factorization in the ring of integers of quadratic number fields by the values of certain Dirichlet L-functions $L(s, \chi)$ at $s = 1$. In particular, the distribution of zeros of L-functions is of special interest with respect to many problems in multiplicative number theory. A first example is the Riemann hypothesis on the non-vanishing of the Riemann zeta-function in the right half of the critical strip and its impact on the distribution of prime numbers. Another example are L-functions $L(s, E)$ attached to elliptic curves E defined over $\mathbb{Q}$. The yet unproved conjecture of Birch and Swinnerton-Dyer claims that $L(s, E)$ has a zero at $s = 1$ whose order is equal to the rank of the Mordell-Weil group of the elliptic curve E.

These notes present recent results in the value-distribution theory of such L-functions with an emphasis on the phenomenon of universality. The starting point of this theory is Bohr's achievement at the first half of the twentieth century. He proved denseness results and first limit theorems for the values of the Riemann zeta-function. Maybe the most remarkable result concerning the value-distribution of $\zeta(s)$ is Voronin's universality theorem from 1975, which roughly states that any non-vanishing analytic function can be approximated uniformly by certain shifts of the zeta-function in the critical strip. More precisely: *let $0 < r < \frac{1}{4}$ and suppose that $g(s)$ is a non-vanishing continuous function on the disc $|s| \leq r$ which is analytic in its interior. Then, for any $\epsilon > 0$, there exists a real number τ such that*

$$\max_{|s| \leq r} \left| \zeta\left(s + \frac{3}{4} + i\tau\right) - g(s) \right| < \epsilon;$$

moreover, the set of these τ has positive lower density:

$$\liminf_{T \to \infty} \frac{1}{T} \operatorname{meas} \left\{ \tau \in [0, T] : \max_{|s| \leq r} \left| \zeta\left(s + \frac{3}{4} + i\tau\right) - g(s) \right| < \epsilon \right\} > 0.$$

This is a remarkable property! We say that $\zeta(s)$ is *universal* since it allows uniform approximation of a large class of functions. Voronin's universality theorem, in a spectacular way, indicates that Riemann's zeta-function is a transcendental function; clearly, rational functions cannot be universal. In some literature the validity of the Riemann hypothesis for abelian varieties (proved by Hasse for elliptic curves and by Weil in the general case) is regarded as evidence for the truth of Riemann's hypothesis for $\zeta(s)$. However, the zeta-function of an abelian variety is a rational function and so its value-distribution is of a rather different type.

The Linnik–Ibragimov conjecture asserts that any Dirichlet series (which has a *sufficiently rich* value-distribution) is universal. Meanwhile we know quite many universal Dirichlet series; for instance, Dirichlet L-functions (Voronin, 1975), Dedekind zeta-functions (Reich, 1980), Lerch zeta-functions (Laurinčikas, 1997), and L-functions associated with newforms (Laurinčikas, Matsumoto and Steuding, 2003). One aim of these notes is to prove an extension of Voronin's universality theorem for a large class of L-functions which covers (at least conjecturally) all known L-functions of number-theoretical significance.

These notes are organized as follows. In the introduction, we give an overview on the value-distribution theory of the classical Riemann zeta-function and Dirichlet L-functions; also we touch some allied zeta-functions which we will not consider in detail in the following chapters. In Chap. 2, we introduce a class $\tilde{S}$ of Dirichlet series, satisfying certain analytic and arithmetic axioms. The members of this class are the main actors in the sequel. Roughly speaking, an L-function in $\tilde{S}$ has a polynomial Euler product and satisfies some hypothesis which may be regarded as some kind of prime number theorem; besides, we require analytic continuation to the left of the half-plane of absolute convergence for the associated Dirichlet series in addition with some growth condition. The axioms defining $\tilde{S}$ are kept quite general and therefore they may appear to be rather abstract and technical; however, as we shall discuss later for many examples (in Chaps. 6, 12 and 13), they hold (or at least they are expected to hold) for all L-functions of number theoretical interest. This abstract setting has the advantage that we can derive a rather general universality theorem.

Our proof of universality, in the main part, relies on Bagchi's probabilistic approach from 1981. For the sake of completeness we briefly present in Chap. 3 some basic facts from probability theory and measure theory. In Chap. 4, we prove along the lines of Laurinčikas' extension of Bagchi's method a limit theorem (in the sense of weakly convergent probability measures) for functions in the class $\tilde{S}$. In the following chapter we give the proof of the main result, a universality theorem for L-functions in $\tilde{S}$. The proof depends on the limit theorem of the previous chapter and the so-called positive density method, recently introduced by Laurinčikas and Matsumoto to tackle L-functions attached to cusp forms. Furthermore, we discuss the phenomenon of discrete universality; here the attribute *discrete* means that the shifts τ are taken from arithmetic progressions. This concept of universality was introduced by Reich in 1980.

In Chap. 6, we introduce the Selberg class S consisting of Dirichlet series with Euler product and a functional equation of Riemann-type (and a bit more). It is a folklore conjecture that the Selberg class consists of all automorphic L-functions. We study basic facts about S and discuss the main conjectures, in particular, the far-reaching Selberg conjectures on primitive elements. We shall see that the class $\tilde{S}$ fits rather well into the setting of the Selberg class S (especially with respect to Selberg's conjectures). Hence, our general universality theorem extends to the Selberg class, unconditionally for many of the classical L-function and conditionally to all elements of S subject to some widely believed but rather deep conjectures. However, the Selberg class is too small with respect to universality; for instance, a Dirichlet L-function to an imprimitive character does not lie in the Selberg class (by lack of an appropriate functional equation) but it is known to be universal. Furthermore, some important L-functions are only conjectured to lie in the Selberg class, and, in spite of this, for some of them we can derive universality unconditionally.

In the following chapter, we consider the value-distribution of Dirichlet series $\mathcal{L}(s)$ with functional equation in the complex plane. Following Levinson's approach from the 1970s, we shall prove asymptotic formulae for the c-values of $\mathcal{L}$, i.e., roots of the equation $\mathcal{L}(s) = c$, and give applications in Nevanlinna theory. In particular, we give an alternative proof of the Riemann–von Mangoldt formula for the elements in the Selberg class.

The main themes of Chap. 8 are almost periodicity and the Riemann hypothesis. Universality has an interesting feedback to classical problems. Bohr observed that the Riemann hypothesis for Dirichlet L-functions associated with non-principal characters is equivalent to almost periodicity in the right half of the critical strip. Applying Voronin's universality theorem, Bagchi was able to extend this result to the zeta-function in proving that if the Riemann zeta-function can approximate itself uniformly in the sense of Voronin's theorem, then Riemann's hypothesis is true, and vice versa. We sketch an extension of Bagchi's theorem to other L-functions.

Chapter 9 deals with the problem of effectivity. The known proofs of universality are ineffective, giving neither bounds for the first approximating shift τ nor for their density (with the exception of particular results due to Garunk-štis, Good, and Laurinčikas). We give explicit upper bounds for the density of universality; more precisely, we prove upper bounds for the frequency with which a certain class of target functions (analytic isomorphisms) can be uniformly approximated. Moreover, we apply effective results from the theory of inhomogeneous diophantine approximation to prove several explicit estimates for the value-distribution in the half-plane of absolute convergence.

In Chap. 10, we discuss further applications of universality, most of them classical, e.g., an extension of Bohr's and Voronin's results concerning the value-distribution inside the critical strip, and the functional independence which covers Ostrowski's solution of the Hilbert problem on the hyper-transcendence of the zeta-function and some of its generalizations. Here a function is called hyper-transcendental, if it does not satisfy any algebraic differential equation. Further, we study the value-distribution of linear combinations of (strongly) universal Dirichlet series. A subtle consequence of this *strong* concept of universality, and a big contrast to L-functions, can be found in the distribution of zeros off the critical line. Very likely a (universal) Dirichlet series satisfying a functional equation of Riemann-type has either *many* zeros to the right of the critical line (as a generic Dirichlet series with periodic coefficients) or *none* (as it is expected for L-functions). This seems to be the heart of many secrets in the value-distribution theory of Dirichlet series.

Chapter 11 deals with Dirichlet series associated with periodic arithmetical functions. In general, these functions do not have an Euler product but they are additively related to Dirichlet L-functions. Consequently, they share certain properties with L-functions, e.g., a functional equation similar to the one for Riemann's zeta-function. We prove universality for a large class of these Dirichlet series; in contrast to L-functions they can approximate uniformly analytic functions having zeros (provided their Dirichlet coefficients are not multiplicative). Moreover, we study joint universality for Hurwitz zeta-functions with rational parameters.

We conclude with joint universality; here *joint* stands for simultaneous uniform approximation. In Chap. 12, we prove a theorem which reduces joint universality for L-functions in $\tilde{\mathcal{S}}$ to a denseness property in a related function space. Of course, we cannot have joint universality for any set of L-functions; for example, $\zeta(s)$ and $\zeta(s)^2$ cannot approximate any given pair of admissible target functions simultaneously. However, we shall prove that in some instances twists of $\mathcal{L} \in \tilde{\mathcal{S}}$ with pairwise non-equivalent characters fulfill this condition (e.g., Dirichlet L functions). In the following chapter we present several further applications. For instance, we prove joint universality for Artin

L-functions (which lie in the Selberg class if and only if the deep Artin conjecture is true). This universality theorem holds unconditionally despite the fact that Artin L-functions might have infinitely many poles in their strip of universality; this was first proved by Bauer in 2003 by a tricky argument.

At the end of these notes an appendix on the history of the general phenomenon of universality in analysis is given. It is known that universality is a quite regularly appearing phenomenon in limit processes, but among all these universal objects only universal Dirichlet series are explicitly known. At the end an index and a list of the notations and axioms which were used are given.

Value-distribution theory for L-functions with emphasis on aspects of universality was treated in the monographs of Karatsuba and Voronin [166] and Laurinčikas [186]. However, after the publication of these books, many new results and applications were discovered; we refer the reader to the surveys of Laurinčikas [196] and of Matsumoto [242] for some of the progress made in the meantime. The content of this book forms an extract of the authors habilitation thesis written at Frankfurt University in 2003. We have added Chaps. 12 and 13 on joint universality and its applications as well as several remarks and comments concerning the progress obtained in the meantime. Unfortunately, we could not include the most current contributions as, for example, the promising work [245] of Mauclaire which relates universality with almost periodicity.

I am very grateful to Springer for publishing these notes; especially, I want to thank Stefanie Zöller and Catriona M. Byrne from Springer, the editors of the series Lecture Notes in Mathematics, and, of course, the anonymous referees for their excellent work, their valuable remarks and corrections. Furthermore, I am grateful to my family, my friends and my colleagues for their interest and support, in particular those from the Mathematics Departments at the universities of Frankfurt, Madrid, and Würzburg. Especially, I would like to thank Ramūnas Garunkštis and Antanas Laurinčikas for introducing me to questions concerning universality, Ernesto Girondo, Aleksander Ivić, Roma Kačinskaitė, Kohji Matsumoto, Georg Johann Rieger, Jürgen Sander, Wolfgang Schwarz, and Jürgen Wolfart for the fruitful discussions, helpful remarks and their encouragement. Last but not least, I would like to thank my wife Rasa.

Jörn Steuding
Würzburg, December 2006

1

Introduction

The grandmother of all zeta-functions is the Riemann zeta-function.

David Ruelle

In this introduction we give some hints for the importance of the Riemann zeta-function for analytic number theory and present first classic results on its amazing value-distribution due to Harald Bohr but also the remarkable universality theorem of Voronin (including a sketch of his proof). Moreover, we introduce Dirichlet L-functions and other generalizations of the zeta-function, discuss their relevance in number theory and comment on their value-distribution. For historical details we refer to Narkiewicz's monograph [277] and Schwarz's surveys [317, 318].

1.1 The Riemann Zeta-Function and the Distribution of Prime Numbers

The Riemann zeta-function is a function of a complex variable $s = \sigma + it$, for $\sigma > 1$ given by

$$\zeta(s) = \sum_{n=1}^{\infty} \frac{1}{n^s} = \prod_p \left(1 - \frac{1}{p^s}\right)^{-1}; \tag{1.1}$$

here and in the sequel the letter p always denotes a prime number and the product is taken over all primes. The Dirichlet series, and the Euler product, converge absolutely in the half-plane $\sigma > 1$ and uniformly in each compact subset of this half-plane. The identity between the Dirichlet series and the Euler product was discovered by Euler [76] in 1737 and can be regarded as an analytic version of the unique prime factorization of integers. The Euler product gives a first glance on the intimate connection between the zeta-function and the distribution of prime numbers. A first immediate consequence is Euler's proof of the infinitude of the primes. Assuming that there were only finitely many primes, the product in (1.1) is finite, and therefore convergent for $s = 1$, contradicting the fact that the Dirichlet series defining $\zeta(s)$ reduces to the divergent harmonic series as $s \to 1+$. Hence, there exist infinitely many prime numbers. This fact is well known since Euclid's elementary proof, but the analytic access gives deeper knowledge on the distribution of the prime numbers. It was the young Gauss [94] who conjectured in 1791 for the number $\pi(x)$ of primes $p \leq x$ the asymptotic formula

$$\pi(x) \sim \mathrm{li}(x), \tag{1.2}$$

where the logarithmic integral is given by

$$\mathrm{li}(x) = \lim_{\epsilon \to 0+} \left\{ \int_0^{1-\epsilon} + \int_{1+\epsilon}^x \right\} \frac{du}{\log u} = \int_2^x \frac{du}{\log u} - 1.04 \ldots ;$$

this integral is a principal value in the sense of Cauchy. Gauss' conjecture states that, in first approximation, the number of primes $\leq x$ is asymptotically $\frac{x}{\log x}$. By elementary means, Chebyshev [54, 55] proved around 1850 that for sufficiently large x

$$0.921 \ldots \leq \pi(x)\frac{\log x}{x} \leq 1.055 \ldots .$$

Furthermore, he showed that if the limit

$$\lim_{x \to \infty} \pi(x)\frac{\log x}{x}$$

exists, the limit is equal to one, which supports relation (1.2).

Riemann was the first to investigate the Riemann zeta-function as a function of a complex variable. In his only one but outstanding paper [310] on number theory from 1859 he outlined how Gauss' conjecture could be proved by using the function $\zeta(s)$. However, at Riemann's time the theory of functions was not developed sufficiently far, but the open questions concerning the zeta-function pushed the research in this field quickly forward. We shall briefly discuss Riemann's memoir. First of all, by partial summation

$$\zeta(s) = \sum_{n \leq N} \frac{1}{n^s} + \frac{N^{1-s}}{s-1} + s \int_N^\infty \frac{[u] - u}{u^{s+1}} \, du; \tag{1.3}$$

here and in the sequel $[u]$ denotes the maximal integer less than or equal to u. This gives an analytic continuation for $\zeta(s)$ to the half-plane $\sigma > 0$ except for a simple pole at $s = 1$ with residue 1. This process can be continued to the left half-plane and shows that $\zeta(s)$ is analytic throughout the whole complex plane except for $s = 1$. Riemann gave the functional equation

$$\pi^{-s/2} \Gamma\left(\frac{s}{2}\right) \zeta(s) = \pi^{-(1-s)/2} \Gamma\left(\frac{1-s}{2}\right) \zeta(1-s), \tag{1.4}$$

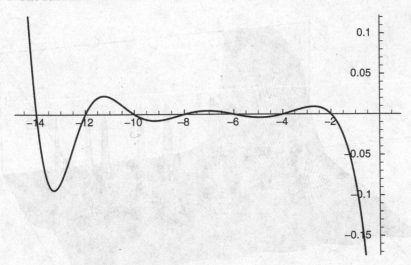

Fig. 1.1. $\zeta(s)$ in the range $s \in [-14.5, 0.5]$

where $\Gamma(s)$ denotes Euler's Gamma-function; it should be noted that
Euler [76] had partial results in this direction (namely, for integral s and
for half-integral s; see [7]). In view of the Euler product (1.1) it is easily seen
that $\zeta(s)$ has no zeros in the half-plane $\sigma > 1$. It follows from the functional
equation and from basic properties of the Gamma-function that $\zeta(s)$ vanishes
in $\sigma < 0$ exactly at the so-called trivial zeros $s = -2n$ with $n \in \mathbb{N}$ (see Fig. 1.1
for the *first* trivial zeros). All other zeros of $\zeta(s)$ are said to be non-trivial,
and we denote them by $\varrho = \beta + i\gamma$. Obviously, they have to lie inside the
so-called critical strip $0 \le \sigma \le 1$, and it is easily seen that they are non-real.
The functional equation (1.4), in addition with the identity

$$\zeta(\overline{s}) = \overline{\zeta(s)},$$

shows some symmetries of $\zeta(s)$. In particular, the non-trivial zeros of $\zeta(s)$ are
distributed symmetrically with respect to the real axis and to the vertical line
$\sigma = \frac{1}{2}$. It was Riemann's ingenious contribution to number theory to point
out how the distribution of these non-trivial zeros is linked to the distribution
of prime numbers. Riemann conjectured that the number $N(T)$ of non-trivial
zeros $\varrho = \beta + i\gamma$ with $0 < \gamma \le T$ (counted according multiplicities) satisfies
the asymptotic formula

$$N(T) \sim \frac{T}{2\pi} \log \frac{T}{2\pi e}.$$

This was proved in 1895 by von Mangoldt [235, 236] who found more precisely

$$N(T) = \frac{T}{2\pi} \log \frac{T}{2\pi e} + \mathrm{O}(\log T). \tag{1.5}$$

Riemann worked with the function $t \mapsto \zeta(\frac{1}{2} + it)$ and wrote that very likely
all roots t are real, i.e., all non-trivial zeros lie on the so-called critical line

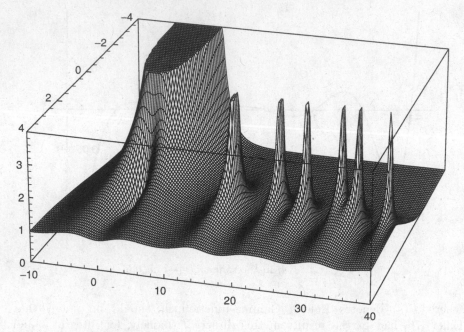

Fig. 1.2. The reciprocal of the absolute value of $\zeta(s)$ for $\sigma \in [-4, 4], t \in [-10, 40]$. The zeros of $\zeta(s)$ appear as poles

$\sigma = \frac{1}{2}$. This is the famous, yet unproved Riemann hypothesis which we rewrite equivalently as

Riemann's Hypothesis. $\zeta(s) \neq 0$ *for* $\sigma > \frac{1}{2}$.

In support of his conjecture, Riemann calculated some zeros; the first one with positive imaginary part is $\varrho = \frac{1}{2} + i14.134\ldots$ (see Fig. 1.2 and also Fig. 8.1).* Furthermore, Riemann conjectured that there exist constants A and B such that

$$\frac{1}{2}s(s-1)\pi^{-s/2}\Gamma\left(\frac{s}{2}\right)\zeta(s) = \exp(A + Bs)\prod_{\varrho}\left(1 - \frac{s}{\varrho}\right)\exp\left(\frac{s}{\varrho}\right).$$

This was shown by Hadamard [113] in 1893 (recall the Hadamard product theorem from the theory of functions). Finally, Riemann conjectured the so-called explicit formula which states that

---

* In 1932, Siegel [329] published an account of Riemann's work on the zeta-function found in Riemann's private papers in the archive of the university library in Göttingen. It became evident that behind Riemann's speculation there was extensive analysis and computation.

$$\pi(x) + \sum_{n=2}^{\infty} \frac{\pi(x^{1/n})}{n} = \mathrm{li}(x) - \sum_{\substack{\varrho = \beta + i\gamma \\ \gamma > 0}} \left(\mathrm{li}(x^{\varrho}) + \mathrm{li}(x^{1-\varrho}) \right) \qquad (1.6)$$

$$+ \int_x^{\infty} \frac{du}{u(u^2 - 1)\log u} - \log 2$$

for any $x \geq 2$ not being a prime power (otherwise a term $\frac{1}{2k}$ has to be added on the left-hand side, where $x = p^k$); the appearing integral logarithm is defined by

$$\mathrm{li}(x^{\beta + i\gamma}) = \int_{(-\infty + i\gamma)\log x}^{(\beta + i\gamma)\log x} \frac{\exp(z)}{z + \delta i \gamma} \, dz,$$

where $\delta = +1$ if $\gamma > 0$ and $\delta = -1$ otherwise. The explicit formula was proved by von Mangoldt [235] in 1895 as a consequence of both product representations for $\zeta(s)$, the Euler product (1.1) on the one hand and the Hadamard product on the other hand.

Riemann's ideas led to the first proof of Gauss' conjecture (1.2), the celebrated prime number theorem, by Hadamard [114] and de la Vallée-Poussin [357] (independently) in 1896. We give a very brief sketch (for the details we refer to Ivić [141]). For technical reasons it is of advantage to work with the logarithmic derivative of $\zeta(s)$ which is for $\sigma > 1$ given by

$$\frac{\zeta'}{\zeta}(s) = - \sum_{n=1}^{\infty} \frac{\Lambda(n)}{n^s},$$

where the von Mangoldt Λ-function is defined by

$$\Lambda(n) = \begin{cases} \log p & \text{if } n = p^k \text{ with } k \in \mathbb{N}, \\ 0 & \text{otherwise.} \end{cases} \qquad (1.7)$$

A lot of information concerning the prime counting function $\pi(x)$ can be recovered from information about

$$\psi(x) := \sum_{n \leq x} \Lambda(n) = \sum_{p \leq x} \log p + \mathrm{O}\left(x^{1/2} \log x\right).$$

Partial summation gives

$$\pi(x) \sim \frac{\psi(x)}{\log x}.$$

First of all, we shall express $\psi(x)$ in terms of the zeta-function. If c is a positive constant, then

$$\frac{1}{2\pi i} \int_{c - i\infty}^{c + i\infty} \frac{x^s}{s} \, ds = \begin{cases} 1 & \text{if } x > 1, \\ 0 & \text{if } 0 < x < 1. \end{cases} \qquad (1.8)$$

This yields the Perron formula: for $x \notin \mathbb{Z}$ and $c > 1$,

$$\psi(x) = -\frac{1}{2\pi i} \int_{c-i\infty}^{c+i\infty} \frac{\zeta'}{\zeta}(s) \frac{x^s}{s} \, ds. \qquad (1.9)$$

Moving the path of integration to the left, we find that the latter expression is equal to the corresponding sum of residues, that are the residues of the integrand at the pole of $\zeta(s)$ at $s = 1$, at the zeros of $\zeta(s)$, and at the pole of the integrand at $s = 0$. The main term turns out to be

$$\operatorname{Res}_{s=1} \left\{ -\frac{\zeta'}{\zeta}(s) \frac{x^s}{s} \right\} = \lim_{s \to 1} (s-1) \left(\frac{1}{s-1} + O(1) \right) \frac{x^s}{s} = x,$$

whereas each non-trivial zero ϱ gives the contribution

$$\operatorname{Res}_{s=\varrho} \left\{ -\frac{\zeta'}{\zeta}(s) \frac{x^s}{s} \right\} = -\frac{x^\varrho}{\varrho}.$$

By the same reasoning, the trivial zeros contribute

$$\sum_{n=1}^{\infty} \frac{x^{-2n}}{2n} = \frac{1}{2} \log \left(1 - \frac{1}{x^2} \right).$$

Incorporating the residue at $s = 0$, this leads to the the *exact* explicit formula

$$\psi(x) = x - \sum_{\varrho} \frac{x^\varrho}{\varrho} - \frac{1}{2} \log \left(1 - \frac{1}{x^2} \right) - \log(2\pi),$$

which is equivalent to Riemann's formula (1.6). Notice that the right-hand side of this formula is not absolutely convergent. If $\zeta(s)$ would have only finitely many non-trivial zeros, the right-hand side would be a continuous function of x, contradicting the jumps of $\psi(x)$ for prime powers x. However, going on it is much more convenient to cut the integral in (1.9) at $t = \pm T$ which leads to the truncated version

$$\psi(x) = x - \sum_{|\gamma| \leq T} \frac{x^\varrho}{\varrho} + O \left(\frac{x}{T} (\log(xT))^2 \right), \qquad (1.10)$$

valid for all values of x. Next we need information on the distribution of the non-trivial zeros. The largest known zero-free region for $\zeta(s)$ was found by Vinogradov [359] and Korobov [173] (independently) who proved

$$\zeta(s) \neq 0 \quad \text{in} \quad \sigma \geq 1 - \frac{c}{(\log |t| + 3)^{1/3} (\log \log(|t| + 3))^{2/3}},$$

where c is some positive absolute constant; the first complete proof due to Richert appeared in Walfisz [366]. In addition with the Riemann–von Mangoldt formula (1.5) one can estimate the sum over the non-trivial zeros in (1.10). Balancing out T and x, we obtain the prime number theorem with the strongest existing remainder term:

Theorem 1.1. *There exists an absolute positive constant C such that for sufficiently large x*

$$\pi(x) = \operatorname{li}(x) + \mathrm{O}\left(x \exp\left(-C\frac{(\log x)^{3/5}}{(\log\log x)^{1/5}}\right)\right).$$

By the explicit formula (1.10) the impact of the Riemann hypothesis on the prime number distribution becomes visible. Von Koch [172] showed that for fixed $\theta \in [\frac{1}{2}, 1)$

$$\pi(x) - \operatorname{li}(x) \ll x^{\theta+\epsilon} \quad \Longleftrightarrow \quad \zeta(s) \neq 0 \quad \text{for} \quad \sigma > \theta; \qquad (1.11)$$

here and in the sequel ϵ stands for an arbitrary small positive constant, not necessarily the same at each appearance. With regard to known zeros of $\zeta(s)$ on the critical line it turns out that an error term with $\theta < \frac{1}{2}$ is impossible. Thus, the Riemann hypothesis states that the prime numbers are *as uniformly distributed as possible!*

Many computations were done to find a counterexample to the Riemann hypothesis. Van de Lune, te Riele and Winter [232] localized the first $1\,500\,000\,001$ zeros, all lying without exception on the critical line; moreover they all are simple! By observations like this it is conjectured, that *all or at least almost all zeros of the zeta-function are simple.* This so-called essential simplicity hypothesis has arithmetical consequences. Cramér [63] showed, assuming the Riemann hypothesis,

$$\frac{1}{\log X}\int_1^X \left(\frac{\psi(x)-x}{x}\right)^2 \mathrm{d}x \sim \sum_\varrho \left|\frac{m(\varrho)}{\varrho}\right|^2, \qquad (1.12)$$

where the sum is taken over distinct zeros and $m(\varrho)$ denotes their multiplicity. The right-hand side is minimal if all the zeros are simple. Going further, Goldston, Gonek and Montgomery [103] observed interesting relations between the essential simplicity hypothesis, mean-values of the logarithmic derivative of $\zeta(s)$, the error term in the prime number theorem, and Montgomery's pair correlation conjecture.

A classical density theorem due to Bohr and Landau states that *most* of the zeros lie *arbitrarily close to* the critical line. Denote by $N(\sigma, T)$ the number of zeros $\varrho = \beta + i\gamma$ of $\zeta(s)$ for which $\beta > \sigma$ and $0 < \gamma \leq T$ (counting multiplicities). Bohr and Landau [34] (see also [224]) proved that

$$N(\sigma, T) \ll T = \mathrm{O}(N(T)) \qquad (1.13)$$

for any fixed $\sigma > \frac{1}{2}$. Hence, almost all zeros of the zeta-function are clustered around the critical line. A refinement of their method led to bounds $N(\sigma, T) \ll T^\xi$ with $\xi < 1$ (see [35, 49]); for instance:

Theorem 1.2. *For any fixed σ with $\frac{1}{2} < \sigma < 1$,*

$$N(\sigma, T) \ll T^{4\sigma(1-\sigma)+\epsilon}.$$

For the various improvements of this density estimate we refer to Ivić [141].

Hardy [116] showed that infinitely many zeros lie on the critical line, and Selberg [321] was the first to prove that a positive proportion of all zeros lies exactly on $\sigma = \frac{1}{2}$. Let $N_0(T)$ denote the number of zeros ϱ of $\zeta(s)$ on the critical line with imaginary part $0 < \gamma \le T$. The idea to use mollifiers to dampen the oscillations of $|\zeta(\frac{1}{2} + it)|$ led Selberg to

$$\liminf_{T \to \infty} \frac{N_0(T + H) - N_0(T)}{N(T + H) - N(T)} > 0,$$

as long as $H \ge T^{1/2+\epsilon}$. Karatsuba [164] improved this result to $H \ge T^{27/82+\epsilon}$ by some technical refinements. The proportion is very small, about 10^{-6} as Min calculated; a later refinement by Zhuravlev gives after all $\frac{2}{21}$ if $H = T$ (cf. [165, p. 36]). However, the localized zeros are not necessarily simple. By an ingenious new method, working with mollifiers of finite length, Levinson [216] localized more than one third of the non-trivial zeros of the zeta-function on the critical line, and as Heath-Brown [122] and Selberg (unpublished) discovered, they are all simple. By optimizing the technique, Levinson himself and others improved the proportion $\frac{1}{3}$ slightly, but more recognizable is Conrey's idea in introducing Kloosterman sums. So Conrey [57] was able to choose a longer mollifier to show that more than two-fifths of the zeros are simple and on the critical line. Bauer [17, 18] improved this proportion slightly. The use of longer mollifiers leads to larger proportions. Farmer [77] observed that if it is possible to take mollifiers of infinite length, then almost all zeros lie on the critical line and are simple. In [339], Steuding found a new approach (combining ideas and methods of Atkinson, Jutila and Motohashi) to treat short intervals $[T, T + H]$, i.e., $H = O(T)$. It was proved that for $H \ge T^{0.552}$ a positive proportion of the zeros of the zeta-function with imaginary parts in $[T, T + H]$ lie on the critical line and are simple.

Recently, Garaev [81] showed that

$$\sum_{\substack{0 < \gamma < T \\ \zeta'(\varrho) \ne 0}} |\varrho \zeta'(\varrho)|^{-1} \gg (\log T)^{1/2};$$

the divergence of the series $\sum_{\varrho} |\varrho \zeta'(\varrho)|^{-1}$ was known before only subject to the truth of the Riemann hypothesis (see [353, p. 374]). Garaev's result was slightly improved by himself [82] and Sleževičienė and Steuding [333] (independently) by a factor $(\log T)^{1/4}$. Furthermore, using the results of [339] for short intervals, in [319] it was proved that

$$\sum_{\substack{T < \gamma < T+H \\ \zeta'(\frac{1}{2}+i\gamma) \ne 0}} \left| \zeta'\left(\frac{1}{2} + i\gamma\right) \right|^{-1} \gg H(\log T)^{-1/4}$$

and

$$\sum_{\substack{T < \gamma < T+H \\ \zeta'(\frac{1}{2}+i\gamma) \ne 0}} \left| \zeta'\left(\frac{1}{2} + i\gamma\right) \right| \ll H(\log T)^{9/4},$$

both valid for $T^{0.552} \le H \le T$. Such estimates measure how close the zeros of the zeta-function are to being simple.

1.2 Bohr's Probabilistic Approach

In the second decade of the twentieth century, Harald Bohr refined former studies on the value-distribution of the Riemann zeta-function by applying diophantine, geometric, and probabilistic methods. Especially, for us, his probabilistic approach is of interest. It seems to be rather difficult to locate concrete values taken by the zeta-function, but it is much easier to study how often the values lie in a given set. As Bohr found out these frequencies follow strict mathematical laws (analogously to quantum mechanics, where it is impossible to locate simultaneously space and time of a particle by Heisenberg's uncertainty relation, but large numbers of particles obey statistical laws).

In the half-plane of absolute convergence $\sigma > 1$, we have

$$0 < |\zeta(s)| \le \zeta(\sigma).$$

Thus the values of $\zeta(s)$ in half-planes $\sigma \ge \sigma_0 > 1$ are lying in the disk of radius $\zeta(\sigma_0)$ centered in the origin. It can be shown that $\zeta(s)$ takes *quite many* of the complex values inside this disk when t varies in $\mathbb{R}$ (very similar to the classic theorem of Bloch in the theory of functions). On the other side $\zeta(\sigma)$ tends to infinity as $\sigma \to 1+$, and indeed Bohr [28] succeeded in proving

Theorem 1.3. *In any strip* $1 < \sigma < 1 + \epsilon$, $\zeta(s)$ *takes any non-zero value infinitely often.*

Somehow it is more natural to study the logarithm of the zeta-function. To define $\log \zeta(s)$ for any $s \in \mathbb{C}$, we choose the principal branch of the logarithm on the intersection of the real axis with the half-plane of absolute convergence. For other points s we define $\log \zeta(\sigma + it)$ to be the value obtained from $\log \zeta(2)$ by continuous variation along the line segments $[2, 2 + it]$ and $[2 + it, \sigma + it]$, provided that the path does not cross a zero or pole of $\zeta(s)$; if it does, then we take $\log \zeta(\sigma + it) = \lim_{\epsilon \to 0+} \log \zeta(\sigma + i(t + \epsilon))$. Of course, for other L-functions we may proceed analogously. For $\sigma > 1$, we have

$$\log \zeta(s) = -\sum_p \log \left(1 - \frac{1}{p^s} \right) = -\sum_p \sum_{k=1}^{\infty} \frac{1}{k p^{sk}}. \tag{1.14}$$

For fixed prime p and fixed $\sigma > 1$, the set of values taken by the inner sum in the series representation on the right-hand side is a convex curve while t runs through $\mathbb{R}$. Adding up all these curves gives, using some facts from the theory of diophantine approximation, information on the values taken by $\log \zeta(s)$ itself. Actually, it follows that $\log \zeta(s)$ takes any complex value in $1 < \sigma < 1 + \epsilon$ which, by exponentiation, leads to the assertion of Theorem 1.3.

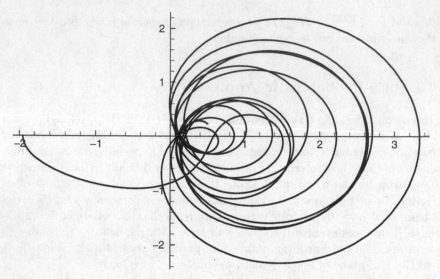

Fig. 1.3. $\zeta(\frac{3}{5} + it)$ for $t \in [0, 60]$. The curve visits any neighbourhood of any given point in the complex plane as t runs through the set of real numbers

The situation inside the critical strip is much more complicated. Here, Bohr studied finite Euler products

$$\zeta_M(s) := \prod_{p \leq M} \left(1 - \frac{1}{p^s}\right)^{-1}. \tag{1.15}$$

As M tends to infinity, these products do not converge any longer in the critical strip but they approximate $\zeta(s)$ in the mean (we will meet this ingenious idea several times later). The value-distribution of finite Euler products is treatable by the theory of diophantine approximation, and by their approximation property this leads to information on the values taken by the zeta-function inside the critical strip. In a series of papers Bohr and his collaborators discovered that the asymptotic behaviour of $\zeta(s)$ is ruled by probability laws on every vertical line to the right of $\sigma = \frac{1}{2}$. In particular, Bohr and Courant [30] proved that for any fixed $\sigma \in (\frac{1}{2}, 1]$ the set of values $\zeta(\sigma + it)$ with $t \in \mathbb{R}$ lies dense in the complex plane (see Fig. 1.3). Later, Bohr refined these results significantly by applying probabilistic methods. Let $\mathcal{R}$ be an arbitrary fixed rectangle in the complex plane whose sides are parallel to the real and the imaginary axes, and let $\mathcal{G}$ be the half-plane $\sigma > \frac{1}{2}$ where all points are removed which have the same imaginary part as, and smaller real part than, one of the possible zeros of $\zeta(s)$ in this region. Then a remarkable limit theorem due to Bohr and Jessen [31, 32] states.

Theorem 1.4. *For any $\sigma > \frac{1}{2}$, the limit*

$$\lim_{T \to \infty} \frac{1}{T} \text{meas} \left\{\tau \in [0, T] : \sigma + i\tau \in \mathcal{G}, \log \zeta(\sigma + i\tau) \in \mathcal{R}\right\}$$

exists.

Here and in the sequel meas A stands for the Lebesgue measure of a measurable set A. The limit value of Theorem 1.4 may be regarded as the *probability* how many values of $\log \zeta(\sigma + it)$ belong to the rectangle $\mathcal{R}$. Hattori and Matsumoto [121] identified this limit distribution, which is too complicated to be written down here. Next, for any complex number c, denote by $N^c(\sigma_1, \sigma_2, T)$ the number of c-values of $\zeta(s)$, i.e., the roots of the equation $\zeta(s) = c$, inside the region $\sigma_1 < \sigma < \sigma_2$, $0 < t \le T$ (counting multiplicities). From the limit theorem mentioned above Bohr and Jessen [32] deduced

Theorem 1.5. *Let c be a complex number $\ne 0$. Then, for any σ_1 and σ_2 satisfying $\frac{1}{2} < \sigma_1 < \sigma_2 < 1$, the limit*

$$\lim_{T \to \infty} \frac{1}{T} N^c(\sigma_1, \sigma_2, T)$$

exists and is positive.

In connection with the density Theorem 1.2 it follows that zeros are indeed *exceptional* values of the zeta-function.

In 1935, Jessen and Wintner [149] proved limit theorems similar to the one above by using more advanced methods from probability theory (infinite convolutions of probability measures). We do not mention further developments of Bohr's ideas by his successors Borchsenius, Jessen and Wintner but refer for more details on Bohr's contribution and results of his collaborators to the monograph of Laurinčikas [186] and the survey of Matsumoto [242]. Bohr's line of investigations appears to have been almost totally abandoned for some decades. Only in 1972, Voronin [361] obtained some significant generalizations of Bohr's denseness result.

Theorem 1.6. *For any fixed numbers $s_1, \ldots, s_n$ with $\frac{1}{2} < \operatorname{Re} s_k < 1$ for $1 \le k \le n$ and $s_k \ne s_\ell$ for $k \ne \ell$, the set*

$$\{(\zeta(s_1 + it), \ldots, \zeta(s_n + it)) : t \in \mathbb{R}\}$$

is dense in $\mathbb{C}^n$. Moreover, for any fixed number s in $\frac{1}{2} < \sigma < 1$,

$$\{(\zeta(s + i\tau), \zeta'(s + i\tau), \ldots, \zeta^{(n-1)}(s + i\tau)) : \tau \in \mathbb{R}\}$$

is dense in $\mathbb{C}^n$.

This result (in addition with Voronin's universality Theorem 1.7) started a second period of intensive research in this field. Comparable limit theorems and related results were obtained, for example, by Laurinčikas [180] for Dirichlet L-functions, by Matsumoto [239] for more general L-functions, and recently by Sleževičienė [331] for powers of the zeta-function, to give only some examples.

We conclude this section with a short view on the value-distribution of the Riemann zeta-function on the critical line which appears to be rather different. It is conjectured but yet unproved that also *the set of values of $\zeta(s)$ taken on the critical line $\sigma = \frac{1}{2}$ is dense in* $\mathbb{C}$. Selberg (unpublished) proved that the values taken by an appropriate normalization of the Riemann zeta-function on the critical line are normally distributed: let $\mathcal{R}$ be an arbitrary fixed rectangle in the complex plane whose sides are parallel to the real and the imaginary axes, then

$$\lim_{T \to \infty} \frac{1}{T} \operatorname{meas} \left\{ t \in (0, T] : \frac{\log \zeta \left(\frac{1}{2} + it \right)}{\sqrt{\frac{1}{2} \log \log T}} \in \mathcal{R} \right\}$$
$$= \frac{1}{2\pi} \iint_{\mathcal{R}} \exp \left(-\frac{x^2 + y^2}{2} \right) \, dx \, dy. \qquad (1.16)$$

The first published proof is due to Joyner [150]. Laurinčikas [184] obtained results near the critical line, comparable to (1.16); more precisely, the weak convergence of a suitably normalized measure to the normal distribution in the region

$$\frac{1}{2} \le \sigma \le \frac{1}{2} + \frac{1}{\log T}$$

as T tends to infinity. Selberg's limit theorem with remainder term includes as a particular case the result

$$\frac{1}{T} \operatorname{meas} \left\{ t \in (0, T] : \left| \zeta \left(\frac{1}{2} + it \right) \right| \le \exp \left(y \sqrt{\frac{1}{2} \log \log T} \right) \right\}$$
$$= \frac{1}{\sqrt{2\pi}} \int_{-\infty}^{y} \exp \left(-\frac{x^2}{2} \right) \, dx + O \left(\frac{(\log \log \log T)^2}{\sqrt{\log \log T}} \right).$$

As Hejhal [130] recently confirmed, this holds uniformly in y. This uniformity is rather useful. Ivić [143] pointed out how to deduce results on small values taken by the zeta-function and results on gaps between the ordinates of consecutive zeros on the critical line. Laurinčikas and Steuding [208] investigated applications to the distribution of large values (with respect to Ω-results on one hand and the Lindelöf hypothesis on the other hand).

1.3 Voronin's Universality Theorem

In 1975, Voronin [363, 364] proved a remarkable universality theorem for $\zeta(s)$ which states, roughly speaking, that any non-vanishing analytic function can be approximated uniformly by certain purely imaginary shifts of the zeta-function in the critical strip.

Theorem 1.7. *Let $0 < r < \frac{1}{4}$ and suppose that $g(s)$ is a non-vanishing continuous function on the disk $|s| \leq r$ which is analytic in the interior. Then, for any $\epsilon > 0$,*

$$\liminf_{T \to \infty} \frac{1}{T} \operatorname{meas} \left\{ \tau \in [0, T] : \max_{|s| \leq r} \left| \zeta \left(s + \frac{3}{4} + i\tau \right) - g(s) \right| < \epsilon \right\} > 0.$$

Thus the set of approximating shifts has positive lower density!

We say that $\zeta(s)$ is *universal* since appropriate shifts approximate uniformly any element of a huge class of functions. What might have been Voronin's intention for his studies which had led him to the discovery of this astonishing universality property? First of all, Voronin's universality theorem can be seen as an infinite dimensional analogue of the second part of Theorem 1.6: the truncated Taylor series of the target function $g(s)$ can be approximated by the truncated Taylor series of a certain shift of zeta. This becomes more clear in Sect. 10.1 when we will deduce the second assertion of Theorem 1.6 from Theorem 1.7 as a particular case in a more general setting (for more details we refer to the nice survey articles of Laurinčikas [198] and Matsumoto [244]). Moreover, in Sect. 10.6 we shall indicate how to deduce the first part from a refined version of Voronin's universality theorem. Another reason for Voronin's investigations could have been Bohr's concept of almost periodicity with which we will get in touch in Sect. 8.2.

We give a brief sketch of Voronin's argument following the book of Karatsuba and Voronin [166]. The Euler product (1.1) is the key to prove the universality theorem in spite of the fact that it does not converge in the region of universality. However, as Bohr [29] observed, an appropriate truncated Euler product (1.15) approximates $\zeta(s)$ in a certain mean-value sense inside the critical strip; this is related to the use of modified truncated Euler products in Voronin's proof (see (1.19) and (1.20)).

It is more convenient to work with series than with products. Therefore, we consider the logarithms of the functions in question. Since $g(s)$ has no zeros in $|s| \leq r$, its logarithm exists and we may define an analytic function $f(s)$ by $g(s) = \exp f(s)$. First we approximate $f(s)$ by the logarithm of a truncated Euler product. Let Ω denote the set of all sequences of real numbers indexed by the prime numbers in ascending order. Further, define for every finite subset M of the set of all primes, every $\omega = (\omega_2, \omega_3, \ldots) \in \Omega$, and all complex s,

$$\zeta_M(s, \omega) = \prod_{p \in M} \left(1 - \frac{\exp(-2\pi i \omega_p)}{p^s} \right)^{-1}.$$

Obviously, $\zeta_M(s, \omega)$ is a non-vanishing analytic function of s in the half-plane $\sigma > 0$. Consequently, its logarithm exists and is equal to

$$\log \zeta_M(s, \omega) = -\sum_{p \in M} \log \left(1 - \frac{\exp(-2\pi i \omega_p)}{p^s} \right);$$

in order to have a definite value we may choose the principal branch of the logarithm. It might be unpleasant that f is only assumed to be continuous on the boundary. However, since $f(s)$ is uniformly continuous in the disk $|s| \leq r$, there exists some $\kappa > 1$ such that $\kappa^2 r < \frac{1}{4}$ and

$$\max_{|s|\leq r} \left| f\left(\frac{s}{\kappa^2}\right) - f(s) \right| < \frac{\epsilon}{2}.$$

The function $f(s/\kappa^2)$ is analytic and bounded on the disk $|s| \leq \kappa r =: R$, and thus belongs to the Hardy space $\mathcal{H}_R^2$, i.e., the Hilbert space consisting of those functions $F(s)$ which are analytic for $|s| < R$ with finite norm

$$\|F\| := \lim_{r \to R-} \iint_{|s|\leq r} |F(s)| \, d\sigma \, dt.$$

Introducing the inner product

$$\langle F, G \rangle = \mathrm{Re} \iint_{|s|\leq R} F(s)\overline{G(s)} \, d\sigma \, dt,$$

$\mathcal{H}_R^2$ becomes a real Hilbert space. Denote by p_k the kth prime number. We consider the series

$$\sum_{k=1}^{\infty} u_k(s, \omega),$$

where

$$u_k(s, \omega) := \log\left(1 - \frac{\exp(-2\pi i \omega_{p_k})}{p_k^{s+(3/4)}}\right)^{-1}.$$

Here comes the first main idea. Riemann proved that any conditionally convergent series of real numbers can be rearranged such that its sum converges to an arbitrary preassigned real number. Pechersky [288] generalized Riemann's theorem to Hilbert spaces (Theorem 5.4). It follows, with the special choice $\omega = \omega_0 = (\frac{1}{4}, \frac{2}{4}, \frac{3}{4}, \ldots)$, that there exists a rearrangement of the series $\sum u_k(s, \omega_0)$ for which

$$\sum_{j=1}^{\infty} u_{k_j}(s, \omega_0) = f\left(\frac{s}{\kappa^2}\right)$$

(the rather difficult and lengthy verification of the conditions of Pechersky's theorem uses classic results of Paley and Wiener and Plancherel from Fourier analysis, a theorem on polynomial approximation due to Markov, and, most importantly, the prime number Theorem 1.1). It might be interesting to notice that in the older paper [362] Voronin had already developed the techniques for proving universality – only a rearrangement theorem for function spaces was not at hand. Recently, Garunkštis [87] found an argument to omit the ineffective rearrangement.

The tail of the rearranged series can be made as small as we please, say of modulus less than $\frac{\epsilon}{2}$. Thus, it turns out that for any $\epsilon > 0$ and any $y > 0$ there exists a finite set M of prime numbers, containing at least all primes $p \leq y$, such that

$$\max_{|s| \leq r} \left| \log \zeta_M \left(s + \frac{3}{4}, \omega_0 \right) - f(s) \right| \leq \epsilon. \tag{1.17}$$

The next and main step in Voronin's proof is to switch from $\log \zeta_M(s)$ to the logarithm of the zeta-function. Of course, $\log \zeta(s)$ has singularities at the zeros of $\zeta(s)$, but since the set of these possibly singularities has measure zero by density Theorem 1.2, they are negligible. Note that for many *higher* L-functions (i.e., functions which satisfy a Riemann-type functional equation with *many* Gamma-factors), which we shall consider later, no density results of this strength are known.

Now we choose $\kappa > 1$ and $\epsilon_1 \in (0,1)$ such that $\kappa r < \frac{1}{4}$ and

$$\max_{|s| \leq r} \left| f\left(\frac{s}{\kappa} \right) - f(s) \right| < \epsilon_1. \tag{1.18}$$

Putting $Q = \{p : p \leq z\}$ and $\mathcal{E} = \{s = \sigma + it : -\kappa r < \sigma \leq 2, |t| \leq 1\}$, one can show, using the approximate functional equation for $\zeta(s)$ (i.e., a representation as a Dirichlet polynomial related to (1.3)), that for any $\epsilon_2 > 0$

$$\int_T^{2T} \iint_{\mathcal{E}} \left| \zeta_Q^{-1} \left(s + \frac{3}{4} + i\tau, \mathbf{0} \right) \zeta \left(s + \frac{3}{4} + i\tau \right) - 1 \right|^2 \, d\sigma \, dt \, d\tau \ll \epsilon_2^4 T, \tag{1.19}$$

provided that z and T are sufficiently large, depending on ϵ_2; here $\mathbf{0} := (0,0,\ldots)$; actually this is *Hilfssatz* 2 from Bohr [29]. Now define

$$\mathcal{A}_T = \left\{ \tau \in [T, 2T] : \iint_{\mathcal{E}+\frac{3}{4}} |\zeta_Q^{-1}(s + i\tau, \mathbf{0})\zeta(s + i\tau) - 1|^2 \, d\sigma \, dt < \epsilon_2^2 \right\}.$$

Then it follows from (1.19) that, for sufficiently large z and T,

$$\text{meas}\, \mathcal{A}_T > (1 - \epsilon_2)T, \tag{1.20}$$

which is surprisingly large. It follows from Cauchy's formula that, for sufficiently small ϵ_2,

$$\max_{|s| \leq r} \left| \log \zeta \left(s + \frac{3}{4} + i\tau \right) - \log \zeta_Q \left(s + \frac{3}{4} + i\tau, \mathbf{0} \right) \right| \ll \epsilon_2, \tag{1.21}$$

provided $\tau \in \mathcal{A}_T$, where the implicit constant depends only on κ. By (1.17) there exists a sequence of finite sets of prime numbers $M_1 \subset M_2 \subset \ldots$ such that $\cup_{k=1}^\infty M_k$ contains all primes and

$$\lim_{k\to\infty} \max_{|s|\leq\kappa r} \left| \log \zeta_{M_k}\left(s + \frac{3}{4}, \omega_0\right) - f\left(\frac{s}{\kappa}\right) \right| = 0. \qquad (1.22)$$

Let $\omega_0 = (\omega_2^{(0)}, \omega_3^{(0)}, \ldots)$. By the continuity of $\log \zeta_M\left(s + \frac{3}{4}, \omega_0\right)$, for any $\epsilon_1 > 0$, there exists a positive δ such that, whenever the inequalities

$$\|\omega_p^{(0)} - \omega_p\| < \delta \qquad \text{for} \quad p \in M_k \qquad (1.23)$$

hold, where $\|z\|$ denotes the minimal distance of z to an integer, then

$$\max_{|s|\leq\kappa r} \left| \log \zeta_{M_k}\left(s + \frac{3}{4}, \omega_0\right) - \log \zeta_{M_k}\left(s + \frac{3}{4}, \omega\right) \right| < \epsilon_1. \qquad (1.24)$$

Let

$$\mathcal{B}_T = \left\{ \tau \in [T, 2T] : \left\| \tau\frac{\log p}{2\pi} - \omega_p^{(0)} \right\| < \delta \right\}.$$

Now we consider

$$\frac{1}{T} \int_{\mathcal{B}_T} \iint_{|s|\leq\kappa r} \left| \log \zeta_Q\left(s + \frac{3}{4} + i\tau, 0\right) - \log \zeta_{M_k}\left(s + \frac{3}{4} + i\tau, 0\right) \right|^2 d\sigma\, dt\, d\tau,$$

respectively,

$$\iint_{|s|\leq\kappa r} \frac{1}{T} \int_{\mathcal{B}_T} \left| \log \zeta_Q\left(s + \frac{3}{4} + i\tau, 0\right) - \log \zeta_{M_k}\left(s + \frac{3}{4} + i\tau, 0\right) \right|^2 d\tau\, d\sigma\, dt.$$

Putting

$$\omega(\tau) = \left(\tau\frac{\log 2}{2\pi}, \tau\frac{\log 3}{2\pi}, \ldots \right), \qquad (1.25)$$

we may rewrite the inner integral as

$$\int_{\mathcal{B}_T} \left| \log \zeta_Q\left(s + \frac{3}{4}, \omega(\tau)\right) - \log \zeta_{M_k}\left(s + \frac{3}{4}, \omega(\tau)\right) \right|^2 d\tau.$$

Next we need Weyl's refinement of Kronecker's approximation theorem. Let $\omega(\tau)$ be any continuous function with domain of definition $[0, \infty)$ and range $\mathbb{R}^N$. Then the curve $\omega(\tau)$ is said to be uniformly distributed mod 1 in $\mathbb{R}^N$ if, for every parallelepiped $\Pi = [\alpha_1, \beta_1] \times \ldots \times [\alpha_N, \beta_N]$ with $0 \leq \alpha_j < \beta_j \leq 1$ for $1 \leq j \leq N$,

$$\lim_{T\to\infty} \frac{1}{T} \text{meas} \left\{ \tau \in (0, T) : \omega(\tau) \in \Pi \mod 1 \right\} = \prod_{j=1}^N (\beta_j - \alpha_j).$$

In some sense, a curve is uniformly distributed mod 1 if the proportion of the fractional values which lie in an arbitrary parallelepiped Π of the unit cube coincides with the volume of Π. In questions about uniform distribution mod 1 one is interested in the fractional part only. For a curve $\omega(\tau)$ in $\mathbb{R}^N$, we define

$$\{\omega(\tau)\} = (\omega_1(\tau) - [\omega_1(\tau)], \ldots, \omega_N(\tau) - [\omega_N(\tau)]),$$

where $[x]$ denotes the integral part of $x \in \mathbb{R}$.

Lemma 1.8. *(i) Let $a_1, \ldots, a_N$ be real numbers, linearly independent over $\mathbb{Q}$, and let γ be a subregion of the N-dimensional unit cube with Jordan content Γ. Then*

$$\lim_{T \to \infty} \frac{1}{T} \operatorname{meas} \{\tau \in (0, T) : (\tau a_1, \ldots, \tau a_N) \in \gamma \bmod 1\} = \Gamma.$$

(ii) Suppose that the curve $\omega(\tau)$ is uniformly distributed mod 1 in $\mathbb{R}^N$. Let $\mathcal{D}$ be a closed and Jordan measurable subregion of the unit cube in $\mathbb{R}^N$ and let Ω be a family of complex-valued continuous functions defined on $\mathcal{D}$. If Ω is uniformly bounded and equicontinuous, then

$$\lim_{T \to \infty} \frac{1}{T} \int_0^T f(\{\omega(\tau)\}) \mathbf{1}_{\mathcal{D}}(\tau) \, \mathrm{d}\tau = \int_{\mathcal{D}} f(x_1, \ldots, x_N) \, \mathrm{d}x_1 \ldots \mathrm{d}x_N$$

uniformly with respect to $f \in \Omega$, where $\mathbf{1}_{\mathcal{D}}(\tau)$ is equal to 1 if $\omega(\tau) \in \mathcal{D} \bmod 1$, and 0 otherwise.

Note that the notion of Jordan content is more restrictive than the notion of Lebesgue measure. But, if the Jordan content exists, then it is also defined in the sense of Lebesgue and equal to it. A proof of Weyl's theorem can be found in his paper [370] or in Karatsuba and Voronin [166].

The unique prime factorization of integers implies the linear independence of the logarithms of the prime numbers over the field of rational numbers. Thus, in some sense, the logarithms of prime numbers behave like random variables. This is the main reason why probabilistic methods can be applied to the zeta-function or, more generally, to Euler products!

By Lemma 1.8(i), the curve $\omega(\tau)$, defined by (1.25), is uniformly distributed mod 1. Application of Lemma 1.8(ii), to the curve $\omega(\tau)$ yields

$$\lim_{T \to \infty} \frac{1}{T} \int_{\mathcal{B}_T} \left| \log \zeta_Q \left(s + \frac{3}{4}, \omega(\tau) \right) - \log \zeta_{M_k} \left(s + \frac{3}{4}, \omega(\tau) \right) \right|^2 \mathrm{d}\tau$$

$$= \int_{\mathcal{D}} \left| \log \zeta_Q \left(s + \frac{3}{4}, \omega \right) - \log \zeta_{M_k} \left(s + \frac{3}{4}, \omega \right) \right|^2 \mathrm{d}\mu,$$

uniformly in s for $|s| \le \kappa r$, where $\mathcal{D}$ is the subregion of the unit cube in $\mathbb{R}^N$ given by the inequalities (1.23) with $N = \sharp M_k$, and $\mathrm{d}\mu$ is the Lebesgue measure. By the definition of $\zeta_M(s, \omega)$ it follows that for $M_k \subset Q$

$$\zeta_Q(s, \omega) = \zeta_{M_k}(s, \omega) \zeta_{Q \setminus M_k}(s, \omega),$$

and thus

$$\int_{\mathcal{D}} \left| \log \zeta_Q \left(s + \frac{3}{4}, \omega \right) - \log \zeta_{M_k} \left(s + \frac{3}{4}, \omega \right) \right|^2 \mathrm{d}\mu$$

$$\ll \operatorname{meas} \mathcal{D} \cdot \int_{[0,1]^N} \left| \log \zeta_{Q \setminus M_k} \left(s + \frac{3}{4}, \omega \right) \right|^2 \mathrm{d}\mu.$$

The latter integral is bounded above by $y_k^{2\kappa r-(1/2)}$ provided that M_k contains all primes $\leq y_k$. It follows that:

$$\frac{1}{T}\int_{\mathcal{B}_T}\iint_{|s|\leq\kappa r}\left|\log\zeta_Q\left(s+\tfrac{3}{4}+\mathrm{i}\tau,\mathbf{0}\right)-\log\zeta_{M_k}\left(s+\tfrac{3}{4}+\mathrm{i}\tau,\mathbf{0}\right)\right|^2\,\mathrm{d}\sigma\,\mathrm{d}t\,\mathrm{d}\tau$$
$$\ll y_k^{2\kappa r-(1/2)}\operatorname{meas}\mathcal{D}.$$

Applying once more Lemma 1.8(i) yields

$$\lim_{T\to\infty}\frac{1}{T}\operatorname{meas}\mathcal{B}_T=\operatorname{meas}\mathcal{D},$$

which implies, for sufficiently large y_k,

$$\operatorname{meas}\left\{\tau\in\mathcal{B}_T:\iint_{|s|\leq\kappa r}\left|\log\zeta_Q\left(s+\frac{3}{4}+\mathrm{i}\tau,\mathbf{0}\right)\right.\right.$$
$$\left.\left.-\log\zeta_{M_k}\left(s+\frac{3}{4}+\mathrm{i}\tau,\mathbf{0}\right)\right|^2\,\mathrm{d}\sigma\,\mathrm{d}t<y_k^{\kappa r-(1/4)}\right\}$$
$$>\frac{\operatorname{meas}\mathcal{D}}{2}T.$$

This gives via Lemma 1.8(ii)

$$\frac{1}{T}\operatorname{meas}\left\{\tau\in\mathcal{B}_T:\max_{|s|\leq\kappa r}\left|\log\zeta_Q\left(s+\frac{3}{4}+\mathrm{i}\tau,\mathbf{0}\right)\right.\right.$$
$$\left.\left.-\log\zeta_{M_k}\left(s+\frac{3}{4}+\mathrm{i}\tau,\mathbf{0}\right)\right|<y_k^{(\kappa r-1/4)/5}\right\}$$
$$>\frac{\operatorname{meas}\mathcal{D}}{2}. \tag{1.26}$$

If we now take $0<\epsilon_2<\frac{1}{2}\operatorname{meas}\mathcal{D}$, then (1.20) implies

$$\frac{1}{T}\operatorname{meas}\mathcal{A}_T\cap\mathcal{B}_T\geq\frac{1}{2}\operatorname{meas}\mathcal{D}-\epsilon_2>0.$$

Thus, in view of (1.18) and (1.22) we may approximate $f(s)$ by $\log\zeta_{M_k}\left(s+\frac{3}{4},\omega_0\right)$ (independent on τ), with (1.24) and (1.26) the latter function by $\log\zeta_Q\left(s+\frac{3}{4}+\mathrm{i}\tau,\mathbf{0}\right)$, and finally with regard to (1.21) by $\log\zeta\left(s+\frac{3}{4}+\mathrm{i}\tau\right)$ on a set of τ with positive measure. Replacing T by $\frac{1}{2}T$, we thus find, for any $\epsilon>0$,

$$\liminf_{T\to\infty}\frac{1}{T}\operatorname{meas}\left\{\tau\in[0,T]:\max_{|s|\leq r}\left|\log\zeta\left(s+\frac{3}{4}+\mathrm{i}\tau\right)-f(s)\right|<\epsilon\right\}>0.$$

Now it is obvious how to deduce Voronin's theorem by taking the exponential.

Voronin called his universality theorem Теорема о кружочках, the *theorem about little disks*. Reich [305] and Bagchi [9] improved Voronin's result significantly in replacing the disk by an arbitrary compact set in the right half of the critical strip with connected complement, and by giving a lucid proof in the language of probability theory. The strongest version of Voronin's theorem has the form:

Theorem 1.9. *Suppose that $\mathcal{K}$ is a compact subset of the strip $\frac{1}{2} < \sigma < 1$ with connected complement, and let $g(s)$ be a non-vanishing continuous function on $\mathcal{K}$ which is analytic in the interior of $\mathcal{K}$. Then, for any $\epsilon > 0$,*

$$\liminf_{T \to \infty} \frac{1}{T} \operatorname{meas} \left\{ \tau \in [0, T] : \max_{s \in \mathcal{K}} |\zeta(s + i\tau) - g(s)| < \epsilon \right\} > 0.$$

The topological restriction on $\mathcal{K}$ is necessary. This follows from basic facts in approximation theory (see the remark to Theorem 5.15). Also the restriction on $g(s)$ to be non-vanishing cannot be removed as we shall show in Sect. 8.1. The domain in which the uniform approximation of admissible target functions takes place is called the strip of universality. In the case of the zeta-function this strip of universality is the open right half of the critical strip; later we will meet examples where the strip of universality is more restricted.

It should be noticed that Voronin's theorem implies that $\zeta(s)^{-1}$ is universal (independent of the truth of the Riemann hypothesis). To see this we observe that given a non-vanishing analytic function $g(s)$ on some admissible set $\mathcal{K}$ also its reciprocal is analytic and non-vanishing. Hence we can approximate the function $g(s)^{-1}$ uniformly by shifts $\zeta(s + i\tau)$. Since the set of such τ has positive lower density and, by the density Theorem 1.2, the set of non-trivial zeros of $\zeta(s)$ is negligible, it follows that we can find τ for which

$$\zeta(s + i\tau)^{-1} \approx g(s) \quad \text{for} \quad s \in \mathcal{K}.$$

Another consequence of the universality for the zeta-function is that its derivatives and also the logarithmic derivative are universal; this was first proved by Bagchi [10] and Laurinčikas [182] by a slight modification of the proof of universality for $\zeta(s)$.

We may interpret the absolute value of an analytic function as an analytic landscape over the complex plane. Then the universality theorem states that any finite analytic landscape can be found (up to an arbitrarily small error) in the analytic landscape of the Riemann zeta-function (see Fig. 1.4 on the next page). On the contrary, Steuding [338] showed that one cannot go to infinity along a rectifiable path inside the critical strip avoiding large values, roughly speaking, one cannot stay in valleys in the analytic landscape all the time (the proof relies on a Phragmén–Lindelöf type theorem for regions with rectifiable boundary and Ahlfors' distortion theorem). Taking into account explicit estimates for extreme values, so-called Ω-results, as for example (9.2) due to Montgomery [260], this can be formulated precisely.

1.4 Dirichlet L-Functions and Joint Universality

A special role in number theory is played by multiplicative arithmetical functions and their associated generating functions. Multiplicative functions respect the multiplicative structure of $\mathbb{N}$: an arithmetic function f is called multiplicative if $f(1) \neq 0$ and

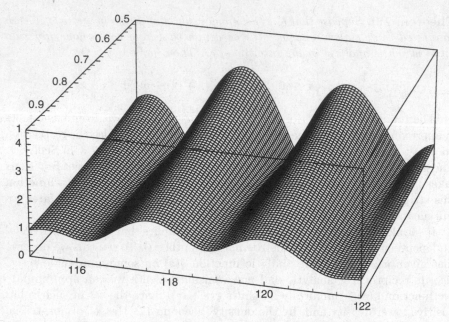

Fig. 1.4. Some dunes at the baltic sea shore or the analytic landscape of $\zeta(s)$ for $\sigma \in [\frac{1}{2}, 1], t \in [115, 122]$

$$f(m \cdot n) = f(m) \cdot f(n)$$

for all coprime integers m, n; if the latter identity holds for all integers, f is said to be completely multiplicative.

Let q be a positive integer. A Dirichlet character χ mod q is a non-vanishing group homomorphism from the group $(\mathbb{Z}/q\mathbb{Z})^*$ of prime residue classes modulo q to $\mathbb{C}$. The character, which is identically one, is called principal, and is denoted by χ_0. By setting $\chi(a) = 0$ on the non-prime residue classes such a character extends via $\chi(n) = \chi(a)$ for $n \equiv a \bmod q$ to a completely multiplicative arithmetic function. For $\sigma > 1$, the Dirichlet L-function $L(s, \chi)$ attached to a character χ mod q is given by

$$L(s, \chi) = \sum_{n=1}^{\infty} \frac{\chi(n)}{n^s} = \prod_p \left(1 - \frac{\chi(p)}{p^s}\right)^{-1}.$$

The zeta-function $\zeta(s)$ may be regarded as the Dirichlet L-function to the principal character χ_0 mod 1. It is possible that for values of n coprime with q the character $\chi(n)$ may have a period less than q. If so, we say that χ is imprimitive, and otherwise primitive; the principal character is not regarded as a primitive character. Every non-principal imprimitive character is induced by a primitive character. Two characters are non-equivalent if they are not induced by the same character. The characters to a common modulus are pairwise non-equivalent. If $\chi \bmod q$ is induced by a primitive character $\chi^* \bmod q^*$, then

$$L(s,\chi) = L(s,\chi^*) \prod_{p|q} \left(1 - \frac{\chi^*(p)}{p^s}\right). \tag{1.27}$$

Being twists of the Riemann zeta-function with multiplicative characters, Dirichlet L-functions share many properties with the zeta-function. For instance, there is an analytic continuation to the complex plane, only with the difference that $L(s,\chi)$ is regular at $s = 1$ if and only if χ is non-principal. Furthermore, L-functions to primitive characters satisfy a functional equation of the Riemann-type; namely,

$$\left(\frac{q}{\pi}\right)^{(s+\delta)/2} \Gamma\left(\frac{s+\delta}{2}\right) L(s,\chi) = \frac{\tau(\chi)}{i^\delta \sqrt{q}} \left(\frac{q}{\pi}\right)^{(1+\delta-s)/2} \Gamma\left(\frac{1+\delta-s}{2}\right) L(1-s,\overline{\chi}),$$

where $\delta := \frac{1}{2}(1 - \chi(-1))$ and

$$\tau(\chi) := \sum_{a \bmod q} \chi(a) \exp\left(\frac{2\pi i a}{q}\right)$$

is the Gauss sum attached to χ. One finds similar zero-free regions (with the exception of possible Siegel zeros on the real line), density theorems, and also for Dirichlet L-functions it is expected that the analogue of the Riemann hypothesis holds; the so-called Generalized Riemann hypothesis states that neither $\zeta(s)$ nor any $L(s,\chi)$ has a zero in the half-plane $\operatorname{Re} s > \frac{1}{2}$.

Dirichlet L-functions were constructed by Dirichlet [69] to tackle the problem of the distribution of primes in arithmetic progressions. The main ingredient in his approach are the orthogonality relations for characters linking prime residue classes with character sums:

$$\frac{1}{\varphi(q)} \sum_{a \bmod q} \chi(a) = \begin{cases} 1 & \text{if } \chi = \chi_0, \\ 0 & \text{otherwise,} \end{cases} \tag{1.28}$$

and its dual variant

$$\frac{1}{\varphi(q)} \sum_{\chi \bmod q} \chi(a) = \begin{cases} 1 & \text{if } a \equiv 1 \bmod q, \\ 0 & \text{otherwise,} \end{cases}$$

valid for a coprime with q, where $\varphi(q)$ is Euler's φ-function which counts the number of prime residue classes $\bmod\, q$. By the latter relation a suitable linear combination of characters can be used as indicator function of prime residue classes modulo q. Using similar techniques as for $\zeta(s)$, one can prove a prime number theorem for arithmetic progressions. Let $\pi(x; a \bmod q)$ denote the number of primes $p \leq x$ in the residue class $a \bmod q$, then, for a coprime with q,

$$\pi(x; a \bmod q) \sim \frac{1}{\varphi(q)} \pi(x). \tag{1.29}$$

This shows that the primes are uniformly distributed in the prime residue classes. Of course, one can prove also an asymptotic formula with error term (the theorem of Page–Siegel–Walfisz gives even an asymptotic formula which is uniform in a small region of values q). Under assumption of the Generalized Riemann hypothesis one has

$$\pi(x; a \bmod q) = \frac{1}{\varphi(q)} \operatorname{li}(x) + \operatorname{O}\left(x^{1/2} \log(qx)\right)$$

for $x \geq 2, q \geq 1$, and a coprime with q, the implicit constant being absolute. There are plenty of results which hold if Riemann's hypothesis is true. Often one can replace this assumption by the celebrated theorem of Bombieri–Vinogradov due to Bombieri [37] and Vinogradov [360] (independently, with a slightly weaker range for Q) which states that, for any $A \geq 1$,

$$\sum_{q \leq Q} \max_{\substack{a \bmod q \\ (a,q)=1}} \max_{y \leq x} \left| \pi(y; a \bmod q) - \frac{1}{\varphi(q)} \operatorname{li}(y) \right| \ll \frac{x}{(\log x)^A} + Q x^{1/2} (\log Qx)^6.$$

This shows that the error term in the prime number theorem (1.29) is, on average over $q \leq x^{1/2}(\log x)^{-A-7}$, of comparable size as predicted by the Riemann hypothesis.

We return to the theme of universality. Voronin [364] proved that a collection of Dirichlet L-functions to non-equivalent characters can uniformly approximate simultaneously non-vanishing analytic functions; in slightly different form this was also established by Gonek [104] and Bagchi [9] (independently; all these sources are unpublished doctoral theses). Again we state the strongest version of this so-called *joint* universality:

Theorem 1.10. *Let* $\chi_1 \bmod q_1, \ldots, \chi_\ell \bmod q_\ell$ *be pairwise non-equivalent Dirichlet characters,* $\mathcal{K}_1, \ldots, \mathcal{K}_\ell$ *be compact subsets of the strip* $\frac{1}{2} < \sigma < 1$ *with connected complements. Further, for each* $1 \leq k \leq \ell$, *let* $g_k(s)$ *be a continuous non-vanishing function on* $\mathcal{K}_k$ *which is analytic in the interior of* $\mathcal{K}_k$. *Then, for any* $\epsilon > 0$,

$$\liminf_{T \to \infty} \frac{1}{T} \operatorname{meas} \left\{ \tau \in [0,T] : \max_{1 \leq k \leq \ell} \max_{s \in \mathcal{K}_k} |L(s + i\tau, \chi_k) - g_k(s)| < \varepsilon \right\} > 0.$$

The proof of this joint universality theorem (in a slightly weaker form) can be found in the monograph of Karatsuba and Voronin [166]. The proof uses the orthogonality relation (1.28). This independence is essential for joint universality. Consider a character $\chi \bmod q$ induced by another character $\chi^* \bmod q^*$. It follows immediately from (1.27) that both $L(s, \chi^*)$ and $L(s, \chi)$ cannot approximate uniformly a given function jointly.

Another type of universality was discovered by Bagchi. In [9], he proved universality for Dirichlet L-functions with respect to the characters; more precisely, if $\mathcal{K}$ is a compact subset of $\frac{1}{2} < \sigma < 1$ with connected complement and $g(s)$ is a non-vanishing continuous function on $\mathcal{K}$, which is analytic in the interior, then, for any sufficiently large prime number p and any $\varepsilon > 0$, there exist a Dirichlet character χ mod p such that

$$\max_{s \in \mathcal{K}} \left| L\left(s + \frac{3}{4}, \chi\right) - g(s) \right| < \varepsilon; \tag{1.30}$$

moreover, the latter inequality holds for more than cp characters χ mod p, where c is a positive constant (recall that there are $\varphi(p) = p - 1$ characters χ mod p). Eminyan [75] showed that if $\mathcal{K} = \{s : |s| \leq r\}$ with $0 < r < \frac{1}{4}$, and $g(s)$ is a non-vanishing continuous function for $s \in \mathcal{K}$, which is analytic in the interior, then, for each prime number p and any $\varepsilon > 0$, there exist a positive integer n and a Dirichlet character χ mod p^n such that (1.30) holds.

We conclude with another interesting type of universality. Any integer $d \neq 0$ with $d \equiv 0, 1 \bmod 4$ is called a discriminant; if $d = 1$ or d is the discriminant of a quadratic number field, d is said to be fundamental. To any fundamental discriminant d we attach a real primitive character χ_d mod $|d|$ defined by the Kronecker symbol $(\frac{d}{\cdot})$. Recently, Mishou and Nagoshi [251] investigated the functional distribution of the associated Dirichlet L-functions $L(s, \chi_d)$ on $\frac{1}{2} < \sigma < 1$ as d varies and proved a universality theorem for $L(s, \chi_d)$ in the d-aspect: let Ω be a simply connected region in the open strip $\mathcal{D} := \{s : \frac{1}{2} < \sigma < 1\}$ which is symmetric with respect to the real axis and let $g(s)$ be a non-vanishing analytic function on Ω which takes positive real values on $\mathcal{D} \cap \mathbb{R}$; then for any compact subset $\mathcal{K}$ of Ω and any positive ϵ,

$$\liminf_{X \to \infty} \frac{\#\{1 \leq d \leq X : \max_{s \in K} |L(s, \chi_d) - g(s)| < \epsilon\}}{\#\{1 \leq d \leq X\}} > 0. \tag{1.31}$$

The condition on the target functions $g(s)$ is natural with respect to the values taken by $L(s, \chi_d)$ on $\Omega \cap \mathbb{R}$. Mishou and Nagoshi [252] also obtained a similar result for prime discriminants. Besides, they also studied the distribution of values $L(1, \chi_d)$. Denote by $\mathcal{D}^+$ the set of positive square-free integers $d > 1$ with $d \equiv 1 \bmod 8$, and by $\mathcal{D}^-$ the set of negative square-free integers $d \equiv 1 \bmod 8$. In [251], they obtained the remarkable result that the sets $\{L(1, \chi_d) : d \in \mathcal{D}^\delta\}$ with $\delta \in \{+, -\}$ are dense in $\mathbb{R}_+$. They further deduced from Dirichlet's class number formula that also

$$\left\{ \frac{h(d) \log \epsilon(d)}{\sqrt{d}} : d \in \mathcal{D}^+ \right\} \quad \text{and} \quad \left\{ \frac{h(d)}{\sqrt{d}} : d \in \mathcal{D}^- \right\}$$

are dense in $\mathbb{R}_+$, where $h(d)$ is the class number of the associated quadratic number field $\mathbb{Q}(\sqrt{d})$ with $d \in \mathcal{D}^\delta$, and $\epsilon(d)$ is the fundamental unit of $\mathbb{Q}(\sqrt{d})$ with $d \in \mathcal{D}^+$.

Nagoshi [275] extended this concept of universality to families of automorphic L-functions, more precisely, to L-functions attached to newforms for congruence subgroups of $\mathsf{SL}_2(\mathbb{Z})$ (which we will introduce in the following section).

1.5 *L*-Functions Associated with Newforms

First we recall some facts from the theory of modular forms; see Iwaniec's book [144] on modular forms and associated *L*-functions for details.

Denote by $\mathbb{H}$ the upper half-plane $\{z := x + iy \in \mathbb{C} : y > 0\}$, and let k and N be positive integers, k being even. The subgroup

$$\Gamma_0(N) := \left\{ \begin{pmatrix} a & b \\ c & d \end{pmatrix} \in \mathsf{SL}_2(\mathbb{Z}) : c \equiv 0 \bmod N \right\}$$

of the full modular group $\mathsf{SL}_2(\mathbb{Z})$ is called Hecke subgroup of level N or congruence subgroup mod N. A holomorphic function $f(z)$ on $\mathbb{H}$ is said to be a cusp form of weight k and level N, if

$$f\left(\frac{az+b}{cz+d}\right) = (cz+d)^k f(z)$$

for all $z \in \mathbb{H}$ and all matrices

$$\begin{pmatrix} a & b \\ c & d \end{pmatrix} \in \Gamma_0(N),$$

and if f vanishes at all cusps. The vanishing of f at the cusps is equivalent with

$$z := x + iy \;\mapsto\; y^k |f(z)|^2$$

is bounded on $\mathbb{H}$. Then f has for $z \in \mathbb{H}$ a Fourier expansion

$$f(z) = \sum_{n=1}^{\infty} c(n) \exp(2\pi i n z). \qquad (1.32)$$

The cusp forms on $\Gamma_0(N)$ of weight k form a finite dimensional complex vector space, denoted by $S_k(\Gamma_0(N))$, with the Petersson inner product, defined by

$$\langle f, g \rangle = \int_{\mathbb{H}/\Gamma_0(N)} f(z)\overline{g(z)} y^k \frac{\mathrm{d}x\,\mathrm{d}y}{y^2}$$

for $f, g \in S_k(\Gamma_0(N))$. Suppose that $M|N$. If $f \in S_k(\Gamma_0(M))$ and $dM|N$, then $z \mapsto f(dz)$ is a cusp form on $\Gamma_0(N)$ of weight k too. The forms which may be obtained in this way from divisors M of the level N with $M \neq N$ span a subspace $S_k^{\mathrm{old}}(\Gamma_0(N))$, called the space of oldforms. Its orthogonal complement with respect to the Petersson inner product is denoted $S_k^{\mathrm{new}}(\Gamma_0(N))$. For $n \in \mathbb{N}$ we define the Hecke operator $T(n)$ by

$$T(n)f = \frac{1}{n} \sum_{ad=n} a^k \sum_{0 \le b < d} f\left(\frac{az+b}{d}\right)$$

for $f \in S_k(\Gamma_0(N))$. The operators $T(n)$ are multiplicative, i.e., $T(mn) = T(m)T(n)$ for coprime m, n, and they encode plenty of arithmetic information about modular forms. The theory of Hecke operators implies the existence of an orthogonal basis of $S_k^{\mathrm{new}}(\Gamma_0(N))$ made of eigenfunctions of the operators $T(n)$ for n coprime with N. By the multiplicity-one principle of Atkin and Lehner [6], the elements f of this basis are in fact eigenfunctions of all $T(n)$, i.e., there exist complex numbers $\lambda_f(n)$ for which $T(n)f = \lambda_f(n)f$ and $c(n) = \lambda_f(n)c(1)$ for all $n \in \mathbb{N}$. Furthermore, it follows that the first Fourier coefficient $c(1)$ of such an f is non-zero. Such a simultaneous eigenfunction is said to be an eigenform. A newform is defined to be an eigenform that does not come from a space of lower level and is normalized to have $c(1) = 1$. The newforms form a finite set which is an orthogonal basis of the space $S_k^{\mathrm{new}}(\Gamma_0(N))$. For instance, Ramanujan's cusp form

$$\sum_{n=1}^{\infty} \tau(n)\exp(2\pi\mathrm{i}nz) := \exp(2\pi\mathrm{i}z)\prod_{n=1}^{\infty}(1 - \exp(2\pi\mathrm{i}nz))^{24} \qquad (1.33)$$

is a normalized eigenform of weight 12 to the full modular group, and hence a newform of level 1. Ramanujan [300] conjectured that the coefficients $\tau(n)$ are multiplicative and satisfy the estimate $|\tau(p)| \leq 2p^{11/2}$ for every prime number p. The multiplicativity was proved by Mordell [261], in particular by the beautiful formula

$$\tau(m)\tau(n) = \sum_{d|(m,n)} d^{11}\tau\left(\frac{mn}{d^2}\right).$$

The estimate was shown by Deligne. More precisely, Deligne [67] proved for the coefficients of any newform f of weight k the estimate

$$|c(n)| \leq n^{(k-1)/2}d(n), \qquad (1.34)$$

where $d(n) := \sum_{d|n} 1$ is the divisor function.

In the 1930s, Hecke [126] started investigations on modular forms and Dirichlet series with a Riemann-type functional equation; his studies were completed by Atkin and Lehner [6] (for newforms). Here we shall focus on newforms. Given a newform f with Fourier expansion (1.32), we define the associated L-function by

$$L(s,f) = \sum_{n=1}^{\infty} \frac{c(n)}{n^s}. \qquad (1.35)$$

In view of the classic bound $d(n) \ll n^\epsilon$ it follows from (1.34) that the series (1.35) converges absolutely for $\sigma > \frac{k+1}{2}$. By the theory of Hecke operators, it turns out that the Fourier coefficients of newforms are multiplicative. Hence, in the half-plane of absolute convergence, there is an Euler product representation and it is given by

$$L(s,f) = \prod_{p|N}\left(1 - \frac{c(p)}{p^s}\right)^{-1}\prod_{p\nmid N}\left(1 - \frac{c(p)}{p^s} + \frac{1}{p^{2s+1-k}}\right)^{-1}. \qquad (1.36)$$

Hecke [126], respectively, Atkin and Lehner [6], proved that $L(s, f)$ has an analytic continuation to an entire function and satisfies the functional equation

$$N^{s/2}(2\pi)^{-s}\Gamma(s)L(s, f) = \omega(-1)^{k/2}N^{(k-s)/2}(2\pi)^{s-k}\Gamma(k-s)L(k-s, f),$$

where $\omega = \pm 1$ is the eigenvalue of the Atkin–Lehner involution $\left(\begin{smallmatrix} 0 & -N \\ 1 & 0 \end{smallmatrix}\right)$ on $S_k(\Gamma_0(N))$. Hecke proved a converse theorem which gives a characterization of these L-functions by their functional equation; this beautiful result generalizes Hamburger's theorem for the Riemann zeta-function [115] (see also [353, Sect. 2.13]).

Laurinčikas and Matsumoto [201] obtained a universality theorem for L-functions attached to normalized eigenforms of the full modular group. Laurinčikas, Matsumoto and Steuding [204, 205] extended this result to newforms:

Theorem 1.11. *Suppose that f is a newform of weight k and level N. Let $\mathcal{K}$ be a compact subset of the strip $\frac{k}{2} < \sigma < \frac{k+1}{2}$ with connected complement, and let $g(s)$ be a continuous non-vanishing function on $\mathcal{K}$ which is analytic in the interior of $\mathcal{K}$. Then, for any $\epsilon > 0$,*

$$\liminf_{T\to\infty} \frac{1}{T} \operatorname{meas} \left\{ \tau \in [0, T] : \max_{s\in\mathcal{K}} |L(s + i\tau, f) - g(s)| < \epsilon \right\} > 0.$$

All known proofs of universality theorems rely on deep knowledge of the coefficients of the underlying Dirichlet series. Rankin's celebrated asymptotic formula [304] for the Fourier coefficients $c(n)$ of newforms f to the full modular group states

$$\sum_{p\leq x} c(p)^2 p^{1-k} \sim \frac{x}{\log x}. \tag{1.37}$$

The main new ingredient in the proof of Theorem 1.11 is the use of an extension of Rankin's formula to newforms to congruence subgroups $\Gamma_0(N)$. The proof is inspired by Rankin's approach; the non-vanishing of the associated Rankin–Selberg L-functions (see Sect. 1.6) on the boundary of the critical strip is essentially due to Moreno [262]. (See also Theorem 6.14.)

The universality of the logarithmic derivative and the first derivative of L-functions to normalized eigenforms for the full modular group were studied by Laurinčikas [197]. Recently, Nagoshi [273, 274] proved universality for L-functions to Maass forms (that are non-holomorphic eigenforms of the Laplace–Beltrami operator on the upper half-plane). Deligne's estimate (1.34) played an essential part in the proofs of universality for L-functions attached to holomorphic eigen- and newforms, respectively, but its analogue, the so-called Ramanujan hypothesis, is not known for Maass forms in general. In [273], Nagoshi used bounds of Kim and Sarnak for the local root numbers of the Euler factors and obtained universality in a restricted range of the right half of the critical strip; in [274], he proved unrestricted universality by use of an asymptotic formula for the fourth moment of these local roots due to

Murty. Remarkably, his argument also yields a proof for Theorem 1.11' without using (1.34). Moreover, Nagoshi [275] proved universality for *L*-functions to certain newforms and their derivatives in the level aspect.

We return to *L*-functions associated with holomorphic modular forms. Laurinčikas and Matsumoto [202] obtained a joint universality theorem for *L*-functions associated with newforms twisted by characters. Let $f \in S_k(\Gamma_0(N))$ be a newform with Fourier expansion (1.32) and let χ be a Dirichlet character mod q where q is coprime with N. The twisted *L*-function is defined by

$$L_\chi(s, f) = \sum_{n=1}^\infty \frac{c(n)}{n^s} \chi(n).$$

As in the non-twisted case (1.35), this Dirichlet series has an Euler product and extends to an entire function.

Theorem 1.12. *Let* $q_1, \ldots, q_n$ *be positive integers coprime with* N *and let* χ_1 *mod* $q_1, \ldots, \chi_n$ *mod* q_n *be pairwise non-equivalent character. Further, for* $1 \leq j \leq n$, *let* g_j *be a continuous function on* $\mathcal{K}_j$ *which is non-vanishing in the interior, where* $\mathcal{K}_j$ *is a compact subset of the strip* $\{s \in \mathbb{C} : \frac{k}{2} < \sigma < \frac{k+1}{2}\}$ *with connected complement. Then, for any* $\epsilon > 0$,

$$\liminf_{T \to \infty} \frac{1}{T} \operatorname{meas} \left\{ \tau \in [0, T] : \max_{1 \leq j \leq n} \max_{s \in \mathcal{K}_j} |L_{\chi_j}(s + i\tau, f) - g_j(s)| < \varepsilon \right\} > 0.$$

The proof relies on a joint limit theorem due to Laurinčikas [191] and some kind of prime number theorem for the coefficients of cusp forms with respect to arithmetic progressions, namely

$$\sum_{\substack{p \leq x \\ p \equiv a \bmod q}} c(p)^2 p^{1-k} \sim \frac{1}{\varphi(q)} \frac{x}{\log x}, \tag{1.38}$$

where a is coprime with q. The key for the proof of the latter result is again the extension of Rankin's asymptotic formula (1.37) to congruence subgroups $\Gamma_0(N)$; the transition to arithmetic progressions relies on analytic properties of the twisted Rankin–Selberg *L*-function.

By Wiles' celebrated proof of the Shimura–Taniyama conjecture for semi-stable modular forms [371] (which was sufficient to prove Fermat's last theorem), and the proof by Breuil, Conrad, Diamond and Taylor of the general case [44], every *L*-function attached to an elliptic curve over the rationals is the *L*-function to some newform of weight 2 for some congruence subgroup. Consequently, Theorem 1.11 proves the universality of *L*-functions associated with elliptic curves. Laurinčikas and Steuding [209] used Theorem 1.12 to give an example of jointly universal *L*-functions associated with elliptic curves. Here one may choose any finite family of elliptic curves of the form

$$E_m : \qquad Y^2 = X^3 - m^2 X \quad \text{with squarefree} \quad m \in \mathbb{N};$$

these curves were first studied by Tunnell [356] with respect to the congruent number problem. For this family one can avoid Wiles' proof of the Shimura–Taniyama–Weil conjecture but show more or less directly that the L-function associated with E_1 corresponds to a newform $f \in S_2(\Gamma_0(32))$ and that the L-function to E_m is a twist of E_1 with the Kronecker symbol $\left(\frac{m}{\cdot}\right)$ (see also [171]).

1.6 The Linnik–Ibragimov Conjecture

It is known that there exists a *rich zoo* of Dirichlet series having some universality property. We list now some more significant examples.

The Dedekind zeta-function of a number field $\mathbb{K}$ over $\mathbb{Q}$ is given by

$$\zeta_{\mathbb{K}}(s) = \sum_{\mathfrak{a}} \frac{1}{N(\mathfrak{a})^s} = \prod_{\mathfrak{p}} \left(1 - \frac{1}{N(\mathfrak{p})^s}\right)^{-1},$$

where the sum is taken over all non-zero integral ideals, the product is taken over all prime ideals of the ring of integers of $\mathbb{K}$, and $N(\mathfrak{a})$ is the norm of the ideal $\mathfrak{a}$. The Riemann zeta-function may be regarded as the Dedekind zeta-function for $\mathbb{Q}$. Universality for the Dedekind zeta-function was first obtained by Voronin [364], Gonek [104], and, in full generality, by Reich [305, 306]; here the strip of universality is restricted to the strip $\max\{\frac{1}{2}, 1 - \frac{1}{d}\} < \sigma < 1$, where d is the degree of $\mathbb{K}$ over $\mathbb{Q}$ (we will later see the reason behind). Reich's universality theorem [306] is *discrete*, i.e., under the same conditions as in Theorem 1.9, for any real $\Delta \neq 0$ and any $\epsilon > 0$, the relation

$$\max_{s \in \mathcal{K}} |\zeta_{\mathbb{K}}(s + \mathrm{i}\Delta n) - g(s)| < \epsilon \tag{1.39}$$

holds for a set of positive integers n with positive lower density. (We shall consider discrete universality more detailed in Sect. 5.7 and Dedekind zeta-functions in Sect. 13.1.) Furthermore, in [305], Reich obtained universality for Euler products formed with Beurling primes under the assumption of certain extra conditions.

Dirichlet series attached to multiplicative functions appear naturally in many aspects of number theory. They were intensively studied by Laurinčikas and Šleževičienė. Laurinčikas [181] succeeded to prove universality for Dirichlet series associated with a multiplicative function f, formally given by

$$\sum_{n=1}^{\infty} \frac{f(n)}{n^s} = \prod_p \left(1 + \sum_{k=1}^{\infty} \frac{f(p^k)}{p^{ks}}\right),$$

subject to certain conditions on f as, for example, the existence of the mean value

$$M(f) = \lim_{x \to \infty} \frac{1}{x} \sum_{n \le x} f(n),$$

the boundedness of f, and a technical condition on the local Euler factors. In [207], Laurinčikas and Šleževičienė introduced another class of Dirichlet series attached to multiplicative functions and proved also universality for its elements, and Šleževičienė [332] obtained joint universality for certain twists by Dirichlet characters.

In [239] Matsumoto introduced the now called Matsumoto zeta-functions and studied their value distribution; he obtained far-reaching generalizations of the classical results of Bohr and his collaborators. For each positive integer m let $g(m)$ and $f(j,m)$ with $1 \le j \le g(m)$ be positive integers, and let $a_m^{(j)}$ with $1 \le j \le g(m)$ be given complex numbers. Further, define the polynomial

$$A_m(X) = \prod_{j=1}^{g(m)} (1 - a_m^{(j)} X^{f(j,m)}).$$

Denoting the mth prime number by p_m, the associated Matsumoto zeta-function is defined by

$$\varphi(s) = \prod_{m=1}^{\infty} A_m(p_m^{-s})^{-1}.$$

Under the conditions $g(m) \ll p_m^\alpha$ and $|a_m^{(j)}| \le p_m^\beta$ the product defining $\varphi(s)$ converges for $\sigma > 1 + \alpha + \beta$ and defines an analytic function in this region. The Matsumoto zeta-function is an interesting generalization of many well-known zeta-functions (e.g., Dedekind zeta-functions). Suppose now that $\varphi(s)$ has an analytic continuation to some half-plane $\sigma > \delta + \alpha + \beta$ with $\frac{1}{2} \le \delta < 1$ (except for at most finitely many poles) with $\varphi(s) \ll |t|^{c_1}$ as $|t| \to \infty$, where c_1 is some positive constant and bounded mean-square. Assuming

$$\left| \sum_{\substack{j=1 \\ f(j,m)=1}}^{g(m)} a_m^{(j)} \right| p_m^{-\alpha-\beta} \ge c_2 > 0, \tag{1.40}$$

and, additionally, some technical condition, Laurinčikas [190] obtained so-called universality with weight, that is the relation

$$\liminf_{T \to \infty} \frac{1}{U} \int_{T_0}^{T} \omega(\tau) \mathbf{1}\left(\tau; \left\{ \tau : \max_{s \in \mathcal{K}} |\varphi(s + i\tau) - f(s)| < \epsilon \right\} \right) d\tau > 0$$

where T_0 is a sufficiently large constant, ω is a positive function of bounded variation, $\mathbf{1}(\tau; A)$ is the indicator function of the set A,

$$U := U(T, \omega) := \int_{T_0}^{T} \omega(\tau) \, d\tau,$$

and $\mathcal{K}$ is a compact subset of the strip $\delta + \alpha + \beta < \sigma < 1 + \alpha + \beta$ with connected complement.

A useful tool in the theory of modular forms are Rankin–Selberg L-functions, introduced by Rankin [302, 303], and Selberg [320] (independently). Given two cusp forms f and g of the full modular group with Fourier series coefficients $a(n)$ and $b(n)$, respectively, the associated Rankin–Selberg L-function is defined by

$$L(s, f \otimes g) = \zeta(2s) \sum_{n=1}^{\infty} \frac{a(n)b(n)}{n^s}.$$

For normalized eigenforms f to the full modular group, Matsumoto [241] proved universality for $L(s, f \otimes f)$. The universality for such L-functions in the level aspect was studied by Mishou and Nagoshi [254].

To mention only some more important examples of universal L-functions (some of them we shall see in Chap. 13 again) this list can be continued with

- Hecke L-functions to grössencharacters (Mishou [249, 250], and additionally in the character aspect by Mishou and Koyama [174]);
- Artin L-functions [16];
- zeta-functions associated with finite abelian groups (Laurinčikas [194, 196]).

There are other interesting examples which even have a *stronger* universality property: they can approximate functions *with* zeros on a set of positive lower density. The first one is, of course, the logarithm of the Riemann zeta-function (or the derivative). This follows immediately from the proof of Voronin's universality Theorem 1.7. Note that $\log \zeta(s)$ has a Dirichlet series representation for $\sigma > 1$ (see (1.14)). The same argument applies to all universal Euler products.

Now we present a completely different example. For $0 < \alpha \leq 1, \lambda \in \mathbb{R}$, the Lerch zeta-function is given by

$$L(\lambda, \alpha, s) = \sum_{n=0}^{\infty} \frac{\exp(2\pi i \lambda n)}{(n + \alpha)^s}. \tag{1.41}$$

This series converges absolutely for $\sigma > 1$. The analytic properties of $L(\lambda, \alpha, s)$ are quite different, if $\lambda \in \mathbb{Z}$ or not. If $\lambda \notin \mathbb{Z}$, the series converges for $\sigma > 0$ and $L(\lambda, \alpha, s)$ can be continued analytically to the whole complex plane. For $\lambda \in \mathbb{Z}$ the Lerch zeta-function becomes the Hurwitz zeta-function

$$\zeta(s, \alpha) = \sum_{m=0}^{\infty} \frac{1}{(m + \alpha)^s}; \tag{1.42}$$

this function has an analytic continuation to $\mathbb{C}$ except for a simple pole at $s = 1$ with residue 1. Denote by $\{\lambda\}$ the fractional part of a real number λ. Setting

$$\lambda^+ = 1 - \{\lambda\} \qquad \text{and} \qquad \lambda^- = \begin{cases} 1 & \text{if } \lambda \in \mathbb{Z}, \\ \{\lambda\} & \text{otherwise}, \end{cases} \tag{1.43}$$

one can prove the functional equation

$$\begin{aligned} L(\lambda, \alpha, 1 - s) = \frac{\Gamma(s)}{(2\pi)^s} \Big(&\exp\Big(2\pi i \Big(\frac{s}{4} - \alpha\lambda^-\Big)\Big) L(-\alpha, \lambda^-, s) \\ &+ \exp\Big(2\pi i \Big(-\frac{s}{4} + \alpha\lambda^+\Big)\Big) L(\alpha, \lambda^+, s)\Big). \end{aligned} \tag{1.44}$$

Twists with additive characters destroy the point symmetry of Riemann-type functional equations. The function $\zeta(s, \alpha)$ was introduced by Hurwitz [137], and $L(\lambda, \alpha, s)$ by Lipschitz [222] and Lerch [215] (independently). Hurwitz and Lerch also gave the first proofs of the corresponding functional equations. For more details on the Lerch zeta-function we refer to the monograph of Garunkštis and Laurinčikas [89].

Gonek [104] and Bagchi [9] (independently) obtained strong universality for the Hurwitz zeta-function $\zeta(s, \alpha)$ if α is transcendental or rational $\neq \frac{1}{2}, 1$ (we will explain this restriction below and in Sects. 11.4). Laurinčikas [189] extended this result by proving that the Lerch zeta-function $L(\lambda, \alpha, s)$ is strongly universal if λ is not an integer and α is transcendental. The joint universality of Lerch zeta-functions $L(\lambda_j, \alpha_j, s)$ was treated by Laurinčikas and Matsumoto [200, 203] and Nakamura [276]; here the parameters α_j have to be algebraically independent over $\mathbb{Q}$. This and the case of joint universality of Hurwitz zeta-functions with rational parameters will be studied in Sect. 11.4.

The list of known zeta-function having the strong universality property can be extended by, for example,

* Estermann's zeta-function, defined by

$$E\Big(s; \frac{k}{\ell}, \alpha\Big) = \sum_{n=1}^{\infty} \frac{\sigma_\alpha(n)}{n^s} \exp\Big(2\pi i n \frac{k}{\ell}\Big),$$

where $\sigma_\alpha(n)$ is the generalized divisor function given by $\sigma_\alpha(n) = \sum_{d|n} d^\alpha$ for any complex α (Garunkštis et al. [90]).
* General Dirichlet series, given by

$$\sum_{n=1}^{\infty} a_n \exp(-s\lambda_n),$$

satisfying certain conditions on the coefficients a_n and the exponents λ_n (Laurinčikas, Schwarz and Steuding [206], respectively, Genys and Laurinčikas [98]). Unfortunately, these conditions are quite restrictive.

For example, it is not known whether Selberg zeta-functions are universal or not.

All examples of strongly universal Dirichlet series do *not* have an Euler product and have *many* zeros in their region of universality; indeed, we shall see (in Chaps. 8, 10 and 11) that the property of approximating analytic functions with zeros is intimately related to the distribution of zeros of the Dirichlet series in question. In particular, Euler products for which the analogue of Riemann's hypothesis is expected to hold should not be capable of approximating functions with zeros.

Roughly speaking, there are two methods to prove universality. First, one can try to mimic Voronin's proof or Bagchi's probabilistic approach (here we do not distinguish between these two since they have a lot in common). This sounds more simple than it actually is, because one has to assure many analytic and arithmetic properties of the function in question. The second way is to find a representation as a linear combination or a product representation of jointly universal functions (we will use this argument in Chaps. 10–13). All known proofs of universality of the first type depend on a certain kind of *independence*. For instance, the logarithms of the prime numbers are linearly independent over $\mathbb{Q}$ (we used this property in the proof of Voronin's universality Theorem 1.7 when we applied Weyl's refinement of Kronecker's approximation theorem, Lemma 1.8). Another example are the numbers $\log(n + \alpha)$ with non-negative integral n which are linearly independent over $\mathbb{Q}$ if α is transcendental. In order to prove universality for the Hurwitz zeta-function (1.42), the first type of proof yields the result aimed at for transcendental α. If α is rational $\neq \frac{1}{2}, 1$, one can find a representation of $\zeta(s, \alpha)$ as a linear combination of non-equivalent Dirichlet L-functions for which we have the joint universality Theorem 1.10; in the cases $\alpha = \frac{1}{2}$ and $\alpha = 1$ the Hurwitz zeta-function has an Euler product representation and is equal to the Riemann zeta-function for $\alpha = 1$, resp., for $\alpha = \frac{1}{2}$,

$$\zeta\left(s, \frac{1}{2}\right) = 2^s L(s, \chi),$$

where χ is the unique character mod 2. In both cases the Hurwitz zeta-function is universal but does not have the strong universality property. It is an interesting open problem whether $\zeta(s, \alpha)$ *is universal or even strongly universal if α is algebraic irrational*. In this case neither of the two methods seem to applicable directly. Laurinčikas and Steuding [210, 211] obtained first limit theorems for Hurwitz zeta-functions with algebraic parameters, which could be a first step towards universality.

It was conjectured by Linnik and Ibragimov (cf. [207]) that *all functions given by Dirichlet series and analytically continuable to the left of the half-plane of absolute convergence are universal*. However, this has to be understood as a program and not as a conjecture. For example, put $a(n) = 1$ if $n = 2^k$ with $k \in \mathbb{N}_0 := \mathbb{N} \cup \{0\}$, and $a(n) = 0$ otherwise. Then

$$\sum_{n=1}^{\infty} \frac{a(n)}{n^s} = \sum_{k=0}^{\infty} \frac{1}{2^{ks}} = (1 - 2^{-s})^{-1},$$

and obviously, this function is far away from being universal. So we have to ask for *natural* conditions needed to prove universality. Dirichlet series having an infinite Euler product seem to be good candidates.

2

Dirichlet Series and Polynomial Euler Products

What is man in nature? Nothing in relation to the infinite, all in relation to nothing, a mean between nothing and everything.

Blaise Pascal

In this chapter, we introduce a class of Dirichlet series satisfying several quite natural analytic axioms in addition with two arithmetic conditions, namely, a polynomial Euler product representation and some kind of prime number theorem. The elements of this class will be the main actors in the sequel; however, for some of the later results we do not need to assume all of these axioms. Further, we shall prove mean-value estimates for the Dirichlet series coefficients of these L-functions as well as asymptotic mean-square formulae on vertical lines in the critical strip. These estimates will turn out to be rather useful in later chapters.

2.1 General Theory of Dirichlet Series

We start with a brief introduction to the general theory of ordinary Dirichlet series. For more details and proofs we refer to Titchmarsh [352, Chap. IX].

An ordinary Dirichlet series is a series of the form

$$A(s) = \sum_{n=1}^{\infty} \frac{a_n}{n^s}, \tag{2.1}$$

where the coefficients a_n are complex numbers and $s = \sigma + it$ is a complex variable. If such a Dirichlet series is convergent for $s = s_0$, then it is uniformly convergent throughout the angular region given by

$$|\arg(s - s_0)| \leq \frac{\pi}{2} - \delta,$$

where δ is any real number satisfying $0 < \delta < \frac{\pi}{2}$. Consequently, if a Dirichlet series converges, it converges in some half-plane $\sigma > \sigma_c$. Obviously, every term of the Dirichlet series (2.1) is analytic and any point s with $\sigma > \sigma_c$ is contained in a domain of uniform convergence. A well-known theorem of Weierstrass states that the limit of a uniformly convergent sequence of analytic functions is analytic. Hence, $\mathcal{A}(s)$ is an analytic function in its half-plane of convergence $\sigma > \sigma_c$. The abscissa of convergence is given by

$$\sigma_c = \limsup_{N \to \infty} \frac{\log \left| \sum_{n=1}^{N} a_n \right|}{\log N} \quad \text{or} \quad \sigma_c = \limsup_{N \to \infty} \frac{\log \left| \sum_{n=N+1}^{\infty} a_n \right|}{\log N}$$

according to whether $\sum_{n=1}^{\infty} a_n$ diverges or converges. Also the region of absolute convergence of a Dirichlet series is a half-plane. The abscissa σ_a of absolute convergence is given by

$$\sigma_a = \limsup_{N \to \infty} \frac{\log \left(\sum_{n=1}^{N} |a_n| \right)}{\log N} \quad \text{or} \quad \sigma_a = \limsup_{N \to \infty} \frac{\log \left(\sum_{n=N+1}^{\infty} |a_n| \right)}{\log N}$$

according to whether $\sum_{n=1}^{\infty} |a_n|$ is divergent or not. It is easily seen that $\sigma_a - \sigma_c \leq 1$ (with equality, for example, for Dirichlet L-functions). If all coefficients a_n are non-negative real numbers, then the real point $s = \sigma_c$ is a singularity of $\mathcal{A}(s)$ (e.g., the zeta-function). A function has at most one Dirichlet series representation, i.e., given two Dirichlet series

$$\mathcal{A}(s) = \sum_{n=1}^{\infty} \frac{a_n}{n^s} \quad \text{and} \quad \mathcal{B}(s) = \sum_{n=1}^{\infty} \frac{b_n}{n^s},$$

both absolutely convergent for $\sigma > \sigma_a$, if $\mathcal{A}(s) = \mathcal{B}(s)$ for each s in an infinite sequence $\{s_k\}$ such that $\operatorname{Re} s_k \to \infty$ as $k \to \infty$, then $a_n = b_n$ for all $n \in \mathbb{N}$. This uniqueness implies the existence of some zero-free half-plane $\sigma > \sigma_0$ with $\sigma_0 \geq \sigma_a$.

Assume that $\mathcal{A}(s)$ is analytic in some strip $\sigma_1 \leq \sigma \leq \sigma_2$ except for at most a finite number of poles. Then $\mathcal{A}(s)$ is said to be of finite order in this strip if there exists a positive constant c such that the estimate

$$\mathcal{A}(s) \ll |t|^c \qquad \text{as} \quad |t| \to \infty \tag{2.2}$$

holds uniformly for $\sigma_1 \leq \sigma \leq \sigma_2$; similarly, one defines the notion of finite order for half-planes $\sigma \geq \sigma_1$. Clearly, a function given by a Dirichlet series is of finite order in its half-plane of convergence. Given σ, define $\mu(\sigma)$ to be the lower bound of all c for which (2.2) holds; this quantity is called the order of $\mathcal{A}(s)$. One can show that the function $\mu(\sigma)$ is convex downwards (in particular, it is continuous). Moreover, $\mu(\sigma) = 0$ for $\sigma > \sigma_a$ and, conversely, the Dirichlet series (2.1) is convergent in the half-plane where $\mathcal{A}(s)$ is regular and $\mu(\sigma) = 0$. If σ_ϵ denotes the abscissa limiting the half-plane where $\mathcal{A}(s)$ is analytic with $\mu(\sigma) = 0$, then $\sigma_c \leq \sigma_\epsilon \leq \sigma_a$.

Now we shall consider the mean-square of $\mathcal{A}(s)$ on vertical lines. In the half-plane of absolute convergence for (2.1) it is easily shown that

$$\frac{1}{2T}\int_{-T}^{T}|\mathcal{A}(\sigma+\mathrm{i}t)|^2\,\mathrm{d}t \sim \sum_{n=1}^{\infty}\frac{|a(n)|^2}{n^{2\sigma}}. \tag{2.3}$$

The convergence of the series on the right-hand side is a matter of course. Carlson [50] proved that the mean-square of $\mathcal{A}(s)$ on vertical lines exists if the left-hand side is bounded as $T\to\infty$.

Theorem 2.1. *Let $\mathcal{A}(s)$ be given by (2.1). Assume that $\mathcal{A}(s)$ is analytic except for at most a finite number of poles, of finite order for $\sigma \geq \sigma_1$, and*

$$\limsup_{T\to\infty}\frac{1}{2T}\int_{-T}^{T}|\mathcal{A}(\sigma+\mathrm{i}t)|^2\,\mathrm{d}t < \infty.$$

Then,

$$\lim_{T\to\infty}\frac{1}{2T}\int_{-T}^{T}|\mathcal{A}(\sigma+\mathrm{i}t)|^2\,\mathrm{d}t = \sum_{n=1}^{\infty}\frac{|a_n|^2}{n^{2\sigma}}$$

for $\sigma > \sigma_1$, and uniformly in any strip $\sigma_1 + \epsilon \leq \sigma \leq \sigma_2$.

Carlson's theorem may be regarded as an analogue of Parseval's theorem for Fourier series. We omit the proof but refer to Carlson [50] and Titchmarsh [352].

In view of Carlson's theorem it makes sense to define a so-called mean-square half-plane by defining its abscissa σ_{m} as the infimum over all σ_1 such that (2.3) holds for all fixed $\sigma > \sigma_1$. One can show that

$$\sigma_{\mathrm{m}} \geq \max\left\{\sigma_{\mathrm{a}} - \frac{1}{2}, \sigma_c\right\} \tag{2.4}$$

and that $\mu(\sigma) \leq \frac{1}{2}$ for any $\sigma > \sigma_{\mathrm{m}}$. In our applications we will normalize the Dirichlet series to be absolutely convergent for $\sigma > \sigma_{\mathrm{a}} = 1$. Inequality (2.4) implies that (at least with present day methods) the strip of universality is at most $\frac{1}{2} < \sigma < 1$; it seems reasonable that we cannot approximate analytic functions uniformly on sets having a non-empty intersection with the half-plane $\sigma \leq \frac{1}{2}$.

2.2 A Class of Dirichlet Series: The Main Actors

Now we introduce several axioms for Dirichlet series; some of them might look a bit technical on first sight but they have been proved to be rather useful in the analytic theory of Dirichlet series.

In the sequel we assume that a function $\mathcal{L}(s)$ has a representation as a Dirichlet series in some half-plane:

$$\mathcal{L}(s) = \sum_{n=1}^{\infty}\frac{a(n)}{n^s}. \tag{2.5}$$

First of all, we pose a hypothesis on the size of the Dirichlet series coefficients.

(i) *Ramanujan hypothesis.* $a(n) \ll n^\epsilon$ for any $\epsilon > 0$, where the implicit constant may depend on ϵ.

The Ramanujan hypothesis implies the absolute convergence of the Dirichlet series (2.5) in the half-plane $\sigma > 1$, and uniform convergence in every compact subset. Thus, $\mathcal{L}(s)$ is analytic for $\sigma > 1$ and it makes sense to ask for analytic continuation.

(ii) *Analytic continuation.* There exists a real number $\sigma_\mathcal{L}$ such that $\mathcal{L}(s)$ has an analytic continuation to the half-plane $\sigma > \sigma_\mathcal{L}$ with $\sigma_\mathcal{L} < 1$ except for at most a pole at $s = 1$.

Actually, we could allow a finite number of poles on the line $\sigma = 1$, however, we do not loose too much by the restriction above.

The next axiom excludes functions which have an extraordinary large order of growth; Dirichlet series satisfying a functional equation are known to grow moderately slowly (by the Phragmén–Lindelöf principle; see Sect. 6.5).

(iii) *Finite order.* There exists a constant $\mu_\mathcal{L} \geq 0$ such that, for any fixed $\sigma > \sigma_\mathcal{L}$ and any $\epsilon > 0$,

$$\mathcal{L}(\sigma + \mathrm{i}t) \ll |t|^{\mu_\mathcal{L} + \epsilon} \quad \text{as} \quad |t| \to \infty;$$

again the implicit constant may depend on ϵ.

The final two axioms are of purely arithmetical nature. Both are rather restrictive; however, they fit well to what we think an L-function should look like. All known examples of L-functions of number theoretical interest are automorphic or at least conjecturally automorphic L-functions, and for all of them it turns out that the local Euler factors for prime p are the inverse of a polynomial.

(iv) *Polynomial Euler product.* There exists a positive integer m and for every prime p, there are complex numbers $\alpha_j(p)$, $1 \leq j \leq m$, such that

$$\mathcal{L}(s) = \prod_p \prod_{j=1}^m \left(1 - \frac{\alpha_j(p)}{p^s} \right)^{-1}.$$

The $\alpha_j(p)$ are called local roots at p. This axiom implies that the Dirichlet series coefficients $a(n)$ are multiplicative (this will be proved in Lemma 2.2). The Ramanujan hypothesis (i), in addition with the polynomial Euler product, implies that the numbers $\alpha_j(p)$ have absolute value less than or equal to one (see Lemma 2.2).

The last axiom may be regarded as some kind of prime number theorem for the coefficients of the Euler product:

(v) *Prime mean-square.* There exists a positive constant κ such that

$$\lim_{x \to \infty} \frac{1}{\pi(x)} \sum_{p \leq x} |a(p)|^2 = \kappa.$$

This asymptotic formula is intimately related to the deep Selberg conjectures for L-functions in the Selberg class (see Chap. 6) and it is known to be satisfied in many cases. This last axiom implies that there are infinitely many primes p for which not all numbers $\alpha_j(p)$ vanish. Consequently, the Euler product is infinite (and it makes sense to ask for universality).

In the sequel we shall not always use all of these axioms; however, we denote the class of Dirichlet series (2.5) satisfying all axioms (i)–(v) by $\tilde{\mathcal{S}}$. In Lemma 2.2 we will show that (iv) $\Rightarrow$ (i) and thus the assumption of the Ramanujan hypothesis is superfluous for the definition of the class $\tilde{\mathcal{S}}$; however, besides $\tilde{\mathcal{S}}$ we shall also consider other classes of Dirichlet series where we assume (i) but not (iv). The class $\tilde{\mathcal{S}}$ is not empty. Obviously, Riemann's zeta-function $\zeta(s)$ is an element of $\tilde{\mathcal{S}}$. Further examples are Dirichlet L-functions, Dedekind zeta-functions, and many more. As a matter of fact, $\tilde{\mathcal{S}}$ is a subset of the large class of Matsumoto zeta-functions, introduced by Matsumoto [239]. The restrictions going beyond (in particular, axioms (ii) and (v)) will turn out to be rather important in the sequel. As already noted, they are motivated by naturally appearing L-functions in number theory.

We give a simple example of an element in $\tilde{\mathcal{S}}$. Assume that f is a newform of weight k to some congruence subgroup $\Gamma_0(N)$ with Fourier expansion (1.32). Writing

$$a(n) = c(n)n^{\frac{1-k}{2}}$$

we find via (1.36) the Euler product representation

$$L\left(s + \frac{k-1}{2}, f\right) = \prod_p \left(1 - \frac{a(p)}{p^s} + \frac{\chi(p)}{p^{2s}}\right)^{-1},$$

where $\chi(p) = 0$ if $p \mid N$, and $\chi(p) = 1$ otherwise. In the latter case, i.e., $p \mid N$, then the corresponding Euler factor can be rewritten as

$$\left(1 - \frac{a(p)}{p^s} + \frac{1}{p^{2s}}\right)^{-1} = \left(1 - \frac{\alpha_1(p)}{p^s}\right)^{-1}\left(1 - \frac{\alpha_2(p)}{p^s}\right)^{-1},$$

where $\alpha_1(p), \alpha_2(p)$ are complex numbers satisfying

$$\alpha_1(p) + \alpha_2(p) = a(p) \quad \text{and} \quad \alpha_1(p)\alpha_2(p) = 1.$$

Deligne's estimate (1.34) translates into

$$|\alpha_1(p)| = |\alpha_2(p)| = 1, \quad \text{i.e.,} \quad \alpha_1(p) = \overline{\alpha_2(p)}.$$

The transformation $s \mapsto s + \frac{k-1}{2}$ maps the critical strip to $0 \le \sigma \le 1$ (independent of the weight). In the sequel we shall assume that L-functions to modular forms are normalized in this way, and we denote them again by $L(s, f)$. Note that the asymptotic formula (1.37) yields the prime number theorem for the coefficients of $L(s, f)$. It follows that L-functions to newforms are in our class $\tilde{\mathcal{S}}$.

2.3 Estimates for the Dirichlet Series Coefficients

Mean-value estimates are very useful tools in analytic number theory. They give information on the size of arithmetical objects. Now we shall prove some elementary estimates for the coefficients $a(n)$ in the Dirichlet series expansion (2.5). First of all, we study the relation between the Ramanujan hypothesis and the size of the local roots of a polynomial Euler product.

Lemma 2.2. *Suppose that $\mathcal{L}(s)$ is given by (2.5) and satisfies axiom (iv). Then $a(n)$ is multiplicative and*

$$a(n) = \prod_{p|n} \sum_{\substack{k_1,\ldots,k_m \geq 0 \\ k_1+\ldots+k_m=\nu(n;p)}} \prod_{j=1}^{m} \alpha_j(p)^{k_j}, \tag{2.6}$$

where $\nu(n;p)$ is the exponent of the prime p in the prime factorization of the integer n. Moreover, if $|\alpha_j(p)| \leq 1$ for $1 \leq j \leq m$ and all primes p, then $a(n) \ll n^\epsilon$ for any $\epsilon > 0$, and vice versa.

Thus, a polynomial Euler product of the form (iv) with fixed degree m together with the Ramanujan hypothesis (i) is equivalent to $|\alpha_j(p)| \leq 1$ for $1 \leq j \leq m$ and all primes p.

Proof. Taking into account the shape of the Euler product (iv) we have the identity

$$\sum_{n=1}^{\infty} \frac{a(n)}{n^s} = \prod_p \prod_{j=1}^{m} \left(1 - \frac{\alpha_j(p)}{p^s}\right)^{-1} = \prod_p \prod_{j=1}^{m} \left(1 + \sum_{k=1}^{\infty} \frac{\alpha_j(p)^k}{p^{ks}}\right),$$

valid for sufficiently large σ. This implies (2.6) and that $a(n)$ is multiplicative.

Now we show that the estimate for the local roots implies the Ramanujan hypothesis. If $|\alpha_j(p)| \leq 1$, then we get

$$|a(n)| \leq \prod_{p|n} \sum_{\substack{k_1,\ldots,k_m \geq 0 \\ k_1+\ldots+k_m=\nu(n;p)}} 1 = d_m(n), \tag{2.7}$$

say. The generalized divisor function $d_m(n)$ appears as Dirichlet coefficients in the representation

$$\zeta(s)^m = \prod_p \left(1 - \frac{1}{p^s}\right)^{-m} = \sum_{n=1}^{\infty} \frac{d_m(n)}{n^s},$$

and is therefore multiplicative (by the same reasoning as for $a(n)$). For p prime and $\nu \in \mathbb{N}$, we have

$$\begin{aligned}
d_m(p^\nu) &= \sharp\{(k_1,\ldots,k_m) \in \mathbb{N}_0^m : k_1 + \ldots + k_m = \nu\} \\
&= \binom{m+\nu-1}{\nu},
\end{aligned} \tag{2.8}$$

and thus

$$d_m(n) = \prod_p \binom{m + \nu(n;p) - 1}{\nu(n;p)}.$$

The function $d_m(n)$ is the m-fold Dirichlet convolution of the function constant 1: writing

$$(f * g)(n) = \sum_{b|n} f(b)g(n/b),$$

we have $d_k = d_{k-1} * 1$. Thus,

$$d_m(n) = \sum_{b|n} d_{m-1}(b) \leq \sum_{b|n} d_{m-1}(n) \leq d_{m-1}(n)d(n),$$

where, $d(n) = d_2(n) = \sum_{d|n} 1$; here the first inequality follows from (2.8). Hence, for $m \geq 2$,

$$d_m(n) \leq d(n)^{m-1}. \tag{2.9}$$

It remains to estimate the divisor function $d(n)$. We start with

$$d(n) \leq \prod_p (1 + \nu(n;p)) \leq \prod_{p \leq x} (1 + \nu(n;p)) \prod_{p > x} 2^{\nu(n;p)}$$

$$\leq \left(1 + \frac{\log n}{\log 2}\right)^x \left(\prod_p p^{\nu(n;p)}\right)^{\log 2/\log n}.$$

Let $n \geq e^3$. Taking $x = \log n (\log \log n)^{-3}$, we obtain

$$d(n) \leq \exp\left(\log 2 \frac{\log n}{\log \log n}\left(1 + O\left(\frac{\log \log \log n}{\log \log n}\right)\right)\right). \tag{2.10}$$

Hence, we obtain

$$|a(n)| < \exp\left((1 + \epsilon)(m - 1)\log 2 \frac{\log n}{\log \log n}\right) \tag{2.11}$$

for any $\epsilon > 0$ and sufficiently large n. This proves that the Ramanujan hypothesis (i) holds provided $|\alpha_j(p)| \leq 1$.

Now assume that the Ramanujan hypothesis (i) is true. Consider the identity between the Dirichlet series and the Euler product representation, respectively, the power series expansion

$$\sum_{k=1}^{\infty} a(p^k) X^k = \prod_{j=1}^{m} (1 - \alpha_j(p)X)^{-1};$$

by the Ramanujan hypothesis, the series is analytic for $|X| < 1$, hence so is the right-hand side which gives the desired bound. This concludes the proof of the lemma.   □

Estimate (2.10) is best possible. To see this note that

$$\limsup_{n \to \infty} \frac{d(n) \log \log n}{\log n} = \log 2$$

(a proof can be found in [120, Sect. 18.1]). Estimate (2.11) is near to be best possible since, if $d(n)$ is replaced by $d_m(n)$, then the latter formula holds with the right-hand side replaced by $\log m$ (for a proof, see [255]).

For our studies on the value-distribution of functions in $\tilde{\mathcal{S}}$ we need to prove several mean-square formulae. It follows immediately from the Ramanujan hypothesis (i) that

$$\sum_{n \leq x} |a(n)|^2 \ll x^{1+\epsilon}. \tag{2.12}$$

This rough bound is sufficient for many applications. However, we can do a bit better by a standard argument. Recall that m is the degree of the polynomial in p^{-s} in the local Euler factor of $\mathcal{L}$, independent of p.

Lemma 2.3. *Suppose $\mathcal{L}(s)$ satisfies axioms (i) and (iv). Then, as $x \to \infty$,*

$$\sum_{n \leq x} |a(n)|^2 \ll x(\log x)^{m^2-1}.$$

Proof. In view of (2.7) it suffices to find a mean-square estimate for $d_m(n)$. By the multiplicativity of $d_m(n)$,

$$d_m(n)^2 = \sum_{d|n} g(d)$$

with some multiplicative function g. By (2.8), we find

$$g(1) = d_m(1)^2 = 1, \quad g(p) = d_m(p)^2 - d_m(1)^2 = m^2 - 1,$$

and by induction

$$g(p^\nu) = d_m(p^\nu)^2 - d_m(p^{\nu-1})^2 \sim \frac{m^{2\nu}}{\nu!},$$

as $\nu \to \infty$. Hence, we obtain

$$\sum_{n \leq x} d_m(n)^2 = \sum_{d \leq x} \sum_{\substack{n \leq x \\ n \equiv 0 \bmod d}} g(d) \leq x \sum_{d \leq x} \frac{g(d)}{d} \leq x \prod_{p \leq x} \left(1 + \sum_{\nu=1}^{\infty} \frac{g(p^\nu)}{p^\nu} \right).$$

In view of the above estimates the right-hand side above equals asymptotically

$$x \prod_{p \leq x} \left(1 + \frac{m^2-1}{p} + \sum_{\nu=2}^{\infty} \frac{m^{2\nu}}{\nu! p^\nu} \right) = x \prod_{p \leq x} \left(1 + \frac{m^2-1}{p} \right) + O(x).$$

A classic estimate due to Mertens,

$$\prod_{p \leq x} \left(1 + \frac{1}{p}\right) \ll \prod_{p \leq x} \left(1 - \frac{1}{p}\right)^{-1} \ll \log x$$

(see [120, Sect. 22.8]), in combination with

$$1 + \frac{m^2 - 1}{p} \leq \left(1 + \frac{1}{p}\right)^{m^2 - 1},$$

gives the estimate of the lemma. □

2.4 The Mean-Square on Vertical Lines

Our next aim is to derive an asymptotic mean-square formulae for $\mathcal{L}(s)$ on some vertical lines to the left of the abscissa of absolute convergence $\sigma = 1$.

Theorem 2.4. *Let $\mathcal{L}(s)$ satisfy the axioms (i)–(iii). Then, for any σ satisfying*

$$\sigma > \max\left\{\frac{1}{2}, 1 - \frac{1 - \sigma_\mathcal{L}}{1 + 2\mu_\mathcal{L}}\right\}, \tag{2.13}$$

we have

$$\lim_{T \to \infty} \frac{1}{2T} \int_{-T}^{T} |\mathcal{L}(\sigma + it)|^2 \, dt = \sum_{n=1}^{\infty} \frac{|a(n)|^2}{n^{2\sigma}}.$$

In particular, the mean-square half-plane of $\mathcal{L}(s)$ has a non-empty intersection with the strip $\frac{1}{2} < \sigma < 1$.

This result is essentially due to Carlson [50, 51]; we follow closely Titchmarsh's proof of the special case of the Riemann zeta-function in [353, Sect. 7.9].

Proof. With regard to Carlson's Theorem 2.1 it suffices to show that

$$\int_{-T}^{T} |\mathcal{L}(\sigma + it)|^2 \, dt \ll T \tag{2.14}$$

for σ satisfying (2.13). For this aim we consider for $\delta > 0$ the Dirichlet series

$$\sum_{n=1}^{\infty} \frac{a(n)}{n^s} \exp(-n\delta).$$

Since $a(n) \ll n^\epsilon$ by (i), this series converges absolutely in the whole complex plane and uniformly in any compact subset. Let $\sigma > 1$ and $c > \sigma$ be a constant. By Mellin's inversion formula,

$$\exp(-\alpha) = \frac{1}{2\pi i} \int_{\beta - i\infty}^{\beta + i\infty} \Gamma(z) \alpha^{-z} \, dz, \tag{2.15}$$

valid for positive real numbers α and β, we get

$$\sum_{n=1}^{\infty} \frac{a(n)}{n^s} \exp(-n\delta) = \sum_{n=1}^{\infty} \frac{a(n)}{n^s} \frac{1}{2\pi i} \int_{c-\sigma-i\infty}^{c-\sigma+i\infty} \Gamma(z)(n\delta)^{-z} \, dz.$$

Hence,

$$\sum_{n=1}^{\infty} \frac{a(n)}{n^s} \exp(-n\delta) = \frac{1}{2\pi i} \int_{c-i\infty}^{c+i\infty} \Gamma(z-s)\mathcal{L}(z)\delta^{s-z} \, dz. \tag{2.16}$$

Here, interchanging summation and integration is allowed in view of the absolute convergence of the Dirichlet series defining $\mathcal{L}(s)$ and Stirling's formula,

$$\log \Gamma(s) = \left(s - \frac{1}{2}\right)\log s - s + \frac{1}{2}\log 2\pi + O\left(\frac{1}{|s|}\right), \tag{2.17}$$

valid for $|\arg s| \leq \pi - \epsilon$ and $|s| \geq 1$, uniformly in σ, as $|s| \to \infty$. By (iii),

$$\mathcal{L}(c + i\tau) \ll |\tau|^{\mu_{\mathcal{L}}+\epsilon} \tag{2.18}$$

for $\sigma > \sigma_{\mathcal{L}}$ as $|\tau| \to \infty$. This and Stirling's formula imply

$$\int_{c-i\infty}^{c+i\infty} \Gamma(z-s)\mathcal{L}(z)\delta^{s-z} \, dz$$

$$\ll \delta^{\sigma-c} \int_{-\infty}^{+\infty} (1 + |\tau|^{\mu_{\mathcal{L}}+\epsilon+c-\sigma-\frac{1}{2}}) \exp\left(-\frac{\pi|\tau|}{2}\right) \, d\tau.$$

Hence, the function on the right-hand side of (2.16) is an analytic function of s in the half-plane $\sigma > \sigma_{\mathcal{L}}$ for any $c > \sigma_{\mathcal{L}}$. Therefore, let $\sigma > \max\{\frac{1}{2}, \sigma_{\mathcal{L}}\}$ and choose $\sigma_1 := \operatorname{Re} z$ such that $\max\{\sigma - 1, \sigma_{\mathcal{L}}\} < \sigma_1 < 1$. We move the path of integration in (2.16) to the line $\operatorname{Re} z = \sigma_1$. We pass the simple pole of $\Gamma(z-s)$ at $z = s$ with residue $\mathcal{L}(z)$. Moreover, by (ii), $\mathcal{L}(z)$ has at most a pole at $z = 1$ of order j, say. The residue coming from the hypothetical pole of $\mathcal{L}(z)$ is a finite sum of terms of the form

$$c\Gamma^{(u)}(s-1)\delta^{s-1}(\log \delta)^v,$$

where c is a certain constant and u and v are non-negative integers with $u + v = j$; by Stirling's formula (2.17), these terms are

$$\ll \delta^{\sigma+\epsilon-1} \exp(-BT),$$

where B is a positive absolute constant. Now let $\delta > |T|^{-B}$. Then, for $\delta > |t|^{-B}$ we deduce from (2.16) by the calculus of residues

$$\mathcal{L}(s) = \sum_{n=1}^{\infty} \frac{a(n)}{n^s} \exp(-n\delta) - \frac{1}{2\pi i} \int_{\sigma_1-i\infty}^{\sigma_1+i\infty} \Gamma(z-s)\mathcal{L}(z)\delta^{s-z} \, dz$$

$$+ O(\exp(-B|t|)). \tag{2.19}$$

Now we integrate the square of this expression with respect to $t = \operatorname{Im} s$ in the range $\frac{1}{2}T \le t \le T$. Note that

$$|\mathcal{L}(\sigma + it)|^2 \ll \left| \sum_{n=1}^{\infty} \frac{a(n)}{n^s} \exp(-n\delta) \right|^2 + \left| \int_{\sigma_1 - i\infty}^{\sigma_1 + i\infty} \Gamma(z - s)\mathcal{L}(z)\delta^{s-z}\,dz \right|^2$$
$$+ O(\exp(-2B|t|)).$$

We find

$$\int_{\frac{1}{2}T}^{T} \left| \sum_{n=1}^{\infty} \frac{a(n)}{n^s} \exp(-n\delta) \right|^2 \, dt$$

$$\ll T \sum_{n=1}^{\infty} \frac{|a(n)|^2}{n^{2\sigma}} \exp(-2n\delta) + \sum_{m=1}^{\infty} \sum_{\substack{n=1 \\ n \ne m}}^{\infty} \frac{a(m)\overline{a(n)}\exp(-(m+n)\delta)}{(mn)^\sigma |\log \frac{m}{n}|}.$$

By the Ramanujan hypothesis (i), the first series on the right-hand side is convergent independent of $\delta > 0$. The double series is bounded by

$$\sum_{n=1}^{\infty} \sum_{m < n} \frac{a(m)\overline{a(n)}\exp(-(m+n)\delta)}{(mn)^\sigma \log \frac{n}{m}}.$$

We split the summation and apply the Ramanujan hypothesis (i) with a suitably small $\epsilon > 0$ to obtain the upper bound

$$\sum_{n=1}^{\infty} \left\{ \sum_{m < \frac{n}{2}} + \sum_{\frac{n}{2} \le m < n} \right\} \frac{\exp(-(m+n)\delta)}{(mn)^{\sigma - \epsilon} \log \frac{n}{m}}.$$

In the first sum, we have $\log \frac{n}{m} > \log 2$, and so it is bounded by

$$\left(\sum_{n=1}^{\infty} \frac{\exp(-n\delta)}{n^{\sigma - \epsilon}} \right)^2 \ll \left(\int_0^{\infty} u^{\epsilon - \sigma} \exp(-u\delta)\,du \right)^2 \ll \delta^{2\sigma - 2 - \epsilon};$$

recall that ϵ denotes an arbitrary small positive quantity, not necessarily the same at each appearance. In the second sum we write $m = n - r$ with $1 \le r \le \frac{n}{2}$. Then $\log \frac{n}{m} = -\log(1 - \frac{r}{n}) > \frac{r}{n}$, and we get in a similar manner the upper bound

$$\sum_{n=1}^{\infty} n^{1 + 2(\epsilon - \sigma)} \exp(-n\delta) \sum_{r \le \frac{n}{2}} \frac{1}{r} \ll \delta^{2\sigma - 2 - \epsilon} \log \frac{1}{\delta}.$$

Hence,

$$\sum_{n=1}^{\infty} \sum_{m < n} \frac{a(m)\overline{a(n)}\exp(-(m+n)\delta)}{(mn)^\sigma \log \frac{n}{m}} \ll \delta^{2\sigma - 2 - \epsilon}.$$

Thus,

$$\int_{\frac{1}{2}T}^{T} \left| \sum_{n=1}^{\infty} \frac{a(n)}{n^s} \exp(-n\delta) \right|^2 dt \ll T + \delta^{2\sigma-2-\epsilon}. \tag{2.20}$$

Let $z = \sigma_1 + i\tau$. Then

$$\int_{\sigma_1-i\infty}^{\sigma_1+i\infty} \Gamma(z-s)\mathcal{L}(z)\delta^{s-z}\, dz \ll \delta^{\sigma-\sigma_1} \int_{-\infty}^{\infty} |\Gamma(z-s)\mathcal{L}(z)|\, d\tau.$$

By the Cauchy–Schwarz inequality this is bounded by

$$\delta^{\sigma-\sigma_1} \left(\int_{-\infty}^{\infty} |\Gamma(z-s)|\, d\tau \cdot \int_{-\infty}^{\infty} |\Gamma(z-s)\mathcal{L}(z)^2|\, d\tau \right)^{1/2}.$$

By Stirling's formula (2.17) the first integral is bounded. For the second one we additionally apply (2.18) and obtain, for $|t| \leq T$,

$$\left\{ \int_{-\infty}^{-2T} + \int_{2T}^{\infty} \right\} |\Gamma(z-s)\mathcal{L}(z)^2|\, d\tau$$

$$\ll \left\{ \int_{-\infty}^{-2T} + \int_{2T}^{\infty} \right\} |\tau|^{2\mu_{\mathcal{L}}+\epsilon+\sigma_1-\sigma-(1/2)} \exp\left(-\frac{\pi|\tau|}{4} \right) d\tau \ll \exp(-BT)$$

with some absolute positive constant B (since $|t-\tau| \geq \frac{1}{2}|\tau|$). Consequently,

$$\int_{\frac{1}{2}T}^{T} \left| \int_{\sigma_1-i\infty}^{\sigma_1+i\infty} \Gamma(z-s)\mathcal{L}(z)\delta^{s-z}\, dz \right|^2 dt$$

$$\ll \delta^{2\sigma-2\sigma_1} \int_{-2T}^{2T} \left(\int_{\frac{1}{2}T}^{T} |\Gamma(z-s)|\, dt \right) |\mathcal{L}(\sigma_1+i\tau)|^2\, d\tau + \delta^{2\sigma-2\sigma_1}$$

$$\ll \delta^{2\sigma-2\sigma_1} \int_{-2T}^{2T} |\mathcal{L}(\sigma_1+i\tau)|^2\, d\tau.$$

By the trivial estimate

$$\int_{-2T}^{2T} |\mathcal{L}(\sigma_1+i\tau)|^2\, d\tau \ll T^{1+2\mu_{\mathcal{L}}+\epsilon},$$

we get

$$\int_{\frac{1}{2}T}^{T} \left| \int_{\sigma_1-i\infty}^{\sigma_1+i\infty} \Gamma(z-s)\mathcal{L}(z)\delta^{s-z}\, dz \right|^2 dt \ll \delta^{2\sigma-2\sigma_1} T^{1+2\mu_{\mathcal{L}}+\epsilon}.$$

This, (2.19), and (2.20) give

$$\int_{\frac{1}{2}T}^{T} |\mathcal{L}(\sigma+it)|^2\, dt \ll T + \delta^{2\sigma-2-\epsilon} + \delta^{2\sigma-2\sigma_1} T^{1+2\mu_{\mathcal{L}}+\epsilon}. \tag{2.21}$$

Using this with $2^{1-n}T$ instead of T and summing up over all $n \in \mathbb{N}$, we find that the same bound holds for the integral with limits 1 and T. The same reasoning applies to the line segment $[-T, -1]$. Thus, putting

$$\delta = T^{-\frac{1+2\mu_{\mathcal{L}}+\epsilon}{2(1-\sigma_1)}},$$

we find, after a short computation, that all terms of the right-hand side of (2.21) are $\ll T$ if

$$\sigma > \max\left\{\frac{1}{2}, \sigma_{\mathcal{L}}, 1 - \frac{1-\sigma_1}{1+2\mu_{\mathcal{L}}+\epsilon}\right\}.$$

Sending $\epsilon \to 0$ and since $\sigma_1 > \max\{\frac{1}{2}, \sigma_{\mathcal{L}}\}$ can be chosen arbitrarily, we find that estimate (2.14) holds for the range (2.13). This proves the theorem. □

We give an example. Using (1.3) with $N = t$, we find for the zeta-function $\mu_\zeta = \frac{1}{2}$ for $\sigma_\zeta = \frac{1}{2}$ which leads to the existence of the mean-square for $\sigma > \frac{3}{4}$. This is far from being optimal. As a matter of fact, the mean-square exists for $\sigma > \frac{1}{2}$ but we may not forget that here we did not use any deeper property of the zeta-function (only the rough estimate via (1.3)). In Sect. 6.5, we will significantly improve the range (2.13) of Theorem 2.4 for L-functions satisfying a Riemann-type functional equation.

Interlude: Results from Probability Theory

Primes play a game of chance.

M. Kac

In this chapter, we briefly present facts from probability theory which will be used later. These results can be found in the monographs of Billingsley [21, 22], Buldygin [45], Cramér and Leadbetter [64], Heyer [133], Laurinčikas [186], and Loève [226]. However, there are two exceptions in this crash course in probability theory. In Sect. 3.3 we present Denjoy's heuristic probabilistic argument for the truth of Riemann's hypothesis. Finally, in Sect. 3.7, we introduce the universe for our later studies on universality, the space of analytic functions, and state some of its properties, following Conway [62] and Laurinčikas [186].

3.1 Weak Convergence of Probability Measures

The notion of weak convergence of probability measures is a useful tool in investigations on the value-distribution of Dirichlet series. This powerful theory was initiated by Kolmogorov, Erdös and Kac and further developed by Doob, Prokhorov, Skorokhod and others. To present the main properties of weakly convergent probability measures we have to introduce the concept of σ-field and the axiomatic setting of probability measures.

Let Ω be a non-empty set. By $\mathcal{P}(\Omega)$ we denote the set of all subsets of Ω. A subset $\mathcal{F}$ of $\mathcal{P}(\Omega)$ is called a field (or algebra) if it satisfies the following axioms:

- $\emptyset, \Omega \in \mathcal{F}$;
- $A^c \in \mathcal{F}$ for $A \in \mathcal{F}$, where A^c denotes the complement of A;
- $\mathcal{F}$ is closed under finite unions and finite intersections, i.e., if $A_1, \ldots, A_n \in \mathcal{F}$, then

$$\bigcup_{j=1}^n A_j \in \mathcal{F} \quad \text{and} \quad \bigcap_{j=1}^n A_j \in \mathcal{F}.$$

$\mathcal{F}$ is called a σ-field (or σ-algebra) if it satisfies the first two axioms above in addition with

- $\mathcal{F}$ is closed under countable unions and countable intersections, i.e., if $\{A_j\}$ is a countable sequence of events in $\mathcal{F}$, then

$$\bigcup_{j=1}^{\infty} A_j \in \mathcal{F} \quad \text{and} \quad \bigcap_{j=1}^{\infty} A_j \in \mathcal{F}.$$

For $\mathcal{C} \subset \mathcal{P}(\Omega)$ we denote by $\sigma(\mathcal{C})$ the smallest σ-field containing $\mathcal{C}$. This σ-field is said to be generated by $\mathcal{C}$.

A non-negative function $\mathbf{P}$ defined on a σ-field $\mathcal{F}$ with the properties:

- $\mathbf{P}(\emptyset) = 0$ and $\mathbf{P}(\Omega) = 1$;
- For every countable sequence $\{A_j\}$ of pairwise disjoint elements of $\mathcal{F}$,

$$\mathbf{P}\left(\bigcup_{j=1}^{\infty} A_j\right) = \sum_{j=1}^{\infty} \mathbf{P}(A_j),$$

is called a probability measure. The triple $(\Omega, \mathcal{F}, \mathbf{P})$ is said to be a probability space. This setting of probability dates back to Kolmogorov who introduced it in the early 1930s.

Let S be a topological space and let $\mathcal{B}(S)$ denote the class of Borel sets of S, i.e., the σ-field generated by the system of all open subsets of the space S. Then each measure on $\mathcal{B}(S)$ is called Borel measure. Usually, we consider probability measures defined on the Borel sets $\mathcal{B}(S)$ of some metric space S. A class $\mathcal{A}$ of sets of S is said to be a determining class (also separating class) in case the measures $\mathbf{P}$ and $\mathbf{Q}$ on $(S, \mathcal{B}(S))$ coincide on the whole of S when $\mathbf{P}(A) = \mathbf{Q}(A)$ for all $A \in \mathcal{A}$.

Given two probability measures $\mathbf{P}_1$ and $\mathbf{P}_2$ on $(S_1, \mathcal{B}(S_1))$ and $(S_2, \mathcal{B}(S_2))$, respectively, there exists a unique measure $\mathbf{P}_1 \times \mathbf{P}_2$ such that

$$(\mathbf{P}_1 \times \mathbf{P}_2)(A_1 \times A_2) = \mathbf{P}_1(A_1)\mathbf{P}_2(A_2)$$

for $A_j \in \mathcal{B}(S_j)$. This measure is a probability measure on $(S, \mathcal{B}(S))$, where $S = S_1 \times S_2$ and $\mathcal{B}(S) = \mathcal{B}(S_1) \times \mathcal{B}(S_2)$, and is said to be the product measure of the measures $\mathbf{P}_1$ and $\mathbf{P}_2$.

In the sequel let $\mathbf{P}_n$ and $\mathbf{P}$ be probability measures on $(S, \mathcal{B}(S))$. We say that $\mathbf{P}_n$ converges weakly to $\mathbf{P}$ as n tends to infinity, and write $\mathbf{P}_n \Rightarrow \mathbf{P}$, if for all bounded continuous functions $f : S \to \mathbb{R}$

$$\lim_{n \to \infty} \int_S f \, d\mathbf{P}_n = \int_S f \, d\mathbf{P}.$$

Since the integrals on the right-hand side completely determine $\mathbf{P}$ (which is a consequence of Lebesgue's dominated convergence theorem), the sequence

$\{\mathbf{P}_n\}$ cannot converge weakly to two different limits at the same time. Further, note that weak convergence depends only on the topology of the underlying space S, not on the metric that generates it.

A set A in S whose boundary ∂A satisfies $\mathbf{P}(\partial A) = 0$ is called a continuity set of $\mathbf{P}$. The Portmanteau theorem provides useful conditions equivalent to weak convergence.

Theorem 3.1. *Let $\mathbf{P}_n$ and $\mathbf{P}$ be probability measures on $(S, \mathcal{B}(S))$. Then the following assertions are equivalent:*

- $\mathbf{P}_n \Rightarrow \mathbf{P}$;
- *For all open sets G,*

$$\liminf_{n\to\infty} \mathbf{P}_n(G) \geq \mathbf{P}(G);$$

- *For all continuity sets A of $\mathbf{P}$,*

$$\lim_{n\to\infty} \mathbf{P}_n(A) = \mathbf{P}(A).$$

This theorem is part of Theorem 2.1 in Billingsley [21]. Next we state a useful criterion for weak convergence.

Lemma 3.2. *We have $\mathbf{P}_n \Rightarrow \mathbf{P}$ if and only if every subsequence $\{\mathbf{P}_{n_k}\}$ contains a subsequence $\{\mathbf{P}_{n_{k_j}}\}$ such that $\mathbf{P}_{n_{k_j}} \Rightarrow \mathbf{P}$.*

This is Theorem 2.3 from Billingsley [21].

Now we consider continuous mappings between metric spaces S_1 and S_2. A function $h : S_1 \to S_2$ is said to be measurable if

$$h^{-1}(\mathcal{B}(S_2)) \subset \mathcal{B}(S_1).$$

Let $h : S_1 \to S_2$ be a measurable function. Then every probability measure $\mathbf{P}$ on $(S_1, \mathcal{B}(S_1))$ induces a probability measure $\mathbf{P}h^{-1}$ on $(S_2, \mathcal{B}(S_2))$ defined by

$$(\mathbf{P}h^{-1})(A) = \mathbf{P} \circ h^{-1}(A) = \mathbf{P}(h^{-1}(A)),$$

where $A \in \mathcal{B}(S_2)$. This measure is uniquely determined. A function $h : S_1 \to S_2$ is continuous if for every open set $G_2 \subset S_2$ the set $h^{-1}(G_2)$ is open in S_1. Continuous mappings transport the property of weak convergence.

Theorem 3.3. *Let $h : S_1 \to S_2$ be a continuous function. If $\mathbf{P}_n \Rightarrow \mathbf{P}$, then also $\mathbf{P}_n h^{-1} \Rightarrow \mathbf{P}h^{-1}$.*

This theorem is a particular case of Theorem 5.1 from Billingsley [21].

A family $\{\mathbf{P}_n\}$ of probability measures on $(S, \mathcal{B}(S))$ is said to be relatively compact if every sequence of elements of $\{\mathbf{P}_n\}$ contains a weakly convergent subsequence. A family $\{\mathbf{P}_n\}$ is called tight if for arbitrary $\varepsilon > 0$ there exists a compact set K such that $\mathbf{P}(K) > 1 - \varepsilon$ for all $\mathbf{P}$ from $\{\mathbf{P}_n\}$. Prokhorov's theorem is a powerful tool in the theory of weak convergence of probability measures; it is given below as Theorem 3.4, the direct half, and as Theorem 3.5, the converse half. These theorems connect relative compactness with the tightness of a family of probability measures.

Theorem 3.4. *If a family of probability measures is tight, then it is relatively compact.*

Theorem 3.5. *Let S be separable (i.e., S contains a countable dense subset) and complete. If a family of probability measures on $(S, \mathcal{B}(S))$ is relatively compact, then it is tight.*

These are Theorems 6.1 and 6.2 from Billingsley [21]. Note that a topological space is said to be separable if it contains a countable dense subset.

In the theory of Dirichlet series we investigate the weak convergence of probability measures $\mathbf{P}_T \Rightarrow \mathbf{P}$, where T is a continuous parameter which tends to infinity. As it is noted in [21], we have $\mathbf{P}_T \Rightarrow \mathbf{P}$, as $T \to \infty$, if and only if $\mathbf{P}_{T_n} \Rightarrow \mathbf{P}$, as $n \to \infty$, for every sequence $\{T_n\}$ with $\lim_{n\to\infty} T_n = \infty$. All theorems on weak convergence analoguous to those stated above remain valid in the case of continuous parameters.

3.2 Random Elements

The theory of weak convergence of probability measures can be paraphrased as the theory of convergence of random elements in distribution.

Let $(\Omega, \mathcal{F}, \mathbf{P})$ be a probability space, $(S, \mathcal{B}(S))$ be a metric space with its class of Borel sets $\mathcal{B}(S)$, and $X : \Omega \to S$ a mapping. If

$$X^{-1}(A) = \{\omega \in \Omega : X(\omega) \in A\} \in \mathcal{F}$$

for every $A \in \mathcal{B}(S)$, then X is called an S-valued random element defined on Ω; if $S = \mathbb{R}$ we say that X is a random variable. The distribution of an S-valued random element X is the probability measure $\mathbf{P}_X$ on $(S, \mathcal{B}(S))$, given by

$$\mathbf{P}_X(A) = \mathbf{P}(X^{-1}(A)) = \mathbf{P}\{\omega \in \Omega : X(\omega) \in A\}$$

for arbitrary $A \in \mathcal{B}(S)$ (in the sequel we will often write $\mathbf{P}$ in place of $\mathbf{P}_X$). We say that a sequence $\{X_n\}$ of random elements converges in distribution to a random element X if the distributions $\mathbf{P}_n$ of the elements X_n converge weakly to the distribution of the element X, and in this case we write

$$X_n \xrightarrow[n\to\infty]{\mathcal{D}} X;$$

if $\mathbf{P}_n \Rightarrow \mathbf{P}_X$, then we also write $X_n \xrightarrow[n\to\infty]{\mathcal{D}} \mathbf{P}_X$.

Let S be a metric space with metric ϱ, and let X_n, Y_n be S-valued random elements defined on $(\Omega, \mathcal{F}, \mathbf{P})$. If X_n and Y_n have a common domain, it makes sense to speak of the distance $\varrho(X_n(\omega), Y_n(\omega))$ for $\omega \in \Omega$. If S is separable, then $\varrho(X_n, Y_n)$ is a random variable. In this case, convergence in distribution of two sequences of random elements X_n and Y_n is related to the distribution of $\varrho(X_n, Y_n)$ (convergence in probability).

Theorem 3.6. *Let S be separable and, for $n \in \mathbb{N}$, let $Y_n, X_{1n}, X_{2n}, \ldots$ be S-valued random elements, all defined on $(\Omega, \mathcal{F}, \mathbf{P})$. Suppose that*

$$X_{kn} \xrightarrow[n\to\infty]{\mathcal{D}} X_k \quad \text{for each } k, \quad \text{and} \quad X_k \xrightarrow[k\to\infty]{\mathcal{D}} X.$$

If for any $\epsilon > 0$

$$\lim_{k\to\infty} \limsup_{n\to\infty} \mathbf{P}\{\varrho(X_{kn}, Y_n) \ge \epsilon\} = 0,$$

then $Y_n \xrightarrow[n\to\infty]{\mathcal{D}} X$.

This is Theorem 4.2 of Billingsley [21].

The mean (expectation value) $\mathbf{E}X$ of a random element X is defined by

$$\mathbf{E}X = \int_{\Omega} X(\omega) \, \mathrm{d}\mathbf{P}$$

if the integral exists in the sense of Lebesgue. A simple but fundamental result on the deviation of a random variable from its expectation value is Chebyshev's (respectively, Markov's) inequality:

Lemma 3.7. *Let X be a real-valued random variable, $h : \mathbb{R} \to [0, \infty)$ be a non-negative function, and $a > 0$. Then*

$$\mathbf{P}\{\omega \in \Omega : h(X) \ge a\} \le \frac{1}{a} \mathbf{E}h(X).$$

A proof can be found, for example, in Billingsley [22]. Taking $h(x) = x^2$, we deduce the classical Chebyshev's inequality:

$$\mathbf{P}\{|X| \ge a\} \le \frac{1}{a^2} \mathbf{E}X^2.$$

We say that some property is valid almost surely if there exists a set $A \in \mathcal{F}$ with $\mathbf{P}(A) = 0$ such that this property is valid for every $\omega \in \Omega \setminus A$. Random variables X and Y are said to be orthogonal if $\mathbf{E}XY = 0$. An important result on almost sure convergence of series of orthogonal random variables is the following

Theorem 3.8. *Assume that the random variables $X_1, X_2, \ldots$ are orthogonal and that*

$$\sum_{n=1}^{\infty} \mathbf{E}|X_n|^2 (\log n)^2 < \infty.$$

Then the series $\sum_{n=1}^{\infty} X_n$ converges almost surely.

Further, we need a similar result for independent random variables. Random variables X and Y are said to be independent if for all $A, B \in \mathcal{B}(S)$

$$\mathbf{P}\{\omega \in \Omega : X \in A, Y \in B\} = \mathbf{P}\{\omega \in \Omega : X \in A\} \cdot \mathbf{P}\{\omega \in \Omega : Y \in B\}.$$

Usually, independence of random variables is defined via the σ-fields generated by the related events; however, for the sake of simplicity, we introduced this notion by the equivalent condition above.

Theorem 3.9. *Assume that the random variables $X_1, X_2, \ldots$ are independent. If the series*

$$\sum_{n=1}^{\infty} \mathbf{E} X_n \quad \text{and} \quad \sum_{n=1}^{\infty} \mathbf{E}(X_n - \mathbf{E} X_n)^2$$

converge, then the series $\sum_{n=1}^{\infty} X_n$ converges almost surely.

Proofs of the last two theorems can be found in Loève [226].

Now we present a criterion on almost sure convergence of series of independent random variables for Hilbert spaces. Let $\mathcal{H}$ be a separable Hilbert space with norm $\|\cdot\|$ and define for an $\mathcal{H}$-valued random element X and a real number c the truncated function

$$X^{(c)} = \begin{cases} X & \text{if } \|X\| \leq c, \\ 0 & \text{if } \|X\| > c. \end{cases}$$

Then

Theorem 3.10. *Let $X_1, X_2, \ldots$ be independent $\mathcal{H}$-valued random elements. If there is a constant $c > 0$ so that the series*

$$\sum_{n=1}^{\infty} \mathbf{E} \|X_n^{(c)} - \mathbf{E} X_n^{(c)}\|^2, \quad \sum_{n=1}^{\infty} \mathbf{E} X_n^{(c)}, \quad \text{and} \quad \sum_{n=1}^{\infty} \mathbf{P} \{\|X_n\| > c\}$$

converge, then the series $\sum_{n=1}^{\infty} X_n$ converges in $\mathcal{H}$ almost surely.

A proof can be found in Buldygin [45].

3.3 Denjoy's Probabilistic Argument for Riemann's Hypothesis

At this point our survey of probability theory will be cut in order to give a heuristic probabilistic argument for the truth of Riemann's hypothesis. The Möbius μ-function is defined by $\mu(1) = 1$, $\mu(n) = 0$ if n has a quadratic divisor $\neq 1$, and $\mu(n) = (-1)^r$ if n is the product of r distinct primes. It is easily seen that $\mu(n)$ is multiplicative and appears as coefficients of the Dirichlet series representation of the reciprocal of the zeta-function:

$$\zeta(s)^{-1} = \prod_p \left(1 - \frac{1}{p^s}\right) = \sum_{n=1}^{\infty} \frac{\mu(n)}{n^s},$$

valid for $\sigma > 1$. Riemann's hypothesis is equivalent to the estimate

$$M(x) := \sum_{n \leq x} \mu(n) \ll x^{(1/2)+\epsilon}.$$

This is related to (1.11); for a proof see, for example, Titchmarsh [353, Sect. 14.25].

Denjoy [68] argued as follows. Assume that $\{X_n\}$ is a sequence of random variables with distribution

$$\mathbf{P}(X_n = +1) = \mathbf{P}(X_n = -1) = \frac{1}{2}.$$

Define

$$S_0 = 0 \quad \text{and} \quad S_n = \sum_{j=1}^{n} X_j,$$

then $\{S_n\}$ is a symmetrical random walk in $\mathbb{Z}^2$ with starting point at 0. A simple application of Chebyshev's inequality yields, for any positive c,

$$\mathbf{P}\{|S_n| \geq cn^{\frac{1}{2}}\} \leq \frac{1}{2c^2},$$

which shows that *large* values for S_n are *rare* events. By the theorem of Moivre–Laplace ([22, Theorem 27.1]) this can be made more precise. It follows that:

$$\lim_{n \to \infty} \mathbf{P}\left\{|S_n| < cn^{\frac{1}{2}}\right\} = \frac{1}{\sqrt{2\pi}} \int_{-c}^{c} \exp\left(-\frac{x^2}{2}\right) \mathrm{d}x.$$

Since the right-hand side above tends to 1 as $c \to \infty$, we obtain

$$\lim_{n \to \infty} \mathbf{P}\left\{|S_n| \ll n^{\frac{1}{2}+\epsilon}\right\} = 1$$

for every $\epsilon > 0$. We observe that this might be regarded as a model for the value-distribution of Möbius μ-function. To say it with the words of Edwards: "Thus these probabilistic assumptions about the values of $\mu(n)$ lead to the conclusion, ludicrous as it seems, that $M(x) = \mathrm{O}(x^{1/2+\epsilon})$ with probability one and hence that the Riemann hypothesis is true with probability one!" (cf. [74]). The law of the iterated logarithm [22, Theorem 9.5] would even gives the stronger estimate

$$\lim_{n \to \infty} \mathbf{P}\left\{|S_n| \ll (n \log \log n)^{1/2}\right\} = 1,$$

which suggests for $M(x)$ the upper bound $(x \log \log x)^{1/2}$. This estimate is pretty close to the so-called weak Mertens hypothesis which states

$$\int_{1}^{X} \left(\frac{M(x)}{x}\right)^2 \mathrm{d}x \ll \log X.$$

Note that this bound implies the Riemann hypothesis and the essential simplicity hypothesis. On the contrary, Odlyzko and te Riele [284] disproved the original Mertens hypothesis [248],

$$|M(x)| < x^{\frac{1}{2}},$$

by showing that

$$\liminf_{x \to \infty} \frac{M(x)}{x^{1/2}} < -1.009 \quad \text{and} \quad \limsup_{x \to \infty} \frac{M(x)}{x^{1/2}} > 1.06; \qquad (3.1)$$

for more details see also the notes to Sect. 14 in Titchmarsh [353].

3.4 Characteristic Functions and Fourier Transforms

There is an intimate relationship between weak convergence and characteristic functions which makes characteristic functions very useful in studying limit distributions.

The characteristic function $\varphi(\tau)$ of a probability measure $\mathbf{P}$ on $(\mathbb{R}^r, \mathcal{B}(\mathbb{R}^r))$ is defined by

$$\varphi(\tau) = \int_{\mathbb{R}^r} \exp(\mathrm{i}\langle \tau, x \rangle) \mathbf{P}(\,\mathrm{d}x),$$

where $\langle \tau, x \rangle$ stands for the inner product of τ and $x \in \mathbb{R}^r$. Notice that the characteristic function uniquely determines the measure it comes from. Let $\{\mathbf{P}_n\}$ be a sequence of probability measures on $(\mathbb{R}^r, \mathcal{B}(\mathbb{R}^r))$ and let $\{\varphi_n(\tau)\}$ be the sequence of the corresponding characteristic functions. Suppose that

$$\lim_{n \to \infty} \varphi_n(\tau) = \varphi(\tau)$$

for all τ, and that $\varphi(\tau)$ is continuous at the point $\mathbf{0} = (0, \ldots, 0)$. Then Lévy's famous continuity theorem (see [22, Sect. 26]) yields the existence of a probability measure $\mathbf{P}$ on $(\mathbb{R}^r, \mathcal{B}(\mathbb{R}^r))$ such that $\mathbf{P}_n \Rightarrow \mathbf{P}$, and $\varphi(\tau)$ is the characteristic function of $\mathbf{P}$. However, later we shall deal with Fourier transforms instead of characteristic functions; their theory is quite similar to the theory of characteristic functions (for details we refer to the first chapter from [186]).

Let $\gamma = \{s \in \mathbb{C} : |s| = 1\}$ and denote by γ^m the cartesian product of m copies of γ. Further, let $\mathbf{P}$ be a probability measure on $(\gamma^m, \mathcal{B}(\gamma^m))$, then the Fourier transform $g(k_1, \ldots, k_m)$ of the measure $\mathbf{P}$ is defined by

$$g(k_1, \ldots, k_m) = \int_{\gamma^m} x_1^{k_1} \ldots x_m^{k_m} \,\mathrm{d}\mathbf{P},$$

where $k_j \in \mathbb{Z}$ and $x_j \in \gamma$ for $1 \le j \le m$. Similarly to the characteristic function, the measure $\mathbf{P}$ is uniquely determined by its Fourier transform. The role of Lévy's continuity theorem in the theory of Fourier transforms is played by

Theorem 3.11. *Let $\{\mathbf{P}_n\}$ be a sequence of probability measures on $(\gamma^m, \mathcal{B}(\gamma^m))$ and let $\{g_n(k_1, \ldots, k_m)\}$ be the sequence of the corresponding Fourier transforms. Suppose that for every vector $(k_1, \ldots, k_m) \in \mathbb{Z}^m$ the limit*

$$g(k_1, \ldots, k_m) = \lim_{n \to \infty} g_n(k_1, \ldots, k_m)$$

exists. Then there is a probability measure $\mathbf{P}$ *on* $(\gamma^m, \mathcal{B}(\gamma^m))$ *such that* $\mathbf{P}_n \Rightarrow$ $\mathbf{P}$, *and* $g(k_1, \ldots, k_m)$ *is the Fourier transform of* $\mathbf{P}$.

Theorem 3.11 is a special case of a more general continuity theorem for probability measures on compact abelian groups. A proof of this result can be found in Heyer [133, Theorem 1.4.2].

3.5 Haar Measure and Characters

Let G be a set equipped with the structures of a group and of a topological space. If the function $h : G \times G \to G$, defined by $h(x, y) = xy^{-1}$, is continuous, then G is called a topological group. A topological group is said to be compact if its topology is compact. In what follows, G is assumed to be compact.

A Borel measure $\mathbf{P}$ on a compact topological group G is said to be invariant if

$$\mathbf{P}(A) = \mathbf{P}(xA) = \mathbf{P}(Ax)$$

for all $A \in \mathcal{B}(G)$ and all $x \in G$, where xA and Ax denote the sets $\{xy : y \in A\}$ and $\{yx : y \in A\}$, respectively. An invariant Borel measure on a compact topological group is called Haar measure.

Theorem 3.12. *On every compact topological group there exists a unique probability Haar measure.*

The uniqueness follows from $m(G) = 1$. For the proof see Hewitt and Ross [132, Chap. IV] or Theorem 5.14 in Rudin [313].

In the sequel, we denote the Haar measure associated with a compact topological group G simply by m; there will be no confusion about the underlying group.

Now assume further that G is a commutative group. A continuous homomorphism $\chi : G \to \mathbb{C}$ is called a character of G. The character of G which is identically 1 is called trivial or principal, and we denote it by χ_0; other characters are said to be non-trivial or non-principal. The characters build up a group $\hat{G}$, the character group. The Fourier transform of a function f defined on G is given by

$$\hat{f}(\chi) = \int_G \chi(g) f(g) \, m(dg),$$

where χ is a character of G. Then $\hat{f}$ is a continuous map defined on $\hat{G}$. The orthogonality relation for characters states

$$\int_G \chi(g) \, m(dg) = \begin{cases} 1 & \text{if } \chi = \chi_0, \\ 0 & \text{if } \chi \neq \chi_0. \end{cases} \tag{3.2}$$

This generalizes the concept of Dirichlet characters.

3.6 Random Processes and Ergodic Theory

To be able to identify later the explicit form of the limit measure in limit theorems, we recall some facts from ergodic theory.

Let $(\Omega, \mathcal{F}, \mathbf{P})$ be a probability space and let $\mathcal{T}$ denote a parameter set. A finite real function $X(\tau, \omega)$ with $\tau \in \mathcal{T}$ and $\omega \in \Omega$ is said to be a random (or stochastic) process if $\omega \mapsto X(\tau, \omega)$ is a random variable for each fixed $\tau \in \mathcal{T}$. For fixed $\omega \in \Omega$, the function $\tau \mapsto X(\tau, \omega)$ is called a sample path of the random process. Let $\tau_1, \ldots, \tau_n$ be an arbitrary set of values of $\mathcal{T}$. Then the family of all common distributions of random variables $X(\tau_1, \omega), \ldots, X(\tau_n, \omega)$, i.e.,

$$\mathbf{P}\{X(\tau_1, \omega) < x_1, \ldots, X(\tau_n, \omega) < x_n\}$$

for all $n \in \mathbb{N}$ and all possible values of τ_j with $1 \leq j \leq n$, is called a family of finite-dimensional distributions of the process $X(\tau, \omega)$. Part of the structure of the random process is specified by its finite-dimensional distributions. However, they do not determine the character of the sample paths (see [22, Sect. 23], for a nice example). Kolmogorov's existence theorem (see Theorem 36.1 in [22]) states that if a family of finite dimensional distributions satisfies certain consistency conditions, then there exists on some probability space a random process having exactly the same finite-dimensional distributions. For instance, a special application of Kolmogorov's existence theorem yields a model for Brownian motion with continuous paths.

Let Y be the space of all finite real-valued functions $y(\tau)$ with $\tau \in \mathbb{R}$. In this case it is known that the family of finite-dimensional distributions of each random process determines a probability measure $\mathbf{P}$ on $(Y, \mathcal{B}(Y))$. Then, on the probability space $(Y, \mathcal{B}(Y), \mathbf{P})$, we define for real u the translation g_u which maps each function $y(\tau) \in Y$ to $y(\tau + u)$. It is easily seen that the translations g_u form a group. A random process $X(\tau, \omega)$ is said to be strongly stationary if all its finite-dimensional distributions are invariant under the translations by u. It is known that if a process $X(\tau, \omega)$ is strongly stationary, then the translation g_u is measure preserving, i.e., for any set $A \in \mathcal{B}(Y)$ and all $u \in \mathbb{R}$ the equality

$$\mathbf{P}(A) = \mathbf{P}(A_u), \quad \text{where} \quad A_u := g_u(A)$$

holds. A set $A \in \mathcal{B}(Y)$ is called an invariant set of the process $X(\tau, \omega)$ if for each u the sets A and A_u differ from each other by a set of zero $\mathbf{P}$-measure. In other words, $\mathbf{P}(A \Delta A_u) = 0$, where Δ denotes the symmetric difference of two sets A and B:

$$A \Delta B := (A \setminus B) \cup (B \setminus A).$$

It is easy to see that all invariant sets of Y form a σ-field which is a sub-σ-field of $\mathcal{B}(Y)$. We say that a strongly stationary process $X(\tau, \omega)$ is ergodic if its σ-field of invariant sets consists only of sets having $\mathbf{P}$-measure equal to 0 or 1. For ergodic processes the Birkhoff–Khintchine theorem gives an expression for the expectation of $X(0, \omega)$ in terms of an integral taken over the sample paths.

Theorem 3.13. *Let $X(\tau, \omega)$ be an ergodic process with $\mathbf{E}|X(\tau, \omega)| < \infty$ and almost surely Riemann-integrable sample paths over every finite interval. Then*

$$\lim_{T \to \infty} \frac{1}{T} \int_0^T X(\tau, \omega)\, \mathrm{d}\tau = \mathbf{E} X(0, \omega)$$

almost surely.

A proof of this theorem can be found in Cramér and Leadbetter [64].

3.7 The Space of Analytic Functions

Let $\mathcal{G}$ be a simply connected region in the complex plane. We denote by $\mathcal{H}(\mathcal{G})$ the space of analytic functions f defined on $\mathcal{G}$ equipped with the topology of uniform convergence on compacta.

In order to introduce an appropriate metric on $\mathcal{H}(\mathcal{G})$ we note

Lemma 3.14. *For any open set $\mathcal{G}$ in the complex plane there exists a sequence of compact subsets K_j of $\mathcal{G}$ with the properties:*

- $K_j \subset K_{j+1}$ *for any $j \in \mathbb{N}$;*
- *If K is compact and $K \subset \mathcal{G}$, then $K \subset K_j$ for some $j \in \mathbb{N}$;*

such that

$$\mathcal{G} = \bigcup_{j=1}^{\infty} K_j$$

The proof is straightforward and can be found in Conway's book [62, Sect. VII.1].

Now, for $f, g \in \mathcal{H}(\mathcal{G})$ let

$$\varrho_j(f, g) = \max_{s \in K_j} |f(s) - g(s)|$$

and put

$$\varrho(f, g) = \sum_{j=1}^{\infty} 2^{-j} \frac{\varrho_j(f, g)}{1 + \varrho_j(f, g)}.$$

This defines a metric on $\mathcal{H}(\mathcal{G})$ which induces the desired topology; of course, the metric ϱ depends on the family $\{K_j\}$. Note that the series above is dominated by $\sum_j 2^{-j}$ and therefore convergent.

Theorem 3.15. *Let $\mathcal{G}$ be a simply connected region in the complex plane. Then $\mathcal{H}(\mathcal{G})$ is a complete separable metric space.*

In Conway [62, Sect. VII.1], it is shown that $\mathcal{H}(\mathcal{G})$ is a complete metric space; the separability, i.e., the existence of a countable dense subset, follows from

Runge's approximation theorem (see [312, Sect. 13]) which states that the set of polynomials is dense in $\mathcal{H}(\mathcal{G})$.

In our later studies, we deal with the supports of $\mathcal{H}(\mathcal{G})$-valued random elements. Let S be a separable metric space and let $\mathbf{P}$ be a probability measure on $(S, \mathcal{B}(S))$. The minimal closed set $S_{\mathbf{P}} \subseteq S$ with $\mathbf{P}(S_{\mathbf{P}}) = 1$ is called the support of $\mathbf{P}$. Note that $S_{\mathbf{P}}$ consists of all $x \in S$ such that for every neighbourhood U of x the inequality $\mathbf{P}(U) > 0$ is satisfied. Let X be a S-valued random element defined on the probability space $(\Omega, \mathcal{F}, \mathbf{P})$. Then the support of the distribution $\mathbf{P}(X \in A)$ for $A \in \mathcal{B}(S)$ is called the support of the random element X. We denote the support of X by S_X.

Theorem 3.16. *Let $\{X_n\}$ be a sequence of independent $\mathcal{H}(\mathcal{G})$-valued random elements, and suppose that the series $\sum_{n=1}^{\infty} X_n$ converges almost everywhere. Then the support of the sum of this series is the closure of the set of all $f \in \mathcal{H}(\mathcal{G})$ which may be written as a convergent series*

$$f = \sum_{n=1}^{\infty} f_n, \quad \text{where} \quad f_n \in S_{X_n}.$$

This is Theorem 1.7.10 of Laurinčikas [186]. The proof follows the lines of an analogous statement for independent real variables due to Lukacs [231]. We only sketch the main ideas. Suppose that the random elements X_n are defined on a probability space $(\Omega, \mathcal{F}, \mathbf{P}^*)$. Put

$$X := \sum_{n=1}^{\infty} X_n = L_N + R_N,$$

where

$$L_N := \sum_{n=1}^{N} X_n \quad \text{and} \quad R_N := \sum_{n=N+1}^{\infty} X_n.$$

Since the series $\sum_{n=1}^{\infty} X_n$ converges almost surely, we have

$$\lim_{N \to \infty} \mathbf{P}^* \{\omega \in \Omega : \varrho(R_N, 0) \geq \epsilon\} = 0.$$

Let

$$\mathbf{P}_N(A) = \mathbf{P}^* \{L_N \in A\} \quad \text{and} \quad \mathbf{P}(A) = \mathbf{P}^* \{X \in A\}$$

for $A \in \mathcal{B}(\mathcal{H}(\mathcal{G}))$. It follows that $\mathbf{P}_n \Rightarrow \mathbf{P}$, which implies

$$S_{\mathbf{P}} \subset \lim S_{\mathbf{P}_N}, \tag{3.3}$$

where $\lim S_{\mathbf{P}_N}$ denotes the set of all $f \in \mathcal{H}(\mathcal{G})$ such that any neighbourhood of f contains at least one g which belongs to $S_{\mathbf{P}_n}$ for almost all $n \in \mathbb{N}$.

To show the converse inclusion, put

$$\mathbf{Q}_N(A) = \mathbf{P}^* \{R_N \in A\}$$

for $A \in \mathcal{B}(\mathcal{H}(\mathcal{G}))$. The distribution of $X = L_N + R_N$ is given by the convolution $\mathbf{P}_N * \mathbf{Q}_N$, defined by

$$(\mathbf{P}_N * \mathbf{Q}_N)(A) = \int_{\mathcal{H}(\mathcal{G})} \mathbf{P}_N(A - g)\mathbf{Q}_N(\,dg).$$

The support of $X = L_N + R_N$ is the closure of the set

$$\{f \in \mathcal{H}(\mathcal{G}) : f = f_1 + f_2, \text{ where } f_1 \in S_{L_N}, f_2 \in S_{R_N}\}. \tag{3.4}$$

For $g \in \lim S_{\mathbf{P}_N}$ let

$$A_\epsilon := \{f \in \mathcal{H}(\mathcal{G}) : \varrho(f,g) < \epsilon\}.$$

It follows that $\mathbf{P}_N(A_\epsilon) = \mathbf{P}^*(L_N \in A_\epsilon) > 0$ and $\mathbf{Q}_N(A_\epsilon) > 0$ for N large enough. This leads to

$$\mathbf{P}(A_{2\epsilon}) \geq \mathbf{P}_N(A_\epsilon)\mathbf{Q}_N(A_\epsilon) > 0.$$

This implies

$$S_X = S_{\mathbf{P}} \supset \lim S_{\mathbf{P}_N}.$$

By (3.3) it follows that the latter inclusion is an equality. In view of (3.4) the support of L_N is the set of all $g \in \mathcal{H}(\mathcal{G})$ which have a representation

$$g = \sum_{n=1}^{N} f_n,$$

where $f_n \in S_{X_n}$. From the definition of $\lim S_{\mathbf{P}_N}$ we deduce that for any $f \in S_X$ there exists a sequence of $g_N \in S_{L_N}$ which converges to f. This yields the assertion of the theorem.

Limit Theorems

Measure what is measurable, and make measurable what is not so.

Galileo Galilei

In this chapter, we prove a limit theorem dealing with weakly convergent probability measures for L-functions from the class $\tilde{S}$ in the space of analytic functions. Throughout this chapter, we assume that $\mathcal{L} \in \tilde{S}$. We remark that we will not make use of axiom (v), so the results hold in a more general context; however, with respect to later applications, there is no need to introduce a further class. We follow the presentation of Laurinčikas [187, 188] (functions in $\tilde{S}$ form a subclass of Matsumoto zeta-functions considered herein). Besides, we refer the interested reader to Laurinčikas' survey [185] and his monograph [186].

4.1 Associated Random Elements and the Main Limit Theorem

Assume that $\mathcal{L} \in \tilde{S}$ and denote by $\mathcal{D}$ the intersection of the corresponding mean-square half-plane $\sigma > \sigma_m$ with the critical strip (see Sect. 2.1):

$$\mathcal{D} := \{ s \in \mathbb{C} : \sigma_m < \sigma < 1 \}. \tag{4.1}$$

In view of Theorem 2.4 we know that $\mathcal{D}$ is not empty. Together with (2.4) we have

$$\frac{1}{2} \le \sigma_m \le \max \left\{ \frac{1}{2}, 1 - \frac{1 - \sigma_{\mathcal{L}}}{1 + 2\mu_{\mathcal{L}}} \right\} < 1, \tag{4.2}$$

where $\sigma_{\mathcal{L}}$ and $\mu_{\mathcal{L}}$ are defined by axioms (ii) and (iii). In particular, as $T \to \infty$,

$$\frac{1}{T} \int_0^T |\mathcal{L}(\sigma + \mathrm{it})|^2 \, \mathrm{dt} \ll 1 \quad \text{for} \quad \sigma \in (\sigma_m, 1). \tag{4.3}$$

Now we attach to $\mathcal{L}(s)$ the probability measure $\mathbf{P}_T$ by setting

$$\mathbf{P}_T(A) = \frac{1}{T}\,\mathrm{meas}\,\{\tau \in [0, T] : \mathcal{L}(s + i\tau) \in A\} \qquad (4.4)$$

for $A \in \mathcal{B}(\mathcal{H}(\mathcal{D}))$, where $\mathcal{H}(\mathcal{D})$ denotes the space of analytic functions on the strip $\mathcal{D}$ (as introduced in Sect. 3.7). We shall show that this measure converges weakly to some probability measure as T tends to infinity. For later purposes it is also important to identify the limit measure.

Denote by γ the compact unit circle $\{s \in \mathbb{C} : |s| = 1\}$ and put

$$\Omega = \prod_p \gamma_p,$$

where $\gamma_p = \gamma$ for each prime p. With product topology and pointwise multiplication this infinite-dimensional torus Ω is a compact topological abelian group; the compactness of Ω follows from Tikhonov's theorem (see [169, Theorem 5.13]). Thus, by Theorem 3.12, the normalized Haar measure m on $(\Omega, \mathcal{B}(\Omega))$ exists. This gives a probability space $(\Omega, \mathcal{B}(\Omega), \mathrm{m})$. Let $\omega(p)$ stand for the projection of $\omega \in \Omega$ to the coordinate space γ_p. The Haar measure m on Ω is the product of the Haar measures m_p on the coordinate spaces γ_p, i.e.,

$$\mathrm{m}\{\omega : \omega \in A\} = \prod_p \mathrm{m}_p\{\omega : \omega(p) \in A_p\},$$

where A_p is the projection of $A \in \mathcal{B}(\Omega)$ onto γ_p. Thus $\{\omega(p) : p\ \text{prime}\}$ is a sequence of independent complex-valued random variables defined on the probability space $(\Omega, \mathcal{B}(\Omega), \mathrm{m})$. Denote by $\mathcal{G}$ the set of all completely multiplicative functions $g : \mathbb{N} \to \gamma$. Further, let $\omega(1) = 1$ and

$$\omega(n) = \prod_p \omega(p)^{\nu(n;p)}, \qquad (4.5)$$

where $\nu(n; p)$ is the exponent of the prime p in the prime factorization of n. This extends the function ω to the set of positive integers as a unimodular, completely multiplicative function. On the other hand, the restriction of any function $g \in \mathcal{G}$ to the set of primes is an element of Ω. Consequently, we may identify the elements of Ω with the functions from $\mathcal{G}$.

Next we are going to attach to

$$\mathcal{L}(s) = \sum_{n=1}^{\infty} \frac{a(n)}{n^s}$$

a random element. For $s \in \mathcal{D}$ and $\omega \in \Omega$, we define

$$\mathcal{L}(s, \omega) = \sum_{n=1}^{\infty} \frac{a(n)\omega(n)}{n^s}. \qquad (4.6)$$

This is equal to $\mathcal{L}(s)$ for $\sigma > 1$ and $\omega(n) \equiv 1$.

Lemma 4.1. *The function* $\mathcal{L}(s,\omega)$ *is an* $\mathcal{H}(\mathcal{D})$-*valued random element on the probability space* $(\Omega, \mathcal{B}(\Omega), \mathrm{m})$.

Proof. Fix a real number $\sigma_1 > \sigma_m$. Define for $n \in \mathbb{N}$

$$\xi_n = \xi_n(\omega) = \frac{a(n)\omega(n)}{n^{\sigma_1}},$$

then $\{\xi_n\}$ is a sequence of complex-valued random variables on $(\Omega, \mathcal{B}(\Omega), \mathrm{m})$. It is easy to compute that

$$\mathbf{E}\xi_m\overline{\xi_n} = \frac{a(m)\overline{a(n)}}{(mn)^{\sigma_1}} \int_\Omega \omega(m)\overline{\omega(n)}\,\mathrm{dm} = \begin{cases} \dfrac{|a(n)|^2}{n^{2\sigma_1}} & \text{if } m = n, \\ 0 & \text{otherwise.} \end{cases}$$

This shows that the sequence $\{\xi_n\}$ is pairwise orthogonal. Since $\sigma_m \geq \frac{1}{2}$ by (2.4), it follows that $2\sigma_1 > 1$. This and the Ramanujan hypothesis (i) imply that the series

$$\sum_{n=1}^\infty \mathbf{E}|\xi_n|^2 (\log n)^2 = \sum_{n=1}^\infty \frac{|a(n)|^2}{n^{2\sigma_1}}(\log n)^2$$

is convergent. It follows from Theorem 3.8 that the series $\sum_{n=1}^\infty \xi_n$ converges almost surely, respectively

$$\sum_{n=1}^\infty \frac{a(n)\omega(n)}{n^{\sigma_1}}$$

converges for almost all $\omega \in \Omega$ with respect to m. Since Dirichlet series converge in half-planes, and uniformly in compact subsets of their half-plane of convergence (see Sect. 2.1), it follows that, for almost all $\omega \in \Omega$, the series (4.6) converges uniformly on compact subsets of the half-plane $\sigma > \sigma_1$. For $k \in \mathbb{N}$ let $\mathcal{A}_k$ be the subset of $\omega \in \Omega$ for which the series (4.6) converges uniformly on compact subsets of the half-plane $\sigma \geq \sigma_m + \frac{1}{k}$. Then $\mathrm{m}(\mathcal{A}_k) = 1$ for all $k \in \mathbb{N}$. Setting

$$\mathcal{A} = \bigcap_{k=1}^\infty \mathcal{A}_k,$$

we get $\mathrm{m}(\mathcal{A}) = 1$, and for $\omega \in \mathcal{A}$ the series (4.6) converges uniformly on compact subsets of $\mathcal{D}$. Since each term of the series in question is an $\mathcal{H}(\mathcal{D})$-valued function, the assertion of the lemma follows. $\qquad\square$

Taking into account the multiplicativity of the coefficients $a(n)$, (2.6), and the decomposition (4.5) of ω, we may rewrite the series (4.6) (formally) as

$$\mathcal{L}(s,\omega) = \prod_p \prod_{j=1}^m \left(1 - \frac{\alpha_j(p)\omega(p))}{p^s}\right)^{-1}$$

$$= \prod_p \left(1 + \sum_{k=1}^\infty \frac{a(p^k)\omega(p)^k}{p^{ks}}\right) \tag{4.7}$$

In the half-plane $\sigma > 1$ both the series (4.6) and the product (4.7) converge absolutely.

Lemma 4.2. *For almost all $\omega \in \Omega$ the product (4.7) converges uniformly on compact subsets of $\mathcal{D}$.*

Proof. Define for each prime p the function

$$x_p = x_p(s, \omega) = \sum_{k=1}^{\infty} \frac{a(p^k)\omega(p)^k}{p^{ks}}. \tag{4.8}$$

Then

$$\mathcal{L}(s, \omega) = \prod_p (1 + x_p(s, \omega)).$$

By the uniform convergence of the x_p-defining series on $\mathcal{D}$ the function $x_p(s, \omega)$ is an $\mathcal{H}(\mathcal{D})$-valued random element for any p. Using the Ramanujan hypothesis (i) we obtain, in a similar way as in the proof of Lemma 4.1, that

$$\sum_p |x_p(s, \omega)|^2$$

converges uniformly on compact subsets of $\mathcal{D}$. To prove the almost sure convergence of the product (4.7), it remains to show that the series

$$\sum_p x_p(s, \omega)$$

converges almost surely. For this, we isolate the leading term in the series expansion of x_p. Put

$$y_p(s) = y_p(s, \omega) = \frac{a(p)\omega(p)}{p^s},$$

then, by the Ramanujan hypothesis (i),

$$|x_p(s, \omega) - y_p(s, \omega)| \leq \sum_{k=2}^{\infty} \frac{|a(p^k)|}{p^{k\sigma}} \ll \sum_{k=2}^{\infty} p^{\epsilon - k\sigma} \ll p^{\epsilon - 2\sigma}.$$

Thus, for sufficiently small ϵ, the series

$$\sum_p |x_p(s, \omega) - y_p(s, \omega)|$$

converges uniformly on compact subsets of $\mathcal{D}$ for all $\omega \in \Omega$. Therefore, we can focus on the series $\sum_p y_p(s, \omega)$. Clearly, the $y_p(s, \omega)$ define a sequence of independent $\mathcal{H}(\mathcal{D})$-valued random elements. It follows from the definition of the projections $\omega(p)$ that $\mathbf{E}y_p(s) = 0$. Furthermore, by the Ramanujan hypothesis (i),

$$\mathbf{E}|y_p(s)|^2 = \frac{|a(p)|^2}{p^{2\sigma}} \ll \frac{1}{p^{2\sigma - \epsilon}}.$$

This implies the convergence of

$$\sum_p \mathbf{E}|y_p(s)|^2(\log p)^2$$

for all $s \in \mathcal{D}$. Thus, by Theorem 3.9 the series $\sum_p y_p(s,\omega)$ converges almost surely for each fixed $s \in \mathcal{D}$. It follows that also $\sum_p x_p(s,\omega)$ converges almost surely for fixed $s \in \mathcal{D}$. In view of convergence properties of Dirichlet series the convergence is uniform on compact subsets of $\mathcal{D}$ for almost all $\omega \in \Omega$. It is clear that the same assertion holds for the product too. This proves the lemma. $\square$

By analytic continuation the identity between the series (4.6) and the product (4.7) holds whenever one of these representations converges uniformly. Now let $\mathbf{P}$ denote the distribution of the random element $\mathcal{L}(s,\omega)$, i.e.,

$$\mathbf{P}(A) = \mathrm{m}\{\omega \in \Omega : \mathcal{L}(s,\omega) \in A\} \tag{4.9}$$

for $A \in \mathcal{B}(\mathcal{H}(\mathcal{D}))$. The main result of this chapter is the limit theorem

Theorem 4.3. *Let $\mathcal{L} \in \tilde{\mathcal{S}}$. The probability measure $\mathbf{P}_T$, defined by (4.4), converges weakly to $\mathbf{P}$ as $T \to \infty$.*

The first limit theorem of this type is due to Bagchi [9]. In the case of the Riemann zeta-function this result might be regarded as an analogue of the limit Theorem 1.4 of Bohr and Jessen on function spaces. To see this, note that Laurinčikas [183] translated their statement into a modern language by defining a probability measure via

$$\mathbf{P}_T(A) = \frac{1}{T}\,\mathrm{meas}\,\{t \in [0,T] : \zeta(\sigma + \mathrm{i}t) \in A\} \quad A \in \mathcal{B}(\mathbb{C}),$$

and proving that, for any fixed $\sigma > \frac{1}{2}$, $\mathbf{P}_T$ converges to a certain probability measure as $T \to \infty$. Meanwhile, functional limit theorems became a useful tool in the value-distribution theory of Dirichlet series, not only with respect to universality.

4.2 Limit Theorems for Dirichlet Polynomials

The first step in the proof of Theorem 4.3 is to establish an analogous result for Dirichlet polynomials. For $N \in \mathbb{N}$ let

$$f_N(s) = \sum_{n=1}^{N} \frac{\alpha(n)}{n^s},$$

where $\alpha(n) \in \mathbb{C}$ for $1 \le n \le N$. Further, put

$$f_N(s,g) = \sum_{n=1}^{N} \frac{\alpha(n)g(n)}{n^s},$$

where $g \in \mathcal{G}$ is a unimodular, completely multiplicative arithmetic function; in particular, $f_N(s, 1) = f_N(s)$, where $1 : \mathbb{N} \to \mathbb{C}$ is the arithmetic function constant 1 (i.e., the principal character $\chi_0 \bmod 1$). Later, we will put $\alpha(n) = a(n)$, where $a(n)$ is the nth Dirichlet coefficient of $\mathcal{L} \in \tilde{\mathcal{S}}$. We define a probability measure $\mathbf{P}_{T,N}^g(A)$, corresponding to $\mathbf{P}_T$, by setting

$$\mathbf{P}_{T,N}^g(A) = \frac{1}{T} \operatorname{meas} \{\tau \in [0, T] : f_N(s + \mathrm{i}\tau, g) \in A\} \tag{4.10}$$

for $A \in \mathcal{B}(\mathcal{H}(\mathcal{D}))$. First, we show that the measure $\mathbf{P}_{T,N}^g$ converges weakly to some measure as T tends to infinity.

Lemma 4.4. *Let $g \in \mathcal{G}$. Then there exists a probability measure $\mathbf{P}_N^g$ on $(\mathcal{H}(\mathcal{D}), \mathcal{B}(\mathcal{H}(\mathcal{D})))$ such that $\mathbf{P}_{T,N}^g$ converges weakly to $\mathbf{P}_N^g$ as $T \to \infty$.*

Proof. Let $p_1, \ldots, p_r$ be the distinct prime divisors of the product

$$\prod_{\substack{n=1 \\ \alpha(n) \neq 0}}^{N} n. \tag{4.11}$$

Moreover, set

$$\Omega_r = \prod_{j=1}^{r} \gamma_{p_j}.$$

Finally, define the function $h_g : \Omega_r \to \mathcal{H}(\mathcal{D})$ by

$$h_g(x_1, \ldots, x_r) = \sum_{n=1}^{N} \frac{\alpha(n)g(n)}{n^s} \prod_{j=1}^{r} x_j^{-\nu(n;p_j)}.$$

Obviously, h_g is continuous on Ω_r. Since $\nu(n; p)$ is the exponent in the prime factorization of n,

$$n = \prod_{j=1}^{r} p_j^{\nu(n;p_j)},$$

h_g satisfies the identity

$$f_N(s + \mathrm{i}\tau, g) = \sum_{n=1}^{N} \frac{\alpha(n)g(n)}{n^s} \prod_{j=1}^{r} p^{-\mathrm{i}\tau\nu(n;p_j)} = h_g(p_1^{\mathrm{i}\tau}, \ldots, p_r^{\mathrm{i}\tau}). \tag{4.12}$$

Now we define a probability measure $\mathbf{H}_T$ on $(\Omega_r, \mathcal{B}(\Omega_r))$ by

$$\mathbf{H}_T(A) = \frac{1}{T} \operatorname{meas} \{\tau \in [0, T] : (p_1^{\mathrm{i}\tau}, \ldots, p_r^{\mathrm{i}\tau}) \in A\}$$

for $A \in \mathcal{B}(\Omega_r)$. For $n_1, \ldots, n_r \in \mathbb{N}$, the Fourier transform h_T of $\mathbf{H}_T$ is given by

$$h_T(n_1, \ldots, n_r) = \int_{\Omega_r} x_1^{n_1} \cdot \ldots \cdot x_r^{n_r} \, d\mathbf{H}_T = \frac{1}{T} \int_0^T \prod_{j=1}^{r} p_j^{\mathrm{i}\tau n_j} \, d\tau.$$

Hence, $h_T(0,\ldots,0) = 1$, and

$$h_T(n_1,\ldots,n_r) = \frac{\exp\left(iT\sum_{j=1}^r n_j \log p_j\right) - 1}{iT\sum_{j=1}^r n_j \log p_j}$$

if at least one $n_j \neq 0$. Since the logarithm of the prime numbers are linearly independent, we find in the latter case

$$\lim_{T\to\infty} h_T(n_1,\ldots,n_r) = 0.$$

Thus

$$\lim_{T\to\infty} h_T(n_1,\ldots,n_r) = \begin{cases} 1 & \text{if } (n_1,\ldots,n_r) = (0,\ldots,0), \\ 0 & \text{otherwise.} \end{cases} \quad (4.13)$$

By Theorem 3.11 the measure $\mathbf{H}_T$ converges weakly to a probability measure on $(\Omega_r, \mathcal{B}(\Omega_r))$ as T tends to infinity; since this measure is uniquely determined by its Fourier transform (4.13), it has to be the Haar measure m on $(\Omega_r, \mathcal{B}(\Omega_r))$. In view of the continuity of the function h_g, formula (4.12), and Theorem 3.3, we deduce that the probability measure $\mathbf{P}_{T,N}^g$ converges weakly to

$$\text{m} \circ h_g^{-1} = \text{m}h_g^{-1}$$

as T tends to infinity. This proves the lemma. $\qquad\square$

Now we shall show that the limiting measure $\mathbf{P}_N^g = \text{m}h_g^{-1}$ is independent of g.

Lemma 4.5. *Let g be a unimodular completely multiplicative function. Then both probability measures $\mathbf{P}_{T,N}^g$ and $\mathbf{P}_{T,N}^1$ converge weakly to the same measure as $T \to \infty$.*

Proof. Let $p_1,\ldots,p_r$ be, as in the last proof, the prime divisors of (4.11). We define the function $h^* : \Omega_r \to \Omega_r$ by

$$h^*(x_1,\ldots,x_r) = (x_1\exp(-i\theta_1),\ldots,x_r\exp(-i\theta_r)),$$

where $\theta_j := \arg g(p_j)$ for $1 \leq j \leq r$. It follows from Lemma 4.1 that the probability measures $\mathbf{P}_{T,N}^g$ and $\mathbf{P}_{T,N}^1$ converge weakly to the measures $\text{m}h_g^{-1}$ and $\text{m}h_1^{-1}$, respectively. A simple computation shows

$$h_1(x_1,\ldots,x_r) = \sum_{n=1}^N \frac{\alpha(n)}{n^s} \prod_{j=1}^r \exp(i\nu(n;p_j)\theta_j)x_j^{-\nu(n;p_j)}$$

$$= \sum_{n=1}^N \frac{\alpha(n)g(n)}{n^s} \prod_{j=1}^r x_j^{-\nu(n;p_j)}$$

$$= h_g(h^*(x_1,\ldots,x_r)).$$

It follows that:
$$mh_1^{-1} = m(h_g(h^*))^{-1} = m(h^*)^{-1}h_g^{-1}.$$

Since the Haar measure m is invariant with respect to translation of points in Ω, we deduce $m(h^*)^{-1} = m$. In addition with the last but one identity we get $mh_1^{-1} = mh_g^{-1}$. The lemma is proved.   □

4.3 An Ergodic Process

A one-parameter family $\{\phi_\tau\}$ of transformations on a compact topological abelian group G is called a one-parameter group of transformations if

$$\phi_{\tau_1+\tau_2}(g) = \phi_{\tau_1}(\phi_{\tau_2}(g)) \quad \text{and} \quad \phi_{-\tau}(g) = \phi_\tau^{-1}(g)$$

for all real numbers τ, τ_1 and τ_2 and all $g \in G$. We consider one-parameter groups of measurable transformations on Ω. Define

$$a_\tau : \mathbb{R} \to \Omega, \quad \tau \mapsto a_\tau = \{p^{-i\tau} : p \text{ prime}\}.$$

This map is continuous, and the image of every neighbourhood of the neutral element of $\mathbb{R}$ is a neighbourhood of the neutral element in Ω. Clearly, the family $\{a_\tau : \tau \in \mathbb{R}\}$ is a one-parameter group. Moreover, we define the one-parameter family $\{\phi_\tau : \tau \in \mathbb{R}\}$ of transformations on Ω by setting

$$\phi_\tau(\omega) = a_\tau\omega \tag{4.14}$$

for $\omega \in \Omega$. Then $\{\phi_\tau : \tau \in \mathbb{R}\}$ defines a one-parameter group of measurable and measure preserving transformations on Ω.

Now we transport the notions from ergodic theory of random processes to one-parameter groups. Let $\{\varphi_\tau : \tau \in \mathbb{R}\}$ be an arbitrary one-parameter group of measurable transformations on G. A set $A \in \mathcal{B}(G)$ is invariant with respect to the group $\{\varphi_\tau : \tau \in \mathbb{R}\}$ if for each τ the sets A and $A_\tau := \varphi_\tau(A)$ differ one from another by a set of zero m-measure, i.e.,

$$m(A\Delta A_\tau) = 0,$$

where m is the Haar measure on G. All invariant sets form a sub-σ-field of $\mathcal{B}(G)$. A one-parameter group is said to be ergodic if its σ-field of invariant sets consists only of sets having Haar measure equal to 0 or 1.

Lemma 4.6. *The one-parameter group $\{\phi_\tau : \tau \in \mathbb{R}\}$ is ergodic.*

Proof. For a positive rational number $\alpha = \frac{a}{b}$ with a, b coprime and $\omega \in \Omega$ we put

$$\omega(\alpha) = \frac{\omega(a)}{\omega(b)}.$$

This extends the function ω from $\mathbb{N}$ to the set of positive rational numbers. Now define $\chi_\alpha : \Omega \to \gamma$ by $\chi_\alpha(\omega) = \omega(\alpha)$. Then, for any fixed positive rational number α, χ_α is a character of Ω. On the other side, since the dual group of Ω has a representation as a direct sum

$$\bigoplus_p \mathbb{Z}_p, \quad \text{where} \quad \mathbb{Z}_p = \mathbb{Z} \quad \text{for any prime } p$$

(see [132, Chap. 6]), for any character χ of Ω there exists a positive rational α such that $\chi = \chi_\alpha$.

Now suppose that $A \in \mathcal{B}(\Omega)$ satisfies $\mathrm{m}(A \Delta A_\tau) = 0$ for any real τ. We have to show that either $\mathrm{m}(A) = 0$ or $\mathrm{m}(A) = 1$. Denote by $\mathbf{1}(\omega; A)$ the indicator function of A, i.e.,

$$\mathbf{1}(\omega; A) = \begin{cases} 1 & \text{if} \quad \omega \in A, \\ 0 & \text{otherwise.} \end{cases}$$

Since the set A is invariant, it follows that for arbitrary $\tau \in \mathbb{R}$ we have

$$\mathbf{1}(a_\tau \omega; A) = \mathbf{1}(\omega; A) \quad \text{for almost all} \quad \omega \in \Omega \qquad (4.15)$$

with respect to the Haar measure m.

First, assume that χ is non-trivial. By the same reasoning as above, there exists a positive rational number α such that $\chi = \chi_\alpha$. Consequently,

$$\chi(a_\tau) = \alpha^{-i\tau},$$

and so we can find a τ_0 for which $\chi(a_{\tau_0}) \neq 1$. In view of (4.15) this implies for the Fourier transform of the indicator function

$$\hat{\mathbf{1}}(\chi; A) = \int_\Omega \chi(\omega)\mathbf{1}(\omega; A)\,\mathrm{m}(\,\mathrm{d}\omega) = \chi(a_{\tau_0}) \int_\Omega \chi(\omega)\mathbf{1}(\omega; A)\,\mathrm{m}(\,\mathrm{d}\omega)$$

$$= \chi(a_{\tau_0})\hat{\mathbf{1}}(\chi; A).$$

Since we have chosen τ_0 such that $\chi(a_{\tau_0}) \neq 1$, it follows that $\hat{\mathbf{1}}(\chi; A) = 0$. Obviously, this argument holds for any non-trivial character of Ω.

Now suppose that χ_0 is the trivial character of Ω, and let us assume that $\hat{\mathbf{1}}(\chi_0; A) = u$. By the orthogonality relation for characters (3.2) we find

$$\hat{u}(\chi) = u \int_\Omega \chi(\omega)\,\mathrm{m}(\,\mathrm{d}\omega) = \begin{cases} u & \text{if} \quad \chi = \chi_0, \\ 0 & \text{otherwise.} \end{cases}$$

Since $\hat{\mathbf{1}}(\chi; A) = 0$ for $\chi \neq \chi_0$, we obtain, for any character χ,

$$\hat{\mathbf{1}}(\chi; A) = \hat{u}(\chi).$$

The function $\mathbf{1}(\omega; A)$ is uniquely determined by its Fourier transform. Hence, it follows that $\mathbf{1}(\omega; A) = u$ for almost all $\omega \in \Omega$. Since $\mathbf{1}(\omega; A)$ is the indicator function of A, we get $u = 0$ or $u = 1$, i.e., either $\mathbf{1}(\omega; A) = 0$ for almost all ω, or $\mathbf{1}(\omega; A) = 1$ for almost all $\omega \in \Omega$. This implies that $\mathrm{m}(A) = 0$ or $\mathrm{m}(A) = 1$, which proves the Lemma. $\square$

Lemma 4.7. *Let* $\sigma \in (\sigma_m, 1)$. *Then, for almost all* $\omega \in \Omega$,

$$\int_0^T |\mathcal{L}(\sigma + \mathrm{i}t, \omega)|^2 \, \mathrm{d}t \ll T.$$

Recall that σ_m is the abscissa of the mean-square half-plane for $\mathcal{L}$.

Proof. We start as in the proof of Lemma 4.1. For $n \in \mathbb{N}$ let

$$\xi_n(\sigma, \omega) = \frac{a(n)\omega(n)}{n^\sigma}.$$

The $\xi_n(\sigma, \omega)$ define a sequence of pairwise orthogonal random variables with second moment

$$\mathbf{E}|\xi_n(\sigma, \omega)|^2 = \frac{|a(n)|^2}{n^{2\sigma}}.$$

Thus the random variable

$$\xi(\sigma, \omega) := \left| \sum_{n=1}^\infty \xi_n(\sigma, \omega) \right|^2 = |\mathcal{L}(\sigma, \omega)|^2$$

has the finite expectation

$$\mathbf{E}\xi(\sigma, \omega) = \sum_{n=1}^\infty \mathbf{E}|\xi_n(\sigma, \omega)|^2 = \sum_{n=1}^\infty \frac{|a(n)|^2}{n^{2\sigma}}; \tag{4.16}$$

the convergence follows from the Ramanujan hypothesis (i) (since $\sigma_m \geq \frac{1}{2}$ and so $2\sigma > 1$). It is obvious that

$$\xi(\sigma, \phi_\tau(\omega)) = |\mathcal{L}(\sigma, a_\tau\omega)|^2 = |\mathcal{L}(\sigma + \mathrm{i}t, \omega)|^2 \tag{4.17}$$

for any transformation (4.14). Since the Haar measure m is invariant, the equality $\mathrm{m}(\phi_\tau(A)) = \mathrm{m}(A)$ is valid for every $A \in \mathcal{B}(\Omega)$ and every $\tau \in \mathbb{R}$. Hence, $|\mathcal{L}(\sigma + \mathrm{i}\tau, \omega)|^2$ is a strongly stationary process. We shall show that it is even ergodic.

Let $\mathbf{Q}$ be the probability measure on $(\Omega, \mathcal{B}(\Omega))$ which is determined by the random process $|\mathcal{L}(\sigma + \mathrm{i}\tau, \omega)|^2$. Further, let A be an invariant set of $|\mathcal{L}(\sigma + \mathrm{i}\tau, \omega)|^2$, i.e.,

$$\mathbf{Q}(A \triangle A_u) = 0,$$

where $A_u = \phi_u(A)$ with transformations ϕ_u defined by (4.14). Now put

$$A' = \{\omega \in \Omega : |\mathcal{L}(\sigma + \mathrm{i}\tau, \omega)|^2 \in A\}$$

and

$$A'_u = \{\omega \in \Omega : |\mathcal{L}(\sigma + \mathrm{i}\tau, \omega)|^2 \in A_u\}.$$

It follows that:
$$A' = \{\omega \in \Omega : |\mathcal{L}(\sigma, a_\tau \omega)|^2 \in A\}$$

and

$$A'_u = \{\omega \in \Omega : |\mathcal{L}(\sigma + i\tau + iu, \omega)|^2 \in A\}$$
$$= \{\omega \in \Omega : |\mathcal{L}(\sigma + i\tau, a_u \omega)|^2 \in A\}.$$

Hence $A'_u = \phi_u(A')$. Since $(A \triangle A_u)' = A' \triangle A'_u$,

$$m(A' \triangle A'_u) = m((A \triangle A_u)') = \mathbf{Q}(A \triangle A_u) = 0,$$

which shows that A' is an invariant set with respect to $\{\phi_\tau : \tau \in \mathbb{R}\}$. However, by Lemma 4.6 the group $\{\phi_\tau : \tau \in \mathbb{R}\}$ is ergodic. Therefore, $m(A') = 0$ or $m(A') = 1$. Hence it follows that $\mathbf{Q}(A) = 0$ or $\mathbf{Q}(A) = 1$, i.e., the process $|\mathcal{L}(\sigma + i\tau, \omega)|^2$ is ergodic.

Clearly, $\xi(\sigma, \phi_\tau(\omega)) \geq 0$. Hence, from (4.16), (4.17) and the Birkhoff–Khintchine Theorem 3.13, we get

$$\lim_{T \to \infty} \frac{1}{T} \int_0^T |\mathcal{L}(\sigma + i\tau, \omega)|^2 \, d\tau = \lim_{T \to \infty} \frac{1}{T} \int_0^T \xi(\sigma, \phi_\tau(\omega)) \, d\tau$$
$$= \mathbf{E}\xi(\sigma, \omega) < \infty$$

for almost all $\omega \in \Omega$. This gives the assertion of the lemma. $\square$

4.4 Approximation in the Mean

Let $\sigma_1 > \sigma_m$, and define for $N \in \mathbb{N}$ and $-\sigma_1 \leq \sigma \leq \sigma_1$

$$\ell_N(s) = \frac{s}{\sigma_1} \Gamma\left(\frac{s}{\sigma_1}\right) N^s.$$

Further, for $\sigma > \sigma_m$, put

$$\mathcal{L}_N(s) = \frac{1}{2\pi i} \int_{\sigma_1 - i\infty}^{\sigma_1 + i\infty} \mathcal{L}(s + z)\ell_N(z) \frac{dz}{z}. \tag{4.18}$$

Taking into account Stirling's formula (2.17) and axiom (iii), the existence of the integral follows. Now we shall show that $\mathcal{L}_N(s)$ approximates $\mathcal{L}(s)$ in an appropriate mean.

Lemma 4.8. *Let K be a compact subset of $\mathcal{D}$. Then*

$$\lim_{N \to \infty} \limsup_{T \to \infty} \frac{1}{T} \int_0^T \max_{s \in K} |\mathcal{L}(s + i\tau) - \mathcal{L}_N(s + i\tau)| \, d\tau = 0.$$

This result should be compared with formula (1.19) in the proof of Voronin's universality Theorem 1.7.

Proof. Let Re $z = \sigma_1$. Since $\sigma + \sigma_1 > 1$ for $s \in K$, the function $\mathcal{L}(s+z)$ has a representation as an absolutely convergent Dirichlet series:

$$\mathcal{L}(s+z) = \sum_{n=1}^{\infty} \frac{a(n)}{n^{s+z}}.$$

Next we twist $\mathcal{L}(s)$ by an arithmetic function b_N. For $n, N \in \mathbb{N}$ let

$$b_N(n) = \frac{1}{2\pi i} \int_{\sigma_1 - i\infty}^{\sigma_1 + i\infty} \frac{\ell_N(z)}{n^z} \frac{dz}{z}.$$

By Stirling's formula (2.17),

$$b_N(n) \ll \frac{1}{n^{\sigma_1}} \int_{-\infty}^{\infty} |\ell_N(\sigma_1 + it)| \, dt \ll \frac{1}{n^{\sigma_1}}.$$

Consequently, the series

$$\sum_{n=1}^{\infty} \frac{a(n) b_N(n)}{n^s},$$

converges absolutely for $\sigma > 1 - \sigma_1$. Hence,

$$\frac{1}{2\pi i} \int_{\sigma_1 - i\infty}^{\sigma_1 + i\infty} \mathcal{L}(s+z) \ell_N(z) \frac{dz}{z} = \sum_{n=1}^{\infty} \frac{a(n)}{n^s} \frac{1}{2\pi i} \int_{\sigma_1 - i\infty}^{\sigma_1 + i\infty} \frac{\ell_N(z)}{n^z} \frac{dz}{z}$$

$$= \sum_{n=1}^{\infty} \frac{a(n) b_N(n)}{n^s};$$

interchanging integration and summation is allowed by absolute convergence. Thus (4.18) yields

$$\mathcal{L}_N(s) = \sum_{n=1}^{\infty} \frac{a(n) b_N(n)}{n^s}.$$

Using Mellin's inversion formula (2.15), we obtain

$$b_N(n) = \frac{1}{2\pi i \sigma_1} \int_{\sigma_1 - i\infty}^{\sigma_1 + i\infty} \Gamma\left(\frac{z}{\sigma_1}\right) \left(\frac{N}{n}\right)^z dz = \exp\left(-\left(\frac{n}{N}\right)^{\sigma_1}\right)$$

Thus we get

$$\mathcal{L}_N(s) = \sum_{n=1}^{\infty} \frac{a(n)}{n^s} \exp\left(-\left(\frac{n}{N}\right)^{\sigma_1}\right), \tag{4.19}$$

where the series converges absolutely for $\sigma > \sigma_m$. Now we move the path of integration in (4.18) to the left. The integrand has a simple pole at $z = 0$, and a pole of order f at $z = 1 - s$ if $\mathcal{L}(s)$ has a pole at $s = 1$ of the same order. Since K is compact in $\mathcal{D}$, we may choose $\epsilon > 0$ such that $\sigma \in [\sigma_m + \epsilon, 1 - \epsilon]$ for any $s \in K$. Putting $\sigma_2 = \sigma_m + \frac{\epsilon}{2}$, we get by the calculus of residues

$$\mathcal{L}_N(s) = \frac{1}{2\pi i} \int_{\sigma_2 - \sigma - i\infty}^{\sigma_2 - \sigma + i\infty} \mathcal{L}(s+z)\ell_N(z)\frac{dz}{z} \tag{4.20}$$
$$+ \{\mathrm{Res}_{z=0} + \mathrm{Res}_{z=1-s}\}\left(\mathcal{L}(s+z)\frac{\ell_N(z)}{z}\right).$$

On behalf of the functional equation for Γ, for the residue at $z = 0$ we find

$$\mathrm{Res}_{z=0}\left(\mathcal{L}(s+z)\frac{\ell_N(z)}{z}\right) = \lim_{z \to 0} \mathcal{L}(s+z)\ell_N(z) = \mathcal{L}(s).$$

If $\mathcal{L}(s)$ has a pole of order f at $s = 1$, then

$$\mathrm{Res}_{z=1-s}\left(\mathcal{L}(s+z)\frac{\ell_N(z)}{z}\right) = \mathrm{Res}_{w=1}\left(\mathcal{L}(w)\frac{\ell_N(w-s)}{w-s}\right)$$
$$= \frac{1}{(f-1)!}\left(\frac{d}{dw}\right)^{f-1}(w-1)^f\mathcal{L}(w)\frac{\ell_N(w-s)}{w-s}\bigg|_{w=1} = \lambda(s),$$

say.

Let $\mathcal{C}$ be a Jordan curve (i.e., a simple closed contour) in $\mathcal{D}$ of length $|\mathcal{C}|$, enclosing the set K, and denote by δ the distance from $\mathcal{C}$ to K. By Cauchy's formula,

$$\max_{s\in K}|\mathcal{L}(s+i\tau) - \mathcal{L}_N(s+i\tau)| \le \frac{1}{2\pi\delta}\int_{\mathcal{C}}|\mathcal{L}(z+i\tau) - \mathcal{L}_N(z+i\tau)|\,|dz|.$$

Thus, for sufficiently large T,

$$\frac{1}{T}\int_0^T \max_{s\in K}|\mathcal{L}(s+i\tau) - \mathcal{L}_N(s+i\tau)|\,d\tau$$
$$\ll \frac{1}{\delta T}\int_{\mathcal{C}}\int_{\mathrm{Im}\,z}^{T+\mathrm{Im}\,z}|\mathcal{L}(\mathrm{Re}\,z+i\tau) - \mathcal{L}_N(\mathrm{Re}\,z+i\tau)|\,d\tau\,|dz|$$
$$\ll \frac{|\mathcal{C}|}{\delta T}\max_{\sigma;\,s\in\mathcal{C}}\int_0^{2T}|\mathcal{L}(\sigma+it) - \mathcal{L}_N(\sigma+it)|\,dt. \tag{4.21}$$

We may choose the contour $\mathcal{C}$ such that for $s \in \mathcal{C}$ the inequalities $\sigma \ge \sigma_m + \frac{3\epsilon}{4}$ and $\delta \ge \frac{\epsilon}{4}$ hold. In view of (4.20)

$$\mathcal{L}(\sigma+it) - \mathcal{L}_N(\sigma+it) \ll \int_{-\infty}^{\infty}|\mathcal{L}(\sigma_2+it+i\tau)||\ell_N(\sigma_2-\sigma+i\tau)|\,d\tau$$
$$+ |\lambda(1-\sigma-it)|.$$

Thus, for the same σ, we find

$$\frac{1}{T} \int_0^{2T} |\mathcal{L}(\sigma + it) - \mathcal{L}_N(\sigma + it)| \, dt$$

$$\ll \frac{1}{T} \int_{-\infty}^{\infty} |\ell_N(\sigma_2 - \sigma + i\tau)| \int_{-|\tau|}^{|\tau|+2T} |\mathcal{L}(\sigma_2 + it)| \, dt \, d\tau$$

$$+ \frac{1}{T} \int_0^{2T} |\lambda(1 - \sigma - it)| \, dt.$$

By the Cauchy–Schwarz inequality and the existence of the mean-square for $\sigma > \sigma_m$,

$$\int_0^T |\mathcal{L}(\sigma + it)| \, dt \ll T^{1/2} \left(\int_0^T |\mathcal{L}(\sigma + it)|^2 \, dt \right)^{1/2} \ll T.$$

This leads in the last but one estimate to

$$\frac{1}{T} \int_0^{2T} |\mathcal{L}(\sigma + it) - \mathcal{L}_N(\sigma + it)| \, dt$$

$$\ll \max_{\sigma; s \in \mathcal{C}} \int_{-\infty}^{\infty} |\ell_N(\sigma_2 - \sigma + i\tau)| \left(1 + \frac{|\tau|}{T} \right) d\tau \qquad (4.22)$$

$$+ \frac{1}{T} \max_{\sigma; s \in K} \int_0^{2T} |\lambda(1 - \sigma - it)| \, dt.$$

Let $I = [\sigma_m - 1 + \frac{3\epsilon}{2}, -\frac{\epsilon}{4}]$. Then, by Stirling's formula (2.17), the latter expression is bounded by

$$\max_{\sigma \in I} \int_{-\infty}^{\infty} |\ell_N(\sigma + i\tau)|(1 + |\tau|) \, d\tau + o(1),$$

as $T \to \infty$. By the definition of $\ell_N(s)$, we obtain

$$\lim_{N \to \infty} \max_{\sigma \in I} \int_{-\infty}^{\infty} |\ell_N(\sigma + i\tau)|(1 + |\tau|) \, d\tau = 0.$$

This in addition with (4.21) and (4.22) proves the lemma.   □

4.5 A Limit Theorem for Absolutely Convergent Series

We define for $\omega \in \Omega$, $s \in \mathcal{D}$ and $N \in \mathbb{N}$

$$\mathcal{L}_N(s, \omega) = \sum_{n=1}^{\infty} \frac{a(n)\omega(n)}{n^s} \exp\left(-\left(\frac{n}{N} \right)^{\sigma_1} \right).$$

Due to the absolute convergence of the series $\mathcal{L}_N(s)$, given by (4.19), $\mathcal{L}_N(s,\omega)$ also converges absolutely for $\sigma > \sigma_m$. Further, we define the probability measures $\mathbf{P}_{T,N}$ and $\mathbf{Q}_{T,N}$ by setting

$$\mathbf{P}_{T,N}(A) = \frac{1}{T} \operatorname{meas}\{\tau \in [0,T] : \mathcal{L}_N(s+i\tau) \in A\}$$

and

$$\mathbf{Q}_{T,N}(A) = \frac{1}{T} \operatorname{meas}\{\tau \in [0,T] : \mathcal{L}_N(s+i\tau,\omega) \in A\}$$

for $A \in \mathcal{B}(\mathcal{H}(\mathcal{D}))$. Now we prove that both measures converge to the same limit as T tends to infinity.

Lemma 4.9. *There exists a probability measure $\mathbf{P}'_N$ on $(\mathcal{H}(\mathcal{D}), \mathcal{B}(\mathcal{H}(\mathcal{D})))$ such that the measures $\mathbf{P}_{T,N}$ and $\mathbf{Q}_{T,N}$ both converge weakly to $\mathbf{P}'_N$ as $T \to \infty$.*

Proof. For $\omega \in \Omega$, $s \in \mathcal{D}$ and $M,N \in \mathbb{N}$ let

$$\mathcal{L}_{N,M}(s) = \sum_{n=1}^{M} \frac{a(n)}{n^s} \exp\left(-\left(\frac{n}{N}\right)^{\sigma_1}\right)$$

and

$$\mathcal{L}_{N,M}(s,\omega) = \sum_{n=1}^{M} \frac{a(n)\omega(n)}{n^s} \exp\left(-\left(\frac{n}{N}\right)^{\sigma_1}\right).$$

Define the probability measures $\mathbf{P}_{T,N,M}$ and $\mathbf{Q}_{T,N,M}$ by putting

$$\mathbf{P}_{T,N,M}(A) = \frac{1}{T} \operatorname{meas}\{\tau \in [0,T] : \mathcal{L}_{N,M}(s+i\tau) \in A\}$$

and

$$\mathbf{Q}_{T,N,M}(A) = \frac{1}{T} \operatorname{meas}\{\tau \in [0,T] : \mathcal{L}_{N,M}(s+i\tau,\omega) \in A\}$$

for $A \in \mathcal{B}(\mathcal{H}(\mathcal{D}))$. By Lemma 4.5 both measures converge weakly to the same probability measure, say $\mathbf{P}_{N,M}$, as T tends to infinity.

First, we show that the family $\{\mathbf{P}_{N,M}\}$ is tight for fixed N. Let θ be a random variable uniformly distributed on $[0,1]$, defined on some probability space $(\mathbb{R}, \mathcal{B}(\mathbb{R}), \mathbf{P}^*)$. Let

$$X_{T,N,M}(s) = \mathcal{L}_{N,M}(s+i\theta T). \tag{4.23}$$

Then, by Lemma 4.4,

$$X_{T,N,M} \xrightarrow[T\to\infty]{\mathcal{D}} X_{N,M}, \tag{4.24}$$

where $X_{N,M} = X_{N,M}(s)$ is an $\mathcal{H}(\mathcal{D})$-valued random element with distribution $\mathbf{P}_{N,M}$. Let $\{K_j\}$ be a sequence of compact subsets of $\mathcal{D}$ satisfying the conditions of Lemma 3.14 with $\mathcal{G} = \mathcal{D}$, i.e.,

- $K_j \subset K_{j+1}$ for any $j \in \mathbb{N}$;
- If K is compact and $K \subset \mathcal{D}$, then $K \subset K_j$ for some $j \in \mathbb{N}$;

and

$$\mathcal{D} = \bigcup_{j=1}^{\infty} K_j.$$

Further, for $j \in \mathbb{N}$ let M_j be a positive parameter, which will be defined later. By Chebyshev's inequality, Lemma 3.7,

$$\mathbf{P}^* \left(\max_{s \in K_j} |X_{T,N,M}(s)| > M_j \right) \le \frac{1}{M_j T} \int_0^T \max_{s \in K_j} |\mathcal{L}_{N,M}(s + i\tau)| \, d\tau < \infty.$$

Hence,

$$\limsup_{T \to \infty} \mathbf{P}^* \left(\max_{s \in K_j} |X_{T,N,M}(s)| > M_j \right)$$

$$\le \frac{1}{M_j} \limsup_{T \to \infty} \frac{1}{T} \int_0^T \max_{s \in K_j} |\mathcal{L}_{N,M}(s + i\tau)| \, d\tau.$$

Since the series for $\mathcal{L}_N(s)$ converges absolutely in $\mathcal{D}$,

$$\sup_{N \in \mathbb{N}} \limsup_{T \to \infty} \frac{1}{T} \int_0^T \max_{s \in K_j} |\mathcal{L}_{N,M}(s + i\tau)| \, d\tau < R_j$$

with some positive constant R_j depending on K_j. Setting

$$M_j = \frac{2^j R_j}{\epsilon}, \tag{4.25}$$

we obtain

$$\limsup_{T \to \infty} \mathbf{P}^* \left(\max_{s \in K_j} |X_{T,N,M}(s)| > M_j \right) < \frac{\epsilon}{2^j}. \tag{4.26}$$

By Theorem 3.3 and (4.24) it follows that

$$\max_{s \in K_j} |X_{T,N,M}(s)| \xrightarrow[T \to \infty]{\mathcal{D}} \max_{s \in K_j} |X_{N,M}(s)|.$$

Thus, we obtain from (4.26)

$$\mathbf{P}^* \left(\max_{s \in K_j} |X_{N,M}(s)| > M_j \right) \le \frac{\epsilon}{2^j}. \tag{4.27}$$

Now, for $j \in \mathbb{N}$, define a function $k_j : \mathcal{H}(\mathcal{D}) \to \mathbb{R}$ by

$$k_j(f) = \max_{s \in K_j} |f(s)|.$$

The function k_j is continuous. Furthermore, let

$$\mathcal{H}_\epsilon = \{ f \in \mathcal{H}(\mathcal{D}) : k_j(f) \le M_j \quad \text{for } j \in \mathbb{N} \}.$$

A family of analytic functions is said to be compact on a region $\mathcal{G}$ if every sequence of this family contains a subsequence which converges uniformly on every compact subset of $\mathcal{G}$. The theorem of Montel (see [62, Sect. VII.2] or [313]) asserts that if a family of analytic functions on $\mathcal{G}$ is uniformly bounded on every compact subset of $\mathcal{G}$, then it is compact. Clearly, the family of functions $\mathcal{H}_\epsilon$ is uniformly bounded on every compact subset $K \subset \mathcal{D}$, and thus it follows that $\mathcal{H}_\epsilon$ is compact. Using (4.27), we obtain

$$\mathbf{P}^*(X_{N,M}(s) \in \mathcal{H}_\epsilon) \geq 1 - \sum_{j=1}^\infty \mathbf{P}^*\left(\max_{s \in K_j} |X_{N,M}(s)| > M_j\right)$$

$$\geq 1 - \epsilon \sum_{j=1}^\infty \frac{1}{2^j} = 1 - \epsilon.$$

Since $\mathbf{P}_{N,M}$ is the distribution of $X_{N,M}$, the inequality

$$\mathbf{P}_{N,M}(\mathcal{H}_\epsilon) \geq 1 - \epsilon$$

holds for all $M, N \in \mathbb{N}$. Therefore the family of probability measures $\{\mathbf{P}_{N,M}\}$ is tight with respect to M, and by Prokhorov's Theorem 3.4 it follows that it is relatively compact.

It is clear that, for $\sigma > \sigma_m$,

$$\lim_{M \to \infty} \mathcal{L}_{N,M}(s) = \mathcal{L}_N(s),$$

and since the series defining $\mathcal{L}_N(s)$ converges absolutely, the convergence is uniform in any half-plane $\sigma \geq \sigma_m + \epsilon$. Thus, for every $\epsilon > 0$,

$$\lim_{M \to \infty} \limsup_{T \to \infty} \frac{1}{T} \text{meas} \{\tau \in [0, T] : \varrho(\mathcal{L}_{N,M}(s + i\tau), \mathcal{L}_N(s + i\tau)) \geq \epsilon\}$$

$$\leq \lim_{M \to \infty} \limsup_{T \to \infty} \frac{1}{\epsilon T} \int_0^T \varrho(\mathcal{L}_{N,M}(s + i\tau), \mathcal{L}_N(s + i\tau)) \, d\tau = 0;$$

the definition of the metric ϱ in the space of analytic function was given in Sect. 3.7. Now let, similarly to (4.23),

$$Y_{T,N}(s) = \mathcal{L}_N(s + i\theta T).$$

Then the latter inequality can be rewritten as

$$\lim_{M \to \infty} \limsup_{T \to \infty} \mathbf{P}^*(\varrho(X_{T,N,M}(s), Y_{T,N}(s)) \geq \epsilon) = 0.$$

Let $\{\mathbf{P}_{N,M_k}\}$ be a subsequence $\{\mathbf{P}_{N,M}\}$ which converges weakly to some measure $\mathbf{P}'_N$. Then,

$$X_{N,M_k} \xrightarrow[M_k \to \infty]{\mathcal{D}} \mathbf{P}'_N.$$

Since $\mathcal{H}(\mathcal{D})$ is separable (see Theorem 3.15), the conditions of Theorem 3.6 are fulfilled. Thus, we get by (4.24)

$$Y_{T,N} \xrightarrow[T\to\infty]{\mathcal{D}} \mathbf{P}'_N.$$

Hence, there exists a measure $\mathbf{P}'_N$ such that $\mathbf{P}_{T,N}$ converges weakly to $\mathbf{P}'_N$ as T tends to infinity, and this limit is independent of the subsequence $\{\mathbf{P}_{N,M_k}\}$. Since the family $\{\mathbf{P}_{N,M}\}$ is relatively compact, by Theorem 3.2, we obtain that $\mathbf{P}_{N,M}$ converges weakly to $\mathbf{P}'_N$ as M tends to infinity:

$$X_{N,M} \xrightarrow[M\to\infty]{\mathcal{D}} \mathbf{P}'_N.$$

With the same argument applied to the random elements $\mathcal{L}_{N,M}(s+i\theta T,\omega)$ and $\mathcal{L}_N(s+i\theta T,\omega)$ we find that $\mathbf{Q}_{T,N}$ converges weakly to $\mathbf{P}'_N$ as T tends to infinity. The assertion of the lemma follows. $\square$

Let Ω' be a subset of Ω such that for $\omega \in \Omega'$ the series (4.6) for $\mathcal{L}(s,\omega)$ converges uniformly on compact subsets of $\mathcal{D}$, and that for $\sigma > \sigma_m$ the estimate

$$\int_0^T |\mathcal{L}(\sigma + it,\omega)|^2 \, dt \ll T.$$

holds. It follows from Lemmas 4.1 and 4.7 that $\mathrm{m}(\Omega') = 1$. By the same argument as in the proof for Lemma 4.8 one can show

Lemma 4.10. *Let K be a compact subset of $\mathcal{D}$. Then, for $\omega \in \Omega'$,*

$$\lim_{N\to\infty} \limsup_{T\to\infty} \frac{1}{T} \int_0^T \max_{s\in K} |\mathcal{L}(s+i\tau,\omega) - \mathcal{L}_N(s+i\tau,\omega)| \, d\tau = 0.$$

4.6 Proof of the Main Limit Theorem

Define a probability measure $\mathbf{Q}_T$ by setting

$$\mathbf{Q}_T(A) = \frac{1}{T} \, \mathrm{meas}\, \{\tau \in [0,T] \,:\, \mathcal{L}(s+i\tau,\omega) \in A\}$$

for $A \in \mathcal{B}(\mathcal{H}(\mathcal{D}))$.

Lemma 4.11. *There exists a probability measure $\mathbf{P}'$ on $(\mathcal{H}(\mathcal{D}), \mathcal{B}(\mathcal{H}(\mathcal{D})))$ such that the measures $\mathbf{P}_T$ (given by (4.4)) and $\mathbf{Q}_T$ both converge weakly to $\mathbf{P}'$ as $T \to \infty$.*

Proof. Lemma 4.9 asserts that the measures $\mathbf{P}_{T,N}$ and $\mathbf{Q}_{T,N}$ both converge weakly with $T \to \infty$ to the same measure $\mathbf{P}'_N$.

First, we prove that the family $\{\mathbf{P}'_N\}$ is tight. We follow the argument given in the proof of Lemma 4.9 and use the notations introduced there.

Again, let θ be a random variable uniformly distributed on $[0,1]$, defined on some probability space $(\mathbb{R}, \mathcal{B}(\mathbb{R}), \mathbf{P}^*)$. Define

$$X_{T,N}(s) = \mathcal{L}_N(s + i\theta T).$$

Then, by Lemma 4.4,

$$X_{T,N} \xrightarrow[T \to \infty]{\mathcal{D}} X_N,$$

where X_N is an $\mathcal{H}(\mathcal{D})$-valued random element with distribution $\mathbf{P}'_N$. Now let $\{K_j\}$ be a sequence of compact subsets of $\mathcal{D}$ satisfying the conditions of Lemma 3.14 with $\mathcal{G} = \mathcal{D}$. Since the series for $\mathcal{L}_N(s)$ converges absolutely for $\sigma > \sigma_m$,

$$\sup_{N \in \mathbb{N}} \limsup_{T \to \infty} \frac{1}{T} \int_0^T \max_{s \in K_j} |\mathcal{L}_N(s + i\tau)| \, d\tau \le R_j < \infty.$$

By Chebyshev's inequality, Lemma 3.7,

$$\mathbf{P}^* \left(\max_{s \in K_j} |X_{T,N}(s)| > M_j \right) \le \frac{1}{M_j T} \int_0^T \max_{s \in K_j} |\mathcal{L}_N(s + i\tau)| \, d\tau.$$

Thus,

$$\limsup_{T \to \infty} \mathbf{P}^* \left(\max_{s \in K_j} |X_{T,N}(s)| > M_j \right) < \frac{\epsilon}{2^j},$$

where M_j and R_j are related by (4.25). We deduce

$$\mathbf{P}^*(X_N(s) \in \mathcal{H}_\epsilon) \ge 1 - \epsilon,$$

respectively,

$$\mathbf{P}'_N(\mathcal{H}_\epsilon) \ge 1 - \epsilon$$

for all $N \in \mathbb{N}$. Since the set $\mathcal{H}_\epsilon$ is compact, the family of probability measures $\{\mathbf{P}'_N\}$ is tight, and by Theorem 3.4 it follows that it is relatively compact.

Applying the Chebyshev inequality once more, by Lemma 4.8, we obtain, for every $\epsilon > 0$,

$$\lim_{N \to \infty} \limsup_{T \to \infty} \frac{1}{T} \operatorname{meas} \{\tau \in [0,T] : \varrho(\mathcal{L}(s + i\tau), \mathcal{L}_N(s + i\tau)) \ge \epsilon\}$$

$$\le \lim_{N \to \infty} \limsup_{T \to \infty} \frac{1}{\epsilon T} \int_0^T \varrho(\mathcal{L}(s + i\tau), \mathcal{L}_N(s + i\tau)) \, d\tau = 0.$$

Now let

$$Y_T(s) = \mathcal{L}(s + i\theta T).$$

Then the latter inequality can be rewritten as

$$\lim_{N \to \infty} \limsup_{T \to \infty} \mathbf{P}^*(\varrho(X_{T,N}(s), Y_T(s)) \ge \epsilon) = 0.$$

Let $\{\mathbf{P}'_{N_j}\}$ be a subsequence $\{\mathbf{P}'_N\}$ which converges weakly to some measure $\mathbf{P}'$. Then

$$X_{N_j} \xrightarrow[j\to\infty]{\mathcal{D}} \mathbf{P}'.$$

This result and Theorem 3.6 imply

$$Y_T \xrightarrow[T\to\infty]{\mathcal{D}} \mathbf{P}',$$

which is equivalent to the weak convergence of $\mathbf{P}_T$ to $\mathbf{P}'$. Taking into account that the family $\{\mathbf{P}_N\}$ is relatively compact, by Lemma 3.2 we get

$$X_N \xrightarrow[N\to\infty]{\mathcal{D}} \mathbf{P}'.$$

Repeating this argument for the random elements $\mathcal{L}_N(s + i\theta T, \omega)$ and $\mathcal{L}(s + i\theta T, \omega)$, Lemma 4.10 shows that $\mathbf{Q}_T$ converges weakly to $\mathbf{P}'$ as T tends to infinity. The lemma is proved. □

It remains to identify the limit of the probability measure $\mathbf{P}_T$ as $T \to \infty$. This will finish the

Proof. By Lemma 4.11 the measures, $\mathbf{P}_T$ and $\mathbf{Q}_T$ both converge weakly to the same limit $\mathbf{P}'$ as T tends to infinity. It remains to show that $\mathbf{P}' = \mathbf{P}$, where $\mathbf{P}$ is defined by (4.9).

Let $A \in \mathcal{B}(\mathcal{H}(\mathcal{D}))$ be a continuity set of $\mathbf{P}'$. By Theorem 3.1 and Lemma 4.11 it follows that:

$$\lim_{T\to\infty} \frac{1}{T} \operatorname{meas} \{\tau \in [0, T] : \mathcal{L}(s + i\tau, \omega) \in A\} = \mathbf{P}'(A). \qquad (4.28)$$

Fixing the set A, we define the random variable θ on $(\Omega, \mathcal{B}(\Omega), \mathrm{m})$ by

$$\theta(\omega) = \begin{cases} 1 & \text{if } \mathcal{L}(s, \omega) \in A, \\ 0 & \text{otherwise.} \end{cases}$$

Then

$$\mathbf{E}\theta = \int_\Omega \theta \, \mathrm{dm} = \mathrm{m}\{\omega : \mathcal{L}(s, \omega) \in A\} = \mathbf{P}(A) < \infty. \qquad (4.29)$$

By Lemma 4.6, and with the same argument as in the proof of Lemma 4.7, we see that $\theta(\phi_\tau(\omega))$ is an ergodic process. Thus, in view of Theorem 3.13,

$$\lim_{T\to\infty} \frac{1}{T} \int_0^T \theta(\phi_\tau(\omega)) \, \mathrm{d}\tau = \mathbf{E}\theta \qquad (4.30)$$

for almost all $\omega \in \Omega$. From the definition of θ and of the one-parameter group $\{\phi_\tau : \tau \in \mathbb{R}\}$ it follows that:

$$\frac{1}{T} \int_0^T \theta(\phi_\tau(\omega)) \, \mathrm{d}\tau = \frac{1}{T} \operatorname{meas} \{\tau \in [0, T] : \mathcal{L}(s + i\tau, \omega) \in A\}.$$

This result, (4.29) and (4.30) imply

$$\lim_{T\to\infty} \frac{1}{T} \operatorname{meas}\{\tau \in [0,T] : \mathcal{L}(s+i\tau,\omega) \in A\} = \mathbf{P}(A)$$

for almost all $\omega \in \Omega$. In view of (4.28) we deduce

$$\mathbf{P}'(A) = \mathbf{P}(A)$$

for any continuity set A of the measure $\mathbf{P}'$. Since the continuity sets form a determining class, we see that the latter equality holds for all $A \in \mathcal{B}(\mathcal{H}(\mathcal{D}))$. This proves the assertion of the theorem.   □

4.7 Generalizations

The proof of limit Theorem 4.3 allows generalizations in various directions. We conclude with the situation of Dirichlet series which do not necessarily have multiplicative coefficients.

Theorem 4.12. *Assume that $\mathcal{A}(s)$ is a meromorphic function in the half-plane $\sigma > \sigma_0$, all poles of $\mathcal{A}(s)$ in this region are included in a compact set, that $\mathcal{A}(s)$ satisfies the estimate*

$$\mathcal{A}(\sigma + it) \ll |t|^\delta$$

for $\sigma \geq \sigma_0$ with some $\delta > 0$, and that

$$\int_0^T |\mathcal{A}(\sigma + it + i\tau)|^2 d\tau \ll T$$

for all $\sigma > \sigma_0$. Further, suppose that $\mathcal{A}(s)$ has a representation as a Dirichlet series

$$\sum_{n=1}^{\infty} \frac{a_n}{n^s},$$

the series being absolutely convergent for $\sigma > \sigma_0 + \frac{1}{2}$, and $\sum_{n \leq x} |a_n|^2 \ll x^{2\sigma_0}$. Then the probability measure, defined by

$$\frac{1}{T} \operatorname{meas}\{\tau \in [0,T] : \mathcal{A}(s + i\tau) \in A\}$$

for $A \in \mathcal{B}(\mathcal{H}(\mathcal{D}_0))$, where $\mathcal{D}_0 := \{s \in \mathbb{C} : \sigma > \sigma_0\}$, converges weakly to the distribution of the random element

$$\Omega \ni \omega \mapsto \sum_{n=1}^{\infty} \frac{a_n \omega(n)}{n^s}$$

for $s \in \mathcal{D}_0$, as $T \to \infty$.

This limit theorem is due to Laurinčikas [187, 192].

There are several further generalizations and extensions, most of them belong to Laurinčikas and his school. For instance, Laurinčikas [188, 190] proved a limit theorem for Matsumoto zeta-functions in the space of *mero-morphic* functions. Other limit theorems for Dirichlet series (e.g., [185]) deal with the distribution of values in the complex plane and may be regarded as probabilistic analogues of Bohr's early results (see Sect. 1.2) and Selberg's normal distribution result on the critical line. For a first overview on the various results we refer to [185]) and Matsumoto [244]; the best introduction to limit theorems for Dirichlet series is Laurinčikas' monograph [186].

In the following section, we present another type of limit theorem.

4.8 A Discrete Limit Theorem

Whereas the preceding limit theorems were all of continuous character, we shall now briefly discuss the case of *discrete* limit theorems. Here the attribute '*discrete*' refers to the value-distribution on arithmetic progressions. We discuss this topic with respect to Matsumoto zeta-functions for which Kačinskaite [153, 154] obtained several discrete limit theorems.

We recall the definition of the Matsumoto zeta-function from Sect. 1.6. For any positive integer m, we are given positive integers $g(m)$ and further $f(j, m)$ as well as complex numbers $a_m^{(j)}$ for $1 \leq j \leq g(m)$. Denoting the mth prime number by p_m, the associated Matsumoto zeta-function is given by

$$\varphi(s) = \prod_{m=1}^{\infty} \prod_{j=1}^{g(m)} \left(1 - a_m^{(j)} p_m^{-sf(j,m)}\right)^{-1}.$$

Under the conditions $g(m) \ll p_m^{\alpha}$ and $|a_m^{(j)}| \leq p_m^{\beta}$ the product converges for $\sigma > 1 + \alpha + \beta$. Suppose that $\varphi(s)$ has an analytic continuation to some half-plane $\sigma > \delta + \alpha + \beta$ with $\frac{1}{2} \leq \delta < 1$ (except for at most finitely many poles) with $\varphi(s) \ll |t|^{c_1}$ as $|t| \to \infty$, where c_1 is some positive constant and bounded mean-square. We shall consider the value-distribution of $\varphi(s)$ on arithmetic progressions $\sigma + in\Delta$ with $n \in \mathbb{N}$ in the strip

$$\delta + \alpha + \beta < \sigma < 1 + \alpha + \beta,$$

where Δ is a positive real number. For this aim we consider the probability measure given by

$$\mathbf{P}_N(A) = \frac{1}{N} \sharp \{1 \leq n \leq N : \varphi(s + in\Delta) \in A\}$$

for any Borel set A of the appropriate function space. Similarly as we did for our L-functions, we consider the random variable

$$\varphi(s,\omega) = \prod_{m=1}^{\infty} \prod_{j=1}^{g(m)} (1 - a_m^{(j)} \omega(p_m)^{f(j,m)} p_m^{-sf(j,m)})^{-1}.$$

We denote by $\mathbf{P}_\varphi$ the distribution of the random element $\varphi(s,\omega)$. Kačinskaitė [153, 154] proved

Theorem 4.13. *Assume that Δ is a real number such that*

$$\exp\left(\frac{2\pi k}{\Delta}\right) \tag{4.31}$$

is irrational for all $k \in \mathbb{N}$. Then, for $\sigma > \alpha + \beta + \frac{1}{2}$, the probability measure $\mathbf{P}_N$ converges weakly to $\mathbf{P}_\varphi$ as $N \to \infty$.

We omit the proof of this theorem but briefly indicate where the assumption on Δ is needed. To obtain (4.13) in the proof of Lemma 4.4 the irrationality of (4.31) for all $k \in \mathbb{N}$ was used. To overcome this unfortunate restriction, Kačinskaitė and Laurinčikas [155] recently investigated the case when there exists a positive integer k for which $\exp\left(\frac{2\pi k}{\Delta}\right)$ is rational. By considering a certain subgroup of the infinite torus Ω, they succeeded in proving a discrete limit theorem in the complex plane also for this case; their argument carries over to spaces of analytic functions too.

The general idea and structure of proofs of discrete limit theorems are rather similar to the continuous case, however, some arguments differ in detail.

5

Universality

Wer die Zetafunktion kennt, kennt die Welt!

Now we shall apply the limit theorem from Chap. 4 to derive information on the value-distribution of L-functions. Our approach follows Bagchi [9], respectively, the refinements of Laurinčikas [186]. Using the so-called positive density method, introduced by Laurinčikas and Matsumoto [200], we prove a universality theorem for functions $\mathcal{L} \in \tilde{\mathcal{S}}$. Here, we shall make use of axiom (v). This result is essentially due to Steuding [345] (under slightly more restrictive conditions).

5.1 Dense Sets in Hilbert Spaces

In this section, we prove some preliminary results from the theory of Hilbert spaces. For more details of the proofs below and the theory of Hilbert spaces in general we refer to Bagchi's thesis [10] and the monographs of Laurinčikas [186], Dunford and Schwartz [72], Duren [73] and Rudin [312].

Our first aim is

Lemma 5.1. *Let $x_1, \ldots, x_n$ be linearly dependent vectors in a complex vector space, and let $a_1, \ldots, a_n$ be complex numbers with $|a_j| \leq 1$ for $1 \leq j \leq n$. Then there exist complex numbers $b_1, \ldots, b_n$ with $|b_j| \leq 1$ for $1 \leq j \leq n$, where at least one $|b_j| = 1$, such that*

$$\sum_{j=1}^{n} a_j x_j = \sum_{j=1}^{n} b_j x_j.$$

Proof. By assumption there are complex numbers $c_1, \ldots, c_n$, not all equal zero, such that

$$\sum_{j=1}^{n} c_j x_j = 0. \tag{5.1}$$

Let

$$K = \{(\alpha_1, \ldots, \alpha_n) \in \mathbb{C}^n : |\alpha_j| \leq 1 \quad \text{for} \quad 1 \leq j \leq n\}$$

and

$$I = \{t \in \mathbb{R} : \mathbf{a} + t\mathbf{c} \in K\},$$

where $\mathbf{a} = (a_1, \ldots, a_n)$ and $\mathbf{c} = (c_1, \ldots, c_n)$. Since $\mathbf{a} \in K$ it follows that $0 \in I$, and so I is not empty. K is convex and thus I is convex, too. Hence, I is an interval. Since K is compact and $\mathbf{c}$ is not the null vector, I is bounded. Denote by t_0 one of the endpoints of I. Let

$$\mathbf{b} = (b_1, \ldots, b_n) = \mathbf{a} + t_0\mathbf{c}.$$

Then $\mathbf{b}$ belongs to the boundary of K, that is $|b_j| \leq 1$ for $1 \leq j \leq n$ and $|b_j| = 1$ for at least one of the j's. By (5.1),

$$\sum_{j=1}^{n} b_j x_j = \sum_{j=1}^{n} a_j x_j + t_0 \sum_{j=1}^{n} c_j x_j = \sum_{j=1}^{n} a_j x_j,$$

which proves the lemma. $\square$

In what follows let $\mathcal{H}$ be a complex Hilbert space, and denote, as usual, its inner product by $\langle x, y \rangle$ and its norm by $\|x\| = \sqrt{\langle x, x \rangle}$. Now we apply the Lemma 5.1 to prove

Lemma 5.2. *Let $x_1, \ldots, x_n$ be points in a complex Hilbert space $\mathcal{H}$ and let $a_1, \ldots, a_n$ be complex numbers with $|a_j| \leq 1$ for $1 \leq j \leq n$. Then there exist complex numbers $b_1, \ldots, b_n$ with $|b_j| = 1$ for $1 \leq j \leq n$, satisfying the inequality*

$$\left\| \sum_{j=1}^{n} a_j x_j - \sum_{j=1}^{n} b_j x_j \right\|^2 \leq 4 \sum_{j=1}^{n} \|x_j\|^2.$$

Proof. The proof is achieved by induction on n. The case $n = 1$ is trivial:

$$\|a_1 x_1 - b_1 x_1\|^2 \leq \|2x_1\|^2 = 4\|x_1\|^2.$$

Now assume that the assertion is true for n. Let $x_1, \ldots, x_{n+1}$ be points in $\mathcal{H}$, and let $a_1, \ldots, a_{n+1}$ be complex numbers with $|a_j| \leq 1$ for $1 \leq j \leq n+1$. Denote by y_{n+1} the orthogonal projection of x_{n+1} into the span of $x_1, \ldots, x_n$. Then $x_1, \ldots, x_n, y_{n+1}$ are linearly dependent. Thus, by Lemma 5.1, there exist complex numbers $c_1, \ldots, c_{n+1}$ with $|c_j| \leq 1$ for $1 \leq j \leq n+1$, and $|c_k| = 1$ for some $1 \leq k \leq n+1$, such that

$$\sum_{j=1}^{n} a_j x_j + a_{n+1} y_{n+1} = \sum_{j=1}^{n} c_j x_j + c_{n+1} y_{n+1}.$$

First, suppose that $k = n + 1$. By the induction hypothesis there exist complex numbers $b_1, \ldots, b_n$ with $|b_j| = 1$ for $1 \le j \le n$ such that

$$\left\| \sum_{j=1}^{n} c_j x_j - \sum_{j=1}^{n} b_j x_j \right\|^2 \le 4 \sum_{j=1}^{n} \|x_j\|^2.$$

Putting $b_{n+1} = c_{n+1}$ we get

$$\sum_{j=1}^{n+1} a_j x_j - \sum_{j=1}^{n+1} b_j x_j = \left(\sum_{j=1}^{n} c_j x_j - \sum_{j=1}^{n} b_j x_j \right) + (a_{n+1} - c_{n+1}) z_{n+1},$$

where $z_{n+1} := x_{n+1} - y_{n+1}$ is orthogonal to $x_1, \ldots, x_n$ (since y_{n+1} is the orthogonal projection of x_{n+1} into the span of $x_1, \ldots, x_n$). It follows that:

$$\left\| \sum_{j=1}^{n+1} a_j x_j - \sum_{j=1}^{n+1} b_j x_j \right\|^2 = \left\| \sum_{j=1}^{n} c_j x_j - \sum_{j=1}^{n} b_j x_j \right\|^2 + |a_{n+1} - c_{n+1}|^2 \|z_{n+1}\|^2$$

$$\le 4 \sum_{j=1}^{n} \|x_j\|^2 + 4 \|z_{n+1}\|^2.$$

Since $\|z_{n+1}\|^2 = \|x_{n+1}\|^2 - \|y_{n+1}\|^2 \le \|x_{n+1}\|^2$ we are done in this case.

Next, suppose that $1 \le k \le n$. Without loss of generality we may assume that $k = 1$. By the induction hypothesis there are complex numbers $b_2, \ldots, b_{n+1}$ with $|b_j| = 1$ for $2 \le j \le n + 1$ such that

$$\left\| \sum_{j=2}^{n} c_j x_j + c_{n+1} y_{n+1} - \sum_{j=2}^{n} b_j x_j - b_{n+1} y_{n+1} \right\|^2 \le 4 \sum_{j=2}^{n} \|x_j\|^2 + 4 \|y_{n+1}\|^2.$$

Putting $b_1 = c_1$ we obtain

$$\sum_{j=1}^{n+1} a_j x_j - \sum_{j=1}^{n+1} b_j x_j = \sum_{j=2}^{n} c_j x_j + c_{n+1} y_{n+1} - \sum_{j=2}^{n} b_j x_j - b_{n+1} y_{n+1}$$

$$+ (a_{n+1} - b_{n+1}) z_{n+1}.$$

In view of the choice of y_{n+1} this leads to

$$\left\| \sum_{j=1}^{n+1} a_j x_j - \sum_{j=1}^{n+1} b_j x_j \right\|^2 = \left\| \sum_{j=2}^{n} c_j x_j + c_{n+1} y_{n+1} - \sum_{j=2}^{n} b_j x_j - b_{n+1} y_{n+1} \right\|^2$$

$$+ |a_{n+1} - b_{n+1}|^2 \cdot \|z_{n+1}\|^2$$

$$\le 4 \sum_{j=2}^{n} \|x_j\|^2 + 4 \|y_{n+1}\|^2 + 4 \|z_{n+1}\|^2.$$

This gives the desired estimate for the second case and thus Lemma 5.2 is proved. □

We have to recall a result from the theory of Hilbert spaces which can be found, for example, in Trenogin [354]. A subset L of a complex Hilbert space $\mathcal{H}$ is called a linear manifold (or subspace) if for all $x, y \in L$ and all complex numbers α, β the linear combination $\alpha x + \beta y$ is also an element of L. Denote by $L^\perp$ the orthogonal complement of L. If L is a linear manifold of $\mathcal{H}$, every element $x \in \mathcal{H}$ has a representation $x = y + z$, where $y \in L$ and $z \in L^\perp$.

Lemma 5.3. *A linear manifold L of a complex Hilbert space $\mathcal{H}$ is dense in $\mathcal{H}$ if and only if $L^\perp = \{0\}$.*

Now we are in the position to state the main result of this section.

Theorem 5.4. *Let $\{x_n\}$ be a sequence in $\mathcal{H}$ satisfying*

- $\sum_{n=1}^\infty \|x_n\|^2 < \infty$,
- *For $0 \neq x \in \mathcal{H}$, the series $\sum_{n=1}^\infty |\langle x_n, x \rangle|$ diverges to infinity.*

Then the set of all convergent series

$$\sum_{n=1}^\infty a_n x_n \quad with \quad |a_n| = 1$$

is dense in $\mathcal{H}$.

This is essentially Pechersky's rearrangement theorem [288].

Proof. First, we prove that there exists a sequence $\{\tilde{\epsilon}_n : n \in \mathbb{N}\}$ with $\tilde{\epsilon}_n = \pm 1$ such that

$$\sum_{n=1}^\infty \tilde{\epsilon}_n x_n < \infty. \tag{5.2}$$

For this purpose, let $\{\epsilon_n : n \in \mathbb{N}\}$ be a sequence of independent random variables defined on some probability space $(\mathbb{R}, \mathcal{B}(\mathbb{R}), \mathbf{P})$ such that

$$\mathbf{P}(\epsilon_n = -1) = \mathbf{P}(\epsilon_n = +1) = \frac{1}{2}.$$

Then $X_n := \epsilon_n x_n$ is an $\mathcal{H}$-valued random element with expectation $\mathbf{E}X_n = 0$. By the first assumption of the theorem, the sequence $\{X_n\}$ is uniformly bounded in norm. Moreover,

$$\sum_{n=1}^\infty \mathbf{E}\|X_n\|^2 = \sum_{n=1}^\infty \|x_n\|^2 < \infty.$$

By the second assumption it follows that any non-zero element $x \in \mathcal{H}$ is not orthogonal to all x_n, and, in view of Lemma 5.3, the span of $\{x_n\}$ lies dense in $\mathcal{H}$. Consequently $\mathcal{H}$ is separable. Define a set of linear combinations of the $\{x_n\}$ by

$$S = \left\{ \sum_{n=1}^{m} r_n x_n : m \in \mathbb{N}, r_n \in \mathbb{Q}(i) \right\},$$

where 'i' denotes the imaginary unit. Next we show that S lies dense in $\mathcal{H}$.

We observe that S is countable. Now let x_0 be an arbitrary element of $\mathcal{H}$. By the above density result, for any positive ϵ, there exists an element from the span of $\{x_n\}$

$$y := \sum_{n=1}^{M} c_n x_n$$

with $c_n \in \mathbb{C}$ such that $\|x_0 - y\| < \epsilon$. On the other side we can find an element

$$z := \sum_{n=1}^{M} r_n x_n$$

of S with coefficients $r_n \in \mathbb{Q}(i)$ for which $\|y - z\| < \epsilon$. This implies $\|x_0 - z\| < 2\epsilon$, so S is dense in $\mathcal{H}$ with respect to the norm of $\mathcal{H}$.

Application of Theorem 3.10 implies that the series $\sum_{n=1}^{\infty} X_n$ converges almost surely. Consequently, the desired sequence exists, and we denote it by $\{\tilde{\epsilon}_n\}$.

Next we have to show that for any $x_0 \in \mathcal{H}$ and any positive ϵ there exists a sequence $\{a_n : |a_n| = 1\}$ such that the series $\sum_{n=1}^{\infty} a_n x_n$ converges and

$$\left\| x_0 - \sum_{n=1}^{\infty} a_n x_n \right\| < \epsilon.$$

By the first assumption of the theorem we can find a positive integer N satisfying

$$\sum_{n=N}^{\infty} \|x_n\|^2 < \frac{\epsilon^2}{36}, \tag{5.3}$$

and, in view of (5.2),

$$\left\| \sum_{n=M}^{\infty} \tilde{\epsilon}_n x_n \right\| < \frac{\epsilon}{3} \tag{5.4}$$

for any $M \geq N$. Define

$$K = \left\{ \sum_{n=N}^{M+N} b_n x_n : |b_n| \leq 1, M \in \mathbb{N} \right\}.$$

We show that K is dense in $\mathcal{H}$.

Suppose that the contrary is true. Then there exists an element $0 \neq x_0 \in \mathcal{H} \setminus \overline{K}$. As in the proof of Lemma 5.1, K is convex, and so is $\overline{K}$ too. Thus, $\{x_0\}$ and $\overline{K}$ are disjoint closed convex subsets of $\mathcal{H}$. The one-point set $\{x_0\}$

is compact. By the separation theorem for linear operators (see [72, Sect. V])
there exists a continuous linear functional f on $\mathcal{H}$ such that

$$\operatorname{Re} f(x) > \operatorname{Re} f(x_0)$$

for any $x \in K$. Clearly, f does not vanish identically. By the Riesz representation theorem (see [312, Theorem 6.19]) there exists an element $0 \neq z_0 \in \mathcal{H}$
for which $f(x) = \langle x, z_0 \rangle$. Putting $c = -\operatorname{Re} f(x_0)$ we get

$$\operatorname{Re} \langle x, z_0 \rangle = \operatorname{Re} f(x) > -c \tag{5.5}$$

for any $x \in K$. Now choose b_n so that $|b_n| = 1$ and $b_n \langle x_n, z_0 \rangle = -|\langle x_n, z_0 \rangle|$.
The element

$$y_M = \sum_{n=N}^{M+N} b_n x_n,$$

is in K. It follows from (5.5) that

$$\operatorname{Re} \langle y_M, z_0 \rangle = - \sum_{n=N}^{M+N} |\langle x_n, z_0 \rangle| > -c$$

for any $M \in \mathbb{N}$. Consequently, the series $\sum_{n=N}^{\infty} |\langle x_n, z_0 \rangle|$ is convergent. Since
$z_0 \neq 0$, this contradicts the second assumption of the theorem. So K is dense
in $\mathcal{H}$.

By the density of K there exists a sequence $\{b_n\}$ with $|b_n| \leq 1$ such that

$$\left\| x_0 - \sum_{n=1}^{N-1} x_n - \sum_{n=N}^{M+N} b_n x_n \right\| < \frac{\epsilon}{3}. \tag{5.6}$$

Lemma 5.2 in combination with (5.3) yields the existence of an element

$$u := \sum_{n=N}^{M+N} a_n x_n \qquad \text{with} \quad |a_n| = 1$$

in $\mathcal{H}$ such that

$$\left\| \sum_{n=N}^{M+N} b_n x_n - \sum_{n=N}^{M+N} a_n x_n \right\|^2 \leq 4 \sum_{n=N}^{M+N} \| x_n \|^2 < \frac{\epsilon^2}{9}.$$

Thus,

$$\left\| u - \sum_{n=N}^{M+N} b_n x_n \right\| < \frac{\epsilon}{3}. \tag{5.7}$$

Now define

$$v = \sum_{n=M+N+1}^{\infty} \tilde{\epsilon}_n x_n.$$

In view of (5.4) we have

$$\|v\| < \frac{\epsilon}{3}. \tag{5.8}$$

Further, let

$$w = \sum_{n=1}^{N-1} x_n + u + v = \sum_{n=1}^{\infty} a_n x_n,$$

where $a_n = 1$ for $1 \leq n \leq N - 1$ and $a_n = \tilde{\epsilon}_n$ for $n > M + N$. Then (5.6), (5.7), and (5.8) imply

$$\|x_0 - w\| \leq \left\| x_0 - \sum_{n=1}^{N-1} a_n x_n - \sum_{n=N}^{M+N} b_n x_n \right\| + \left\| \sum_{n=N}^{M+N} b_n x_n - \sum_{n=N}^{M+N} a_n x_n \right\|$$

$$+ \left\| \sum_{n=M+N+1}^{\infty} a_n x_n \right\|$$

$$< \epsilon,$$

which proves the theorem. $\square$

For our later purpose, we prove now a lemma which is not really related to the topic of this section, however, its proof is very similar to a certain aspect in the proof of the previous theorem.

Lemma 5.5. *Let* $\{z_n\}$ *be a sequence of complex numbers such that the series* $\sum_{n=1}^{\infty} |z_n|^2$ *converges. If* $\{\epsilon_n\}$ *is a sequence of independent random variables on some probability space* $(\mathbb{R}, \mathcal{B}(\mathbb{R}), \mathbf{P})$ *such that*

$$\mathbf{P}(\epsilon_n = -1) = \mathbf{P}(\epsilon_n = +1) = \frac{1}{2}$$

for any $n \in \mathbb{N}$, *then the series* $\sum_{n=1}^{\infty} \epsilon_n z_n$ *converges almost surely. In particular, there exists a sequence* $\{\tilde{\epsilon}_n\}$ *with* $\tilde{\epsilon}_n \in \{\pm 1\}$ *such that*

$$\sum_{n=1}^{\infty} \tilde{\epsilon}_n z_n$$

converges.

Proof. Let $X_n = \epsilon_n z_n$, then $\{X_n\}$ is a sequence of independent complex random variables with expectation $\mathbf{E} X_n = 0$. By the assumption of the lemma

$$\sum_{n=1}^{\infty} \mathbf{E}(\mathrm{Re}\, X_n - \mathbf{E}(\mathrm{Re}\, X_n))^2 = \sum_{n=1}^{\infty} \mathbf{E}(\mathrm{Re}\, X_n)^2 \leq \sum_{n=1}^{\infty} |z_n|^2 < \infty.$$

In view of Theorem 3.9 the series $\sum_{n=1}^{\infty} \mathrm{Re}\, X_n$ converges almost surely, respectively, the series $\sum_{n=1}^{\infty} \epsilon_n \mathrm{Re}\, z_n$ converges for almost all sequences $\{\epsilon_n = \pm 1\}$. By the same reasoning for the imaginary part of X_n, it follows that $\sum_{n=1}^{\infty} \epsilon_n \mathrm{Im}\, z_n$ converges for almost all sequences $\{\epsilon_n\}$ with $\epsilon_n \in \{\pm 1\}$ as well. This yields the assertion of the lemma. $\square$

5.2 Application to the Space of Analytic Functions

Next we shall apply Theorem 5.4 to an appropriate function space. We already dealt with Hardy spaces in the proof of Voronin's universality theorem in Sect. 1.3. However, the general situation is a bit more complicated; for details we refer to Duren [73], Laurinčikas [186, Sects. 6.2 and 6.3] and Rudin [312, Chap. 17].

Let D be a simply connected domain in the complex plane with at least two boundary points. The associated Hardy space $\mathcal{H}^2(D)$ consists of those functions $f(s)$ which are analytic in D such that the subharmonic function $|f(s)|^2$ has a harmonic majorant in D. If $F(s)$ is the smallest harmonic majorant for $|f(s)|^2$ and s_0 is a fixed point in D, then the norm of $f \in \mathcal{H}^2(D)$ is defined by

$$\|f\| = (F(s_0))^{\frac{1}{2}}.$$

With this norm $\mathcal{H}^2(D)$ becomes a complex Hilbert space.

Theorem 5.6. *For any given $f \in \mathcal{H}^2(D)$ there exists a Borel measure μ_f with support concentrated in ∂D such that if $g \in \mathcal{H}^2(D)$ has a continuous extension to ∂D, then the inner product of $\mathcal{H}^2(D)$ can be expressed by*

$$\langle f, g \rangle = \int_{\partial D} g \, \mathrm{d}\mu_f.$$

Now we are in the position to apply Theorem 5.4 to the Hardy space $\mathcal{H}^2(D)$ in order to derive a denseness result for the larger space of analytic function $\mathcal{H}(D)$. Recall that γ denotes the unit circle in the complex plane.

Theorem 5.7. *Let D be a simply connected domain in the complex plane. Suppose that the sequence $\{f_n\}$ in $\mathcal{H}(D)$ satisfies the following assumptions:*

- *If μ is a complex Borel measure on $(\mathbb{C}, \mathcal{B}(\mathbb{C}))$ with compact support contained in D such that*

$$\sum_{n=1}^{\infty} \left| \int_{\mathbb{C}} f_n \, \mathrm{d}\mu \right| < \infty,$$

  then

$$\int_{\mathbb{C}} s^r \, \mathrm{d}\mu(s) = 0 \quad \text{for any} \quad r \in \mathbb{N}_0.$$

- *The series $\sum_{n=1}^{\infty} f_n$ converges in $\mathcal{H}(D)$.*
- *For every compact $K \subset D$,*

$$\sum_{n=1}^{\infty} \max_{s \in K} |f_n(s)|^2 < \infty.$$

Then the set of all convergent series

$$\sum_{n=1}^{\infty} b(n) f_n \quad \text{with} \quad b(n) \in \gamma$$

is dense in $\mathcal{H}(D)$.

This is Theorem 6.3.10 of Laurinčikas [186], respectively, Lemma 5.29 of Bagchi [9].

Proof. Let K be a given compact subset of D. We may choose a simply connected domain $\mathcal{G}$ for which $K \subset \mathcal{G}$, the closure $\overline{\mathcal{G}}$ of $\mathcal{G}$ is a compact subset of D, and the boundary of $\mathcal{G}$ is a regular Jordan curve. In view of Theorem 5.6, for any $f_n \in \mathcal{H}^2(D)$ there exists a complex Borel measure μ_{f_n} on the Hilbert space $\mathcal{H}^2(\mathcal{G})$ such that the norm of f_n is given by

$$\|f_n\|^2 = \langle f_n, f_n \rangle = \int_{\partial \mathcal{G}} f_n \, \mathrm{d}\mu_{f_n}.$$

Consequently,

$$\|f_n\|^2 \leq \max_{s \in \overline{\mathcal{G}}} |f_n(s)| \int_{\partial \mathcal{G}} |\mathrm{d}\mu_{f_n}| \ll \max_{s \in \overline{\mathcal{G}}} |f_n(s)|^2.$$

Hence, by the third assumption,

$$\sum_{n=1}^{\infty} \|f_n\|^2 < \infty. \tag{5.9}$$

Now assume that $g \in \mathcal{H}^2(\mathcal{G})$ satisfies

$$\sum_{n=1}^{\infty} |\langle f_n, g \rangle| < \infty.$$

With the Borel measure μ_g assigned to g (according Theorem 5.6) it follows that:

$$\sum_{n=1}^{\infty} \left| \int_{\partial \mathcal{G}} f_n \, \mathrm{d}\mu_g \right| < \infty.$$

By the first assumption of the theorem we get, for every $r \in \mathbb{N}_0$,

$$\langle g(s), s^r \rangle = \int_{\partial \mathcal{G}} s^r \, \mathrm{d}\mu_g(s) = 0.$$

It follows that g is orthogonal to all polynomials. One can show that the set of polynomials is dense in $\mathcal{H}^2(\mathcal{G})$ (see [73]), and thus, by Lemma 5.3, $g(s)$ is identically zero. Hence, for any $g(s) \not\equiv 0$,

$$\sum_{n=1}^{\infty} |\langle f_n, g \rangle| = \infty.$$

This and (5.9) show in view of Theorem 5.4 that the set of all convergent series

$$\sum_{n=1}^{\infty} \beta_n f_n \quad \text{with} \quad |\beta_n| = 1$$

is dense in $\mathcal{H}^2(\mathcal{G})$. It remains to extend this result to the larger space $\mathcal{H}(D)$.

Now let $f \in \mathcal{H}(D)$ and $\epsilon > 0$. By Theorem 5.4 there exists a sequence $\{\beta_n : |\beta_n| = 1\}$ such that the series

$$\sum_{n=1}^{\infty} \beta_n f_n(s)$$

converges on D (in the topology of $\mathcal{H}^2(D)$). This convergence is uniform on every compact subset K of D and we may assume that

$$\max_{s \in K} \left| \sum_{n=1}^{\infty} \beta_n f_n(s) - f(s) \right| < \frac{\epsilon}{4}.$$

Hence, we can find a number M for which

$$\max_{s \in K} \left| \sum_{n \leq M} \beta_n f_n(s) - f(s) \right| < \frac{\epsilon}{2},$$

and with regard to the second assumption of the theorem

$$\max_{s \in K} \left| \sum_{n > M} f_n(s) \right| < \frac{\epsilon}{2}.$$

Now let

$$b(n) = \begin{cases} \beta_n & \text{if } n \leq M, \\ 1 & \text{otherwise.} \end{cases}$$

Then the series $\sum_{n=1}^{\infty} b(n) f_n(s)$ converges in $\mathcal{H}(D)$ and by the above inequalities

$$\max_{s \in K} \left| \sum_{n=1}^{\infty} b(n) f_n(s) - f(s) \right|$$

$$\leq \max_{s \in K} \left| \sum_{n \leq M} \beta(n) f_n(s) - f(s) \right| + \max_{s \in K} \left| \sum_{n > M} f_n(s) \right| < \epsilon.$$

This proves the theorem.   □

5.3 Entire Functions of Exponential Type

Now we state some basic facts from the theory of entire functions.

A function $f(s)$ which is analytic in the closed angular region $|\arg s| \leq \theta_0$ with $0 < \theta_0 \leq \pi$ is said to be of exponential type if

$$\limsup_{r \to \infty} \frac{\log |f(r \exp(i\theta))|}{r} < \infty$$

for $|\theta| \leq \theta_0$, uniformly in θ.

Lemma 5.8. *Let μ be a complex Borel measure on $(\mathbb{C}, \mathcal{B}(\mathbb{C}))$ with compact support contained in the half-plane $\sigma > \sigma_0$. Moreover, for $s \in \mathbb{C}$, define the function*

$$f(s) = \int_{\mathbb{C}} \exp(sz) \, d\mu(z).$$

Then f is an entire function of exponential type. If $f(s)$ does not vanish identically, then

$$\limsup_{r \to \infty} \frac{\log |f(r)|}{r} > \sigma_0.$$

This is Lemma 6.4.10 of Laurinčikas [186]. It is easily seen that the function $f(s)$, defined in the lemma, is indeed an entire function of exponential type. The proof of the lemma relies on the Borel transform $\tilde{f}$ of f which, for complex s that are not contained in the support of μ, is given by

$$\tilde{f}(s) = \int_{\mathbb{C}} \frac{d\mu(z)}{s - z}.$$

The Borel transform is analytic everywhere with the exception of a neighbourhood of $s = 0$. This shows that the closed convex hull of the set of singularities of the Borel transform, the so-called conjugate indicator diagram of f, is contained in the convex hull of the support of μ. By the assumption of the lemma, this is a subset of the half-plane $\sigma > \sigma_0$. This leads via the so-called Phragmén–Lindelöf indicator function and a theorem of Boas to the desired inequality; for the details we refer to Chap. 5 in Boas' monograph [26] and Sect. 6.4 in Laurinčikas [186].

Pólya investigated the rate of growth of functions of exponential type on arithmetic progressions. This was extended by Bernstein [19] to more general divergent sequences. We state a variant of Bernstein's theorem.

Theorem 5.9. *Let $f(s)$ be an entire function of exponential type, and let $\{\xi_m\}$ be a sequence of complex numbers. Moreover, assume that there are positive real constants λ, η and ω such that*

- $\limsup_{y \to \infty} \frac{\log |f(\pm iy)|}{y} \leq \lambda$,
- $|\xi_m - \xi_n| \geq \omega |m - n|$,
- $\lim_{m \to \infty} \frac{\xi_m}{m} = \eta$,
- $\lambda \eta < \pi$.

Then

$$\limsup_{m \to \infty} \frac{\log |f(\xi_m)|}{|\xi_m|} = \limsup_{r \to \infty} \frac{\log |f(r)|}{r}.$$

The case $\eta = 1$ follows directly from Bernstein's theorem [19]. The general case can be deduced by applying the case $\eta = 1$ to the function $F(s) = f(\eta s)$ and an appropriate scaling of the other appearing quantities. A proof can be found in Laurinčikas' book [186, Sect. 6.4], or in Boas [26, Chap. 10]; notice that the proof in [26] contains an error as was pointed out by Bagchi [10].

5.4 The Positive-Density Method

It follows from the prime number Theorem 1.1 that there exist constants c_1 and $c_2 > 0$ such that

$$\sum_{p \leq x} \frac{1}{p} = \log\log x + c_1 + \mathrm{O}\left(\exp\left(-c_2\sqrt{\log x}\right)\right). \tag{5.10}$$

This asymptotic formula plays an essential role in Bagchi's probabilistic proof of Voronin's universality theorem (as a substitute of Voronin's use of the prime number theorem in his approach). However, for many L-functions, given by a Dirichlet series

$$\sum_{n=1}^{\infty} \frac{a(n)}{n^s},$$

an analogous formula for

$$\sum_{p \leq x} \frac{|a(p)|}{p},$$

is not known! Laurinčikas and Matsumoto [200] overcame this difficulty by using sets of prime numbers of positive density within the set of all primes in combination with (5.10). This is manifested in a mean-square formula for the coefficients $a(p)$ (which are easier obtainable). This is the essential idea of the so-called positive-density method which we shall use now. For more details about the origin of this nice argument see Matsumoto [242].

In what follows, we assume that $\mathcal{L} \in \tilde{\mathcal{S}}$. Recall that $\mathcal{D}$ denotes the intersection of the mean-square half-plane $\sigma > \sigma_m$ for $\mathcal{L}(s)$ with the critical strip (see (4.1)); we have already noticed that the strip $\mathcal{D}$ is not empty (see (4.2)). For $s \in \mathcal{D}$, $b(p) \in \gamma = \{z \in \mathbb{C} : |z| = 1\}$, and any prime p, we define the function

$$g_p(s, b(p)) = \log \prod_{j=1}^{m} \left(1 - \frac{\alpha_j(p)b(p)}{p^s}\right)^{-1} \tag{5.11}$$

$$= -\sum_{j=1}^{m} \log\left(1 - \frac{\alpha_j(p)b(p)}{p^s}\right).$$

We observe the similarities between $\exp(g_p(s, b(p)))$ and the corresponding Euler factor in the Euler product representations of $\mathcal{L}(s)$ and of the associated random element (4.6). Next we shall prove

Theorem 5.10. *Let* $\mathcal{L} \in \tilde{\mathcal{S}}$. *The set of all convergent series*

$$\sum_{p} g_p(s, b(p)) \quad \text{with} \quad b(p) \in \gamma$$

is dense in $\mathcal{H}(\mathcal{D})$, *where* γ *is the unit circle in* $\mathbb{C}$.

This denseness theorem plays a crucial role in the proof of the universality theorem for $\tilde{\mathcal{S}}$. We observe that the statement of the theorem is not true if $\mathcal{D}$ is replaced, for example, by the half-plane of absolute convergence (since the terms $g_p(s, b(p))$ tend to zero as $\sigma \to \infty$).

Proof. Define

$$\tilde{g}_p(s) = g_p(s, 1) = -\sum_{j=1}^{m} \log \left(1 - \frac{\alpha_j(p)}{p^s} \right).$$

For a parameter $N \in \mathbb{N}$ which will be chosen later, define

$$\hat{g}_p(s) = \begin{cases} \tilde{g}_p(s) & \text{if} \quad p > N, \\ 0 & \text{if} \quad p \leq N. \end{cases}$$

We claim that there exists a sequence $\{\hat{b}(p) : \hat{b}(p) \in \gamma\}$ such that the series

$$\sum_p \hat{b}(p)\hat{g}_p(s) \tag{5.12}$$

converges in $\mathcal{H}(\mathcal{D})$.

In order to prove this claim we consider the identity between the Dirichlet series and the Euler product representation of $\mathcal{L} \in \tilde{\mathcal{S}}$,

$$\tilde{g}_p(s) = \sum_{j=1}^{m} \sum_{k=1}^{\infty} \frac{\alpha_j(p)^k}{kp^{ks}} = \frac{a(p)}{p^s} + r_p(s), \tag{5.13}$$

say, where

$$r_p(s) = \sum_{k=2}^{\infty} \sum_{j=1}^{m} \frac{a_j(p)^k}{kp^{ks}} \ll \sum_{k=2}^{\infty} \frac{1}{kp^{k\sigma}}$$

(since all $\alpha_j(p)$ are of modulus ≤ 1 by Lemma 2.2). Hence

$$\sum_p r_p(s) \ll \sum_p \frac{1}{p^{2\sigma}} < \zeta(2\sigma) \tag{5.14}$$

and so the series $\sum_p r_p(s)$ converges uniformly on compact subsets of $\mathcal{D}$. Now let $\{\sigma_j\}$ be a strictly decreasing sequence of real numbers which tends to σ_m as $j \to \infty$. Taking into account Lemma 5.5, for each j, there exists a sequence $\{\epsilon_p = \pm 1\}$ of independent random variables such that the series

$$\sum_p \frac{a(p)\epsilon_p}{p^{\sigma_j}}$$

converges almost surely. Thus, we can find a sequence $\{\hat{b}(p) : \hat{b}(p) \in \gamma\}$ for which

$$\sum_p \frac{a(p)\hat{b}(p)}{p^{\sigma_j}}$$

converges for any j. Since Dirichlet series converge uniformly on compact subsets in the half-plane of their convergence, it follows that the series

$$\sum_p \frac{a(p)\hat{b}(p)}{p^s} \tag{5.15}$$

converges on any compact subset of $\mathcal{D}$. Together with the convergence of $\sum_p r_p(s)$, this proves our claim that the series (5.12) converges in $\mathcal{H}(\mathcal{D})$.

Now we shall show that the set of all convergent series

$$\sum_p \tilde{b}(p)\hat{g}_p(s) \quad \text{with} \quad \tilde{b}(p) \in \gamma \tag{5.16}$$

is dense in $\mathcal{H}(\mathcal{D})$. Obviously, it is sufficient to show that the set of all convergent series

$$\sum_p \tilde{b}(p)f_p(s) \quad \text{with} \quad \tilde{b}(p) \in \gamma \tag{5.17}$$

is dense in $\mathcal{H}(\mathcal{D})$, where $f_p(s) := \hat{b}(p)\hat{g}_p(s)$. To prove this we apply Theorem 5.7. Taking into account (5.12) we have already verified the second assumption, namely that the series $\sum_p f_p(s)$ converges in $\mathcal{H}(\mathcal{D})$. The third assumption is easily shown by the estimate

$$\sum_p |f_p(s)|^2 \ll \sum_p \frac{1}{p^{2\sigma}},$$

since the series on the right converges on any compact subset of $\mathcal{D}$ (in a similar manner as in (5.14).

In order to verify the first assumption of Theorem 5.7, let μ be a complex Borel measure on $(\mathbb{C}, \mathcal{B}(\mathbb{C}))$ with compact support contained in $\mathcal{D}$ such that

$$\sum_p \left| \int_{\mathbb{C}} f_p(s)\, \mathrm{d}\mu(s) \right| < \infty. \tag{5.18}$$

Define

$$h_p(s) = \frac{a(p)\hat{b}(p)}{p^s}.$$

Then, by (5.13),

$$\sum_p |f_p(s) - h_p(s)| = \sum_{p \le N} \frac{|a(p)|}{p^\sigma} + \sum_{p > N} |r_p(s)| \ll 1$$

uniformly on compact subsets of $\mathcal{D}$. Hence, (5.18) implies

$$\sum_p \left| \int_{\mathbb{C}} h_p(s)\, \mathrm{d}\mu(s) \right| < \infty. \tag{5.19}$$

In view of the polynomial Euler product representation axiom (iv), respectively, Lemma 2.2, for any prime p we have $|a(p)| \leq m$ (where m is the degree of the polynomial defining the local Euler factors). Hence we may define angles $\phi_p \in [0, \frac{\pi}{2}]$ by setting

$$|a(p)| = \left| \sum_{j=1}^{m} \alpha_j(p) \right| = m \cos \phi_p. \tag{5.20}$$

Notice that $\cos \phi_p$ is non-negative. We rewrite (5.19) as

$$\sum_p |\varrho(\log p)| \cos \phi_p < \infty, \tag{5.21}$$

where

$$\varrho(z) := \int_{\mathbb{C}} \exp(-sz) \, d\mu(s).$$

We shall show that $\varrho(z)$ vanishes identically; however, for this purpose it is of advantage to consider an appropriate subseries of (5.21).

Fix a number ϕ with

$$0 < \phi < \min\left\{1, \frac{\sqrt{\kappa}}{m}\right\};$$

here κ is the quantity appearing in axiom (v) on the prime mean-square for $\mathcal{L} \in \tilde{\mathcal{S}}$. Define

$$\mathbb{P}^\phi = \{p : p \text{ prime and } \cos \phi_p > \phi\}.$$

Then (5.21) yields

$$\sum_{p \in \mathbb{P}^\phi} |\varrho(\log p)| < \infty. \tag{5.22}$$

To deduce the vanishing of $\varrho(z)$ we apply Theorem 5.9. We choose a sufficiently large positive constant M such that the support of μ is contained in the region $\{s \in \mathbb{C} : \sigma_m < \sigma < 1, |t| < M\}$. Then, by the definition of $\varrho(z)$,

$$|\varrho(\pm iy)| \leq \exp(My) \int_{\mathbb{C}} |d\mu(s)|$$

for $y > 0$. Therefore,

$$\limsup_{y \to \infty} \frac{\log |\varrho(\pm iy)|}{y} \leq M,$$

and the first condition of Theorem 5.9 is valid with $\lambda = M$. Now fix a number η with $0 < \eta < \frac{\pi}{M}$, and define

$$A = \left\{ n \in \mathbb{N} : \exists r \in \left(\left(n - \frac{1}{4}\right)\eta, \left(n + \frac{1}{4}\right)\eta \right] \text{ with } |\varrho(r)| \leq \exp(-r) \right\}.$$

We want to show that A has natural density 1 (i.e., $\varrho(z)$ is small).

For $n \in \mathbb{N}$, let

$$\alpha = \alpha(n) := \exp\left(\left(n - \frac{1}{4}\right)\eta\right) \quad \text{and} \quad \beta = \beta(n) := \exp\left(\left(n + \frac{1}{4}\right)\eta\right)$$

(notice that α and β depend on n). We observe that for $n \notin A$ we have

$$|\varrho(\log p)| > \frac{1}{p}$$

for all primes $p \in (\alpha, \beta]$. Thus, it follows that:

$$\sum_{\substack{p \in \mathbb{P}^\phi \\ \alpha < p \leq \beta}} |\varrho(\log p)| \geq \sum_{n \notin A} \sum_{\substack{p \in \mathbb{P}^\phi \\ \alpha < p \leq \beta}} |\varrho(\log p)| \geq \sum_{n \notin A} \sum_{\substack{p \in \mathbb{P}^\phi \\ \alpha < p \leq \beta}} \frac{1}{p}.$$

Hence, in view of (5.22),

$$\sum_{n \notin A} \sum_{\substack{p \in \mathbb{P}^\phi \\ \alpha < p \leq \beta}} \frac{1}{p} < \infty. \tag{5.23}$$

Let

$$\pi_\phi(x) = \sharp\{p \leq x : p \in \mathbb{P}^\phi\}.$$

Then, for $\alpha < u \leq \beta$, we obtain

$$\sum_{\alpha < p \leq u} (\cos \phi_p)^2 \leq \sum_{\substack{p \in \mathbb{P}^\phi \\ \alpha < p \leq u}} 1 + \phi^2 \sum_{\substack{p \notin \mathbb{P}^\phi \\ \alpha < p \leq u}} 1 \tag{5.24}$$

$$= (1 - \phi^2)(\pi_\phi(u) - \pi_\phi(\alpha)) + \phi^2(\pi(u) - \pi(\alpha)).$$

By axiom (v) on the mean-square of the coefficients,

$$\sum_{p \leq x} (\cos \phi_p)^2 = \frac{1}{m^2} \sum_{p \leq x} |a(p)|^2 \sim \frac{\kappa}{m^2} \pi(x). \tag{5.25}$$

Let δ be a small positive constant. Substituting (5.25) in (5.24), gives

$$\pi_\phi(u) - \pi_\phi(\alpha) \geq \left(\frac{\frac{\kappa}{m^2} - \phi^2}{1 - \phi^2} + \mathrm{O}(1)\right)(\pi(u) - \pi(\alpha))$$

for $u \geq \alpha(1 + \delta)$, as $n \to \infty$. Thus, we get by partial summation

$$\sum_{\substack{p \in \mathbb{P}^\phi \\ \alpha < p \leq \beta}} \frac{1}{p} = \int_\alpha^\beta \frac{\mathrm{d}\pi_\phi(u)}{u} \geq \left(\frac{\frac{\kappa}{m^2} - \phi^2}{1 - \phi^2} + \mathrm{O}(1)\right)\int_\alpha^\beta \frac{\mathrm{d}\pi(u)}{u}$$

$$\geq \left(\frac{\frac{\kappa}{m^2} - \phi^2}{1 - \phi^2} + \mathrm{O}(1)\right) \sum_{\alpha(1+\delta) < p \leq \beta} \frac{1}{p}, \tag{5.26}$$

as $n \to \infty$. From (5.10) we deduce

$$\sum_{\alpha(1+\delta)<p\leq\beta} \frac{1}{p} = \log \frac{\log \beta}{\log(\alpha(1+\delta))} + \mathrm{O}\left(\exp(-c_2\sqrt{n})\right)$$

$$= \left(\frac{1}{2} - \frac{\log(1+\delta)}{\eta}\right)\frac{1}{n} + \mathrm{O}\left(\frac{1}{n^2}\right).$$

This gives in formula (5.26)

$$\sum_{\substack{p\in\mathbb{P}^\phi \\ \alpha<p\leq\beta}} \frac{1}{p} \geq \frac{\frac{\kappa}{m^2} - \phi^2}{1 - \phi^2}\left(\frac{1}{2} - \frac{\log(1+\delta)}{\eta}\right)\frac{1}{n} + \mathrm{O}\left(\frac{1}{n^2}\right),$$

as $n \to \infty$. Hence, it follows from (5.23) that

$$\sum_{n\notin A} \frac{1}{n} < \infty \qquad\qquad (5.27)$$

and so A has natural density 1.

Let $A = \{a_k : k \in \mathbb{N}\}$ with $a_1 < a_2 < \dots$. Then (5.27) implies

$$\lim_{k\to\infty} \frac{a_k}{k} = 1. \qquad\qquad (5.28)$$

By the definition of the set A, there exists a sequence $\{\xi_k\}$ such that

$$\left(a_k - \frac{1}{4}\right)\eta < \xi_k \leq \left(a_k + \frac{1}{4}\right)\eta \quad \text{and} \quad |\varrho(\xi_k)| \leq \exp(-\xi_k).$$

Hence, from (5.28) it follows that:

$$\lim_{k\to\infty} \frac{\xi_k}{k} = \eta \qquad \text{and} \qquad \limsup_{k\to\infty} \frac{\log|\varrho(\xi_k)|}{\xi_k} \leq -1.$$

Applying Theorem 5.9, we obtain

$$\limsup_{r\to\infty} \frac{\log|\varrho(r)|}{r} \leq -1. \qquad\qquad (5.29)$$

However, by Lemma 5.8, if $\varrho(z)$ does not vanish identically, then

$$\limsup_{r\to\infty} \frac{\log|\varrho(r)|}{r} > 0,$$

contradicting (5.29). Therefore $\varrho(z) \equiv 0$. Differentiating this identity r-times with respect to z, we obtain

$$0 = (-1)^r \int_{\mathbb{C}} s^r \exp(-sz)\,\mathrm{d}\mu(s).$$

Putting $z = 0$ we get

$$\int_{\mathbb{C}} s^r \, d\mu(s) = 0$$

for $r \in \mathbb{N}_0$. Thus the first assumption of Theorem 5.7 is also satisfied and we obtain the denseness of all convergent series (5.16). It remains to show that this does not change when we add terms $\sum_{p \leq N} \tilde{g}_p(s)$.

Let $f \in \mathcal{H}(\mathcal{D})$, K be a compact subset of $\mathcal{D}$, and $\epsilon > 0$. Further, choose the parameter N such that

$$\max_{s \in K} \left(\sum_{p > N} \sum_{\nu=2}^{\infty} \frac{1}{\nu p^{\nu \sigma}} \right) < \frac{\epsilon}{4m}, \tag{5.30}$$

where m is still the degree of the local Euler factors of $\mathcal{L}(s)$. By the denseness of all convergent series (5.16) in $\mathcal{H}(\mathcal{D})$ we see that there exists a sequence $\{\tilde{b}(p) : \tilde{b}(p) \in \gamma\}$ such that

$$\max_{s \in K} \left| f(s) - \sum_{p \leq N} \tilde{g}_p(s) - \sum_{p > N} \tilde{b}(p) \tilde{g}_p(s) \right| < \frac{\epsilon}{2}. \tag{5.31}$$

Setting

$$b(p) = \begin{cases} 1 & \text{if } p \leq N, \\ \tilde{b}(p) & \text{otherwise}, \end{cases}$$

then (5.30) and (5.31) imply

$$\max_{s \in K} \left| f(s) - \sum_{p} g_p(s, b(p)) \right| = \max_{s \in K} \left| f(s) - \sum_{p \leq N} \tilde{g}_p(s) - \sum_{p > N} g_p(s, b(p)) \right|$$

$$\leq \max_{s \in K} \left| f(s) - \sum_{p \leq N} \tilde{g}_p(s) - \sum_{p > N} \tilde{b}(p) \tilde{g}_p(s) \right|$$

$$+ \max_{s \in K} \left| \sum_{p > N} \tilde{b}(p) \tilde{g}_p(s) - \sum_{p > N} g_p(s, b(p)) \right|$$

$$< \frac{\epsilon}{2} + 2m \max_{s \in K} \left(\sum_{p > N} \sum_{\nu=2}^{\infty} \frac{1}{\nu p^{\nu \sigma}} \right) < \epsilon;$$

for the last but one inequality we have used that $|a(p)| \leq m$ by Lemma 2.2. Since $f(s)$, K and ϵ were arbitrary, the theorem is proved. $\square$

5.5 The Support of the Limit Measure

Now we are going to apply limit Theorem 4.3. We restrict our studies to the space of functions analytic on bounded open rectangles. For this purpose, we define, for an arbitrary fixed real number $M > 0$,

$$\mathcal{D}_M = \{s = \sigma + it \in \mathbb{C} : \sigma_m < \sigma < 1, |t| < M\}. \tag{5.32}$$

Obviously, $\mathcal{D}_M \subset \mathcal{D}$ by (4.1). Therefore, we obtain by the induced topology that for $s \in \mathcal{D}_M$ the function $\mathcal{L}(s, \omega)$, defined by (4.6), is an $\mathcal{H}(\mathcal{D}_M)$-valued random element on the probability space $(\Omega, \mathcal{B}(\Omega), \mathrm{m})$. Now denote by $\mathbf{Q}$ the distribution of $\mathcal{L}(s, \omega)$ on $(\mathcal{H}(\mathcal{D}_M), \mathcal{B}(\mathcal{H}(\mathcal{D}_M)))$. Further, define a probability measure $\mathbf{Q}_T$ by setting

$$\mathbf{Q}_T(A) = \lim_{T \to \infty} \frac{1}{T} \operatorname{meas} \{\tau \in [0, T] : \mathcal{L}(s + i\tau) \in A\} \tag{5.33}$$

for $A \in \mathcal{B}(\mathcal{H}(\mathcal{D}_M))$. Then we deduce from Theorem 4.3

Corollary 5.11. *Let* $\mathcal{L} \in \tilde{S}$. *Then* $\mathbf{Q}_T$ *converges weakly to* $\mathbf{Q}$ *as* $T \to \infty$.

Next we examine the support of the measure $\mathbf{Q}_T$:

Lemma 5.12. *The support of the measure* $\mathbf{Q}_T$ *is the set*

$$S_M := \{\varphi \in \mathcal{H}(\mathcal{D}_M) : \varphi(s) \neq 0 \text{ for } s \in \mathcal{D}_M, \text{ or } \varphi(s) \equiv 0\}.$$

In order to prove this lemma we make use of Hurwitz's classical theorem on uniformly convergent sequences of functions and their zeros.

Theorem 5.13. *Let* $\mathcal{G}$ *be a region and* $\{f_n\}$ *be a sequence of functions analytic on* $\mathcal{G}$ *which converges uniformly on* $\mathcal{G}$ *to some function* f. *Suppose that* $f(s) \not\equiv 0$, *then an interior point* s_0 *of* $\mathcal{G}$ *is a zero of* $f(s)$ *if and only if there exists a sequence* $\{s_n\}$ *in* $\mathcal{G}$ *which tends to* s_0 *as* $n \to \infty$, *and* $f_n(s_n) = 0$ *for all sufficiently large* n.

The proof of Hurwitz's theorem relies in the main part on Rouché's theorem (see Theorem 8.1) and can be found in the monographs of Titchmarsh [352, Sect. 3.4.5] or Conway [62, Sect. VII.2].

Now we can give the

Proof of Lemma 5.12. Recall the definition (5.11) of the functions $g_p(s, \omega(p))$. Since $\{\omega(p)\}$ is a sequence of independent random variables on the probability space $(\Omega, \mathcal{B}(\Omega), \mathrm{m})$, it follows that $\{g_p(s, \omega(p))\}$ is a sequence of independent $\mathcal{H}(\mathcal{D}_M)$-valued random elements. The support of each $\omega(p)$ is the unit circle γ, and therefore the support of the random elements $g_p(s, \omega(p))$ is the set

$$\{\varphi \in \mathcal{H}(\mathcal{D}_M) : \varphi(s) = g_p(s, b) \quad \text{with} \quad b \in \gamma\}.$$

Consequently, by Theorem 3.16, the support of the $\mathcal{H}(\mathcal{D}_M)$-valued random element

$$\log \mathcal{L}(s,\omega) = \sum_p g_p(s,\omega(p))$$

is the closure of the set of all convergent series $\sum_p g_p(s,b(p))$. By Theorem 5.10 the set of these series is dense in $\mathcal{H}(\mathcal{D}_M)$. The map

$$\exp : \mathcal{H}(\mathcal{D}_M) \to \mathcal{H}(\mathcal{D}_M), \quad f \mapsto \exp(f),$$

is a continuous function sending $\log \mathcal{L}(s,\omega)$ to $\mathcal{L}(s,\omega)$ and $\mathcal{H}(\mathcal{D}_M)$ to $S_M \setminus \{0\}$. Therefore, the support $S_{\mathcal{L}}$ of $\mathcal{L}(s,\omega)$ contains $S_M \setminus \{0\}$. On the other hand, the support of $\mathcal{L}(s,\omega)$ is closed. By Hurwitz's Theorem 5.13 it follows that

$$\overline{S_M \setminus \{0\}} = S_M.$$

Thus, $S_M \subset S_{\mathcal{L}}$. In view of Lemma 2.2 the functions

$$\exp(g_p(s,\omega(p))) = \prod_{j=1}^m \left(1 - \frac{\alpha_j(p)\omega(p)}{p^s} \right)$$

are non-zero for $s \in \mathcal{D}_M$ and $\omega \in \Omega$. Hence, $\mathcal{L}(s,\omega)$ is an almost surely convergent product of non-vanishing factors. If we apply Hurwitz's Theorem 5.13 again, we conclude that $\mathcal{L}(s,\omega) \in S_M$ almost surely. Therefore $S_{\mathcal{L}} \subset S_M$. The proof is finished.   □

5.6 The Universality Theorem

Now we are in the position to prove the main result of this chapter, a generalization of Voronin's universality theorem to the class $\tilde{\mathcal{S}}$.

Theorem 5.14. *Let $\mathcal{L} \in \tilde{\mathcal{S}}$, $\mathcal{K}$ be a compact subset of the strip $\mathcal{D} = \{s \in \mathbb{C} : \sigma_m < \sigma < 1\}$ with connected complement, and let $g(s)$ be a non-vanishing continuous function on $\mathcal{K}$ which is analytic in the interior of $\mathcal{K}$. Then, for any $\epsilon > 0$*

$$\liminf_{T \to \infty} \frac{1}{T} \operatorname{meas} \left\{ \tau \in [0,T] : \max_{s \in \mathcal{K}} |\mathcal{L}(s + i\tau) - g(s)| < \epsilon \right\} > 0.$$

So L-functions in $\tilde{\mathcal{S}}$ are universal and their strip of universality is at least the intersection of their mean-square half-plane $\sigma > \sigma_m$ with the open right half of the critical strip. Recall that, by (4.2), this strip of universality $\mathcal{D}$ contains at least

$$\max \left\{ \frac{1}{2}, 1 - \frac{1 - \sigma_{\mathcal{L}}}{1 + 2\mu_{\mathcal{L}}} \right\} < \sigma < 1,$$

where $\sigma_{\mathcal{L}}$ and $\mu_{\mathcal{L}}$ are defined by axioms (ii) and (iii).

Proof. Since $\mathcal{K}$ is a compact subset of $\mathcal{D}$, there exists a number M such that $\mathcal{K} \subset \mathcal{D}_M$, where $\mathcal{D}_M$ is defined by (5.32).

First, we suppose that $g(s)$ has a non-vanishing analytic continuation to $\mathcal{D}_M$. By Lemma 5.12 the function $g(s)$ is contained in the support $S_{\mathcal{L}}$ of the random element $\mathcal{L}(s,\omega)$. Denote by Φ the set of functions $\varphi \in \mathcal{H}(\mathcal{D}_M)$ such that

$$\max_{s \in \mathcal{K}} |\varphi(s) - g(s)| < \epsilon.$$

Since by Corollary 5.11 the measure $\mathbf{Q}_T$ converges weakly to $\mathbf{Q}$, as $T \to \infty$, and the set Φ is open, it follows from Theorem 3.1 and from properties of the support that

$$\liminf_{T \to \infty} \frac{1}{T} \operatorname{meas} \left\{ \tau \in [0,T] : \max_{s \in \mathcal{K}} |\mathcal{L}(s + i\tau) - g(s)| < \epsilon \right\}$$

$$= \liminf_{T \to \infty} \mathbf{Q}_T(\Phi) \geq \mathbf{Q}(\Phi) > 0. \tag{5.34}$$

This proves the theorem in the case of functions $g(s)$ which have a non-vanishing analytic continuation to $\mathcal{D}_M$. We remark that analytic functions on $\mathcal{D}$ can be approximated uniformly on compact sets; here the restriction on $\mathcal{K}$ to have connected complement is not necessary.

Now let $g(s)$ be as in the statement of the theorem. In this case we have to apply a complex analogue of Weierstrass' approximation theorem, the theorem of Mergelyan on the approximation of analytic functions by polynomials.

Theorem 5.15. *Let $\mathcal{K}$ be a compact subset of $\mathbb{C}$ with connected complement. Then any continuous function $g(s)$ on $\mathcal{K}$ which is analytic in the interior of $\mathcal{K}$ is uniformly approximable on $\mathcal{K}$ by polynomials in s.*

This generalizes a classic result of Runge and gives the final solution of the problem of uniform approximation by polynomials on compacta. The ingenious proof relies on Green's formula and Riemann's mapping theorem and can be found, for example, in Mergelyan's article [246], Rudin [312] or Walsh's monograph [367, Sect. A.1]. The restriction to compact sets with connected complement is natural. A necessary and sufficient condition that every function analytic on a closed bounded point set C can be uniformly approximated on C by polynomials is that C should not separate the plane, or equivalently, C should be the complement of an infinite region. For instance, the function $g(s) = \frac{1}{s-\alpha}$ cannot be approximated uniformly on C by polynomials if C is an annulus enclosing α.

We continue with the proof of Theorem 5.14. By Mergelyan's approximation Theorem 5.15 there exists a sequence of polynomials $G_n(s)$ which converges uniformly on $\mathcal{K}$ to $g(s)$ as $n \to \infty$. Since $g(s)$ is non-vanishing on $\mathcal{K}$, we have $G_m(s) \neq 0$ on $\mathcal{K}$ for a sufficiently large m, and

$$\max_{s \in \mathcal{K}} |g(s) - G_m(s)| < \frac{\epsilon}{4}. \tag{5.35}$$

Since the polynomial $G_m(s)$ has only finitely many zeros, there exists a region $\mathcal{G}$ whose complement is connected such that $\mathcal{K} \subset \mathcal{G}$ and $G_m(s) \neq 0$ on $\mathcal{G}$. Hence there exists a continuous branch $\log G_m(s)$ on $\mathcal{G}$, and $\log G_m(s)$ is analytic in the interior of $\mathcal{G}$. Thus, by Mergelyan's Theorem 5.15, there exists a sequence of polynomials $F_n(s)$ which converges uniformly on $\mathcal{K}$ to $\log G_m(s)$ as $n \to \infty$. Hence, for sufficiently large k,

$$\max_{s \in \mathcal{K}} |G_m(s) - \exp(F_k(s))| < \frac{\epsilon}{4}.$$

From this and from (5.35) we obtain

$$\max_{s \in \mathcal{K}} |g(s) - \exp(F_k(s))| < \frac{\epsilon}{2}. \tag{5.36}$$

From (5.34) we deduce

$$\liminf_{T \to \infty} \frac{1}{T} \operatorname{meas} \left\{ \tau \in [0, T] : \max_{s \in \mathcal{K}} |\mathcal{L}(s + i\tau) - \exp(F_k(s))| < \frac{\epsilon}{2} \right\} > 0.$$

In combination with (5.36) this proves Theorem 5.14.   □

A close look on the proof of Theorem 5.14 shows that omitting a finite number of Euler factors in the Euler product does not influence the universality property. It is an interesting question to which extent *infinitely* many Euler factors can be omitted. A first approach to answer this problem was given by Schwarz, Steuding and Steuding [319] by using the notion of related arithmetical functions.

We shall compare the universality Theorem 5.14 with the result of Laurinčikas [190] for Matsumoto zeta-functions (see also Sect. 1.6). Both theorems apply to polynomial Euler products; however, the form of the Euler product (iv) is more restrictive than the one of Matsumoto zeta-functions. Furthermore, Theorem 5.14 relies on axiom (v) while Laurinčikas' assumes (1.40); in our case his condition can be translated as follows: for all primes p, $|a(p)|$ is bounded from below by some positive constant. This is in general a rather strong assumption. For example, Lehmer [212] conjectured that Ramanujan's τ-function $\tau(n)$, given implicitly as the Fourier coefficients of the modular discriminant (1.33), is non-zero for any $n \in \mathbb{N}$. This is still unproved (also in the form $\tau(p) \neq 0$ for prime p). Laurinčikas [190] also claims that (1.40) can be replaced by the condition that, for $\epsilon > 0$,

$$\frac{1}{\pi(x)} \sum_{\substack{p \leq x \\ |a(p)| < \epsilon}} 1 \ll x^{-\delta},$$

where $\delta \geq \frac{1}{2}$ (again in our notation). Also this condition is rather restrictive. For example, in the case of the L-function associated with the modular discriminant one would need a quantitative version of the Sato–Tate conjecture (see (6.22)) in order to verify this assumption. Nevertheless, we cannot exclude the possibility of a number-theoretical relevant Matsumoto zeta-function which is not a polynomial Euler product of fixed degree.

Reviewing our lengthy proof of Theorem 5.14 we may understand universality as a kind of ergodicity on function spaces. On the other side, all proofs of universality results, known so far, depend on some arithmetical conditions; but *is universality really an arithmetic phenomenon or not?* This question was raised by Matsumoto [242]. It seems reasonable that the universality of Dirichlet series is a common phenomenon in analysis (see the appendix), that it is related to Julia rays in value-distribution theory and to ergodical dynamical systems as well, but no link is known so far.

5.7 Discrete Universality

Now we consider the phenomenon of *discrete* universality, that means that the shifts τ are restricted to arithmetic progressions. This concept of universality was introduced by Reich [306] (see (1.39)). It is remarkable that this still leads to universality results with positive lower density for the shifts τ. The precise statement of Reich's discrete universality theorem for Dedekind zeta-functions is

Theorem 5.16. *Let $\mathbb{K}$ be an algebraic number field of degree d over $\mathbb{Q}$, let $\mathcal{K}$ be a compact subset of the strip $\max\{\frac{1}{2}, 1 - \frac{1}{d}\} < \sigma < 1$ with connected complement, and assume that $g(s)$ is a non-vanishing continuous function on $\mathcal{K}$ which is analytic in its interior. Then, for any real $\Delta \neq 0$ and any $\epsilon > 0$,*

$$\liminf_{N \to \infty} \frac{1}{N} \sharp \left\{ n \leq N : \max_{s \in \mathcal{K}} |\zeta_{\mathbb{K}}(s + i\Delta n) - g(s)| < \epsilon \right\} > 0.$$

In Sect. 4.8 we presented a discrete limit theorem for Matsumoto zeta-functions due to Kačinskaitė, where Δ was assumed to satisfy a certain arithmetical property. We also mentioned the work of Kačinskaitė and Laurinčikas [155] which removes this restriction. Interestingly, this problem was solved by Reich in a different way. (For more details about Dedekind zeta-functions see Sect. 13.1.)

Bagchi [9, 10] combined Reich's result with ideas from Gonek [104] in order to obtain joint discrete universality for Dirichlet L-functions.

Theorem 5.17. *Let $\chi_1 \bmod q_1, \ldots, \chi_\ell \bmod q_\ell$ be pairwise non-equivalent Dirichlet characters, $\mathcal{K}_1, \ldots, \mathcal{K}_\ell$ be compact subsets of $\frac{1}{2} < \sigma < 1$ with connected complements. Further, for each $1 \leq j \leq \ell$ let $g_j(s)$ be a continuous non-vanishing function on $\mathcal{K}_j$ which is analytic in the interior of $\mathcal{K}_j$. Then, for any $\Delta \neq 0$ and any $\epsilon > 0$,*

$$\liminf_{N \to \infty} \frac{1}{N} \sharp \left\{ n \leq N : \max_{1 \leq j \leq \ell} \max_{s \in \mathcal{K}_j} |L(s + i\Delta n, \chi_j) - g_j(s)| < \epsilon \right\} > 0.$$

Bagchi considered only characters to a fixed modulus q (in which case the characters χ mod q are trivially non-equivalent) and he assumed the compact sets $\mathcal{K}_j$ to be simply connected and locally path connected; however, without big effort his result (Theorem 5.3.4 in [9]) can easily be extended to the form given above.

It should be remarked that these results do not contain the continuous analogues (as, for example, Theorem 5.14 in the latter case) nor that they can be deduced from Theorem 5.14 (to our present knowledge). Nevertheless, the methods of proof are rather similar. In fact, they depend on an appropriate limit theorem as Theorem 4.13; the remaining arguments are essentially the same.

Matsumoto [243] asked for discrete universality for L-functions associated with cusp forms and Matsumoto zeta-functions. Garbaliauskienė and Laurinčikas [84] proved a discrete version of Theorem 1.11 for L-functions to elliptic curves; Laurinčikas, Matsumoto and Steuding [205] obtained discrete universality for the more general case of L-functions to newforms.

6

The Selberg Class

What is a zeta-function (or an L-function)? We know one when we
see one.   M.N. Huxley

In 1989, Selberg defined a rather general class S of Dirichlet series having
an Euler product, analytic continuation and a functional equation of Riemann-
type, and formulated some fundamental conjectures concerning them. His aim
was to study the value-distribution of linear combinations of L-functions. In
the meantime, this so-called Selberg class became an important object of
research, but still it is not understood very well. In this chapter, we shall
investigate universality for functions in the Selberg class. Therefore, we only
present results on this class which are related to our studies; for detailed
surveys on the Selberg class, we refer to Kaczorowski and Perelli [160],
Perelli [290], and M.R. Murty and V.K. Murty [270].

6.1 Definition and First Properties

Many authors, for example, Lekkerkerker [214], Perelli [289], and
Matsumoto [239] have introduced classes of Dirichlet series to find common
patterns in their value-distribution. However, the most successful class seems
to be the class introduced by Selberg [323]. The Selberg class S consists of
Dirichlet series

$$\mathcal{L}(s) = \sum_{n=1}^{\infty} \frac{a(n)}{n^s}$$

satisfying the following hypotheses:

(1) *Ramanujan hypothesis.* $a(n) \ll n^\epsilon$ for any $\epsilon > 0$, where the implicit constant may depend on ϵ.
(2) *Analytic continuation.* There exists a non-negative integer k such that
$(s-1)^k \mathcal{L}(s)$ is an entire function of finite order.

(3) *Functional equation.* $\mathcal{L}(s)$ satisfies a functional equation of type

$$\Lambda_{\mathcal{L}}(s) = \omega \overline{\Lambda_{\mathcal{L}}(1 - \bar{s})},$$

where

$$\Lambda_{\mathcal{L}}(s) := \mathcal{L}(s) Q^s \prod_{j=1}^{f} \Gamma(\lambda_j s + \mu_j)$$

with positive real numbers Q, λ_j, and complex numbers μ_j, ω with $\operatorname{Re} \mu_j \geq 0$ and $|\omega| = 1$.

(4) *Euler product.* $\mathcal{L}(s)$ has a product representation

$$\mathcal{L}(s) = \prod_{p} \mathcal{L}_p(s),$$

where

$$\mathcal{L}_p(s) = \exp\left(\sum_{k=1}^{\infty} \frac{b(p^k)}{p^{ks}}\right)$$

with suitable coefficients $b(p^k)$ satisfying $b(p^k) \ll p^{k\theta}$ for some $\theta < \frac{1}{2}$.

The Euler product hypothesis implies that the coefficients $a(n)$ are multiplicative, and that each Euler factor has the Dirichlet series representation

$$\mathcal{L}_p(s) = \sum_{k=0}^{\infty} \frac{a(p^k)}{p^{ks}}, \tag{6.1}$$

absolutely convergent for $\sigma > 0$. Differentiation of the identity between the two representations for the local Euler factors leads to the relation

$$\log p \sum_{k=1}^{\infty} \frac{k a(p^k)}{p^{ks}} = -\mathcal{L}_p'(s) = -\mathcal{L}_p(s) \left(\sum_{\ell=1}^{\infty} \frac{b(p^\ell)}{p^{\ell s}}\right)'$$

$$= \log p \sum_{m=1}^{\infty} \frac{1}{p^{ms}} \sum_{m=k+\ell} a(p^k) b(p^\ell) \ell.$$

Comparing the coefficients, we obtain $a(p) = b(p)$ and

$$b(p^\ell) = a(p^\ell) - \frac{1}{\ell} \sum_{k=1}^{\ell-1} k a(p^{\ell-k}) b(p^k).$$

Moreover, it turns out that each Euler factor is absolutely convergent in the half-plane $\sigma > 0$ and non-vanishing for $\sigma > \theta$.

Axioms (1) and (2) imply that any element $\mathcal{L}(s)$ of the Selberg class is analytic in the whole complex plane except for a possible pole at $s = 1$ (by the same reasoning as for $\tilde{S}$ in Sect. 2.2).

The structure of the Selberg class is of special interest. The degree of $\mathcal{L} \in \mathcal{S}$ is defined by

$$d_{\mathcal{L}} = 2 \sum_{j=1}^{f} \lambda_j. \tag{6.2}$$

Although the data of the functional equation are not unique, the degree is well-defined. If $N_{\mathcal{L}}(T)$ counts the number of zeros of $\mathcal{L} \in \mathcal{S}$ in the rectangle $0 \le \sigma \le 1, |t| \le T$ (according to multiplicities), by standard contour integration one can show

$$N_{\mathcal{L}}(T) \sim \frac{d_{\mathcal{L}}}{\pi} T \log T, \tag{6.3}$$

in analogy to the Riemann–von Mangoldt formula (1.5) for Riemann's zeta-function; we shall give a more precise asymptotic formula in Theorem 7.7. It is conjectured that *all* $\mathcal{L} \in \mathcal{S}$ *have integral degree*. Slightly stronger is the

Strong λ-conjecture. *Let* $\mathcal{L} \in \mathcal{S}$. *All* λ_j *appearing in the Gamma-factors of the functional equation can be chosen to be equal to* $\frac{1}{2}$.

The constant function 1 is the only element of $\mathcal{S}$ of degree zero. Recently, Kaczorowski and Perelli [162] showed that all functions $\mathcal{L} \in \mathcal{S}$ with degree $0 < d_{\mathcal{L}} < \frac{5}{3}$ have degree equal to one. For our purpose the following slightly weaker result is sufficient.

Theorem 6.1. *If* $\mathcal{L} \in \mathcal{S}$ *and* $0 \le d_{\mathcal{L}} < 1$, *then* $\mathcal{L}(s) \equiv 1$.

For $0 < d_{\mathcal{L}} < 1$, the statement is implicitly contained in the work of Richert [308], Bochner [27] and Vignéras [358]. We give a sketch of a simple proof following Conrey and Ghosh [59] and Molteni [256].

Proof. Let B be a constant such that $a(n) \ll n^B$. By Perron's formula,

$$\sum_{n \le x} a(n) = \frac{1}{2\pi i} \int_{c-i\infty}^{c+i\infty} \mathcal{L}(s) \frac{x^s}{s} ds + O\left(\frac{x^{c+B}}{T}\right),$$

where $c > 1$ is a constant. Shifting the path of integration to the left, yields, by the Phragmén–Lindelöf principle (see Sect. 6.5), the asymptotic formula

$$\sum_{n \le x} a(n) = xP(\log x) + O\left(x^{(1+B)\frac{d_{\mathcal{L}}-1}{d_{\mathcal{L}}+1}+\epsilon}\right),$$

where $P(x)$ is a computable polynomial according to the principal part of the Laurent expansion of $\mathcal{L}(s)$ at $s = 1$. By subtraction, this implies

$$a(n) \ll n^{(1+B)\frac{d_{\mathcal{L}}-1}{d_{\mathcal{L}}+1}+\epsilon}, \tag{6.4}$$

where the implicit constant depends on B. For $d_{\mathcal{L}} < 1$ the exponent is negative, and we may choose B arbitrarily large. Then $\mathcal{L}(s)$ is uniformly bounded in every right half-plane. This is a contradiction for $\mathcal{L} \in \mathcal{S}$ with positive degree since the functional equation implies a certain order of growth. This shows that $\mathcal{S}$ is free of elements having degree $0 < d < 1$.

It remains to consider the case that $d_{\mathcal{L}} = 0$. Then the functional equation takes the form:

$$Q^s \mathcal{L}(s) = \omega Q^{1-s} \overline{\mathcal{L}(1 - \bar{s})}$$

(there are no Gamma-factors). By (6.4) the $a(n)$ are so small that the Dirichlet series for $\mathcal{L}(s)$ converges in the whole complex plane. Thus, we may rewrite the functional equation as

$$\sum_{n=1}^{\infty} a(n) \left(\frac{Q^2}{n} \right)^s = \omega \sum_{n=1}^{\infty} \frac{\overline{a(n)}}{n} n^s. \tag{6.5}$$

We may regard this as an identity between absolutely convergent Dirichlet series. Thus, if $a(n) \neq 0$, then Q^2/n is an integer. In particular, $q := Q^2 \in \mathbb{N}$. Moreover, since q has only finitely many divisors, it follows that $\mathcal{L}(s)$ is a Dirichlet polynomial. If $q = 1$, then $\mathcal{L}(s) \equiv 1$ and we are done with the case $d_{\mathcal{L}} = 0$. Hence, we may assume that $q > 1$.

Since the Dirichlet coefficients $a(n)$ are multiplicative, we have $a(1) = 1$ and via (6.5)

$$a(1)Q^{2s} = \omega Q^{-1} \overline{a(Q^2)} Q^{2s};$$

thus, $|a(q)| = Q$. In particular, there exists a prime p such that the exponent ν of p in the prime factorization of q is positive and, by the multiplicativity of the $a(n)$'s,

$$|a(p^\nu)| \geq p^{\nu/2}.$$

Now consider the logarithm of the corresponding Euler factor:

$$\log \left(1 + \sum_{m=1}^{\nu} \frac{a(p^m)}{p^{ms}} \right) = \sum_{k=1}^{\infty} \frac{b(p^k)}{p^{ks}}.$$

Viewing this as a power series in $X = p^{-s}$, we write

$$\log P(X) = \sum_{k=1}^{\infty} B_k X^k \quad \text{with} \quad B_k = b(p^k).$$

Since $a(1) = 1$, we find

$$P(X) = 1 + \sum_{m=1}^{\nu} a(p^m) X^m = \prod_{j=1}^{\nu} (1 - C_j X) \quad \text{with} \quad B_k = -\frac{1}{k} \sum_{j=1}^{\nu} C_j^k.$$

Now

$$\prod_{j=1}^{\nu} |C_j| = |a(p^\nu)| \geq p^{\nu/2},$$

and thus the maximum of the values $|C_j|$ is greater than or equal to $p^{1/2}$. We have

$$\lim_{k\to\infty} |b(p^k)|^{1/k} = \lim_{k\to\infty} \left| \frac{1}{k} \sum_{j=1}^{\nu} C_j^k \right|^{1/k} = \max_{1\le j \le \nu} |C_j|;$$

by our foregoing observations the right-hand side is greater than or equal to $p^{1/2}$. This is a contradiction to the condition $b(p^k) \ll p^{k\theta}$ with some $\theta < \frac{1}{2}$ in the axiom on the Euler product. Hence, $q = 1$ and $\mathcal{L}(s) \equiv 1$. This proves the theorem.   □

By the work of Kaczorowski and Perelli [161], it is known that the functions of degree one in the Selberg class are the Riemann zeta-function and shifts $L(s + i\theta, \chi)$ of Dirichlet L-functions attached to primitive characters χ with $\theta \in \mathbb{R}$; parts of this classification are already contained in Bochner's extension of Hamburger's theorem [27] and in subsequent work of Gérardin and Li [99] and Piatetski-Shapiro and Rhagunathan [294] on converse theorems. However, for higher degree there is no complete classification so far. Examples of degree two are L-functions associated with holomorphic newforms f; here the notion *normalized* means that $a(n) = c(n)n^{(1-k)/2}$ where the $c(n)$ are the Fourier coefficients of f (as already indicated in Sect. 2.2). Normalized L-functions attached to non-holomorphic newforms are expected to lie in $\mathcal{S}$ but the Ramanujan hypothesis is not yet verified. The Rankin–Selberg L-function of any normalized eigenform is an element of the Selberg class of degree 4. Other examples are Dedekind zeta-functions to number fields $\mathbb{K}$; their degree is equal to the degree of the field extension $\mathbb{K}/\mathbb{Q}$.

In view of the Euler product representation, it is clear that any element $\mathcal{L}(s)$ of the Selberg class does not vanish in the half-plane of absolute convergence $\sigma > 1$. This gives rise to the notions of critical strip and critical line. The zeros of $\mathcal{L}(s)$ located at the poles of gamma-factors appearing in the functional equation are called trivial. They all lie in $\sigma \le 0$, and it is easily seen that they are located at

$$s = -\frac{k + \mu_j}{\lambda_j} \quad \text{with} \quad k \in \mathbb{N}_0 \quad \text{and} \quad 1 \le j \le f. \tag{6.6}$$

All other zeros are said to be non-trivial. In general, we cannot exclude the possibility that $\mathcal{L}(s)$ has a trivial zero and a non-trivial one at the same point. It is expected that for every function in the Selberg class the analogue of the Riemann hypothesis holds.

Grand Riemann Hypothesis. *If $\mathcal{L} \in \mathcal{S}$, then $\mathcal{L}(s) \ne 0$ for $\sigma > \frac{1}{2}$.*

The zero-distribution is essential for the Selberg class. This is also manifested in the defining axioms. Following Conrey and Ghosh [59] and Kaczorowski and Perelli [160], we motivate the axioms defining $\mathcal{S}$. We have already seen that the Ramanujan hypothesis implies the regularity of $\mathcal{L}(s)$ in $\sigma > 1$. Further, we note:

- The assumption that there be at most one pole, and that this one is located at $s = 1$, is natural in the theory of L-functions. It seems that the point $s = 1$ is the only possible pole for an automorphic L-function and that such a pole is always related to the simple pole of the Riemann zeta-function in the sense that the quotient with an appropriate power of $\zeta(s)$ is another L-function which is entire (examples for this scenario are Dedekind zeta-functions).
- The restriction $\operatorname{Re} \mu_j \geq 0$ in the functional equation comes from the theory of Maass waveforms. If we assume the existence of an arithmetic subgroup of $\mathsf{SL}_2(\mathbb{R})$ together with a Maass cusp form that corresponds to an exceptional eigenvalue, and if we further suppose that all local roots are sufficiently small (more precisely, that the Ramanujan–Petersson conjecture holds), then the L-function associated with the Maass cusp form has a functional equation where the μ_j satisfy $\operatorname{Re} \mu_j < 0$, but this L-function violates Riemann's hypothesis.
- Finally, consider the axiom concerning the Euler product. It is well-known that the existence of an Euler product is a necessary (but not sufficient) condition for Riemann's hypothesis. On first sight, the condition $\theta < \frac{1}{2}$ seems to be a little bit unnatural. However, for $\theta = \frac{1}{2}$, there are examples violating the Riemann hypothesis: the function

$$(1 - 2^{1-s})\zeta(s) = \sum_{n=1}^{\infty} \frac{(-1)^{n-1}}{n^s} \tag{6.7}$$

has zeros off the critical line. Moreover, as we have seen in the proof of Theorem 6.1, the bound for θ rules out non-trivial Dirichlet polynomials from $\mathcal{S}$ as for example

$$(1 - 2^{a-s})(1 - 2^{b-s}) \quad \text{with} \quad a + b = 1.$$

Some words of warning: It might turn out that the axioms of the Selberg class are too restrictive; e.g., the form of the functional equation with the condition on the μ_j's to have positive real part might exclude relevant L-functions.

6.2 Primitive Functions and the Selberg Conjectures

The Selberg class is multiplicatively closed. A function $\mathcal{L} \in \mathcal{S}$ is called primitive if it cannot be factored as a product of two elements non-trivially, i.e., the equation

$$\mathcal{L} = \mathcal{L}_1 \mathcal{L}_2 \quad \text{with} \quad \mathcal{L}_1, \mathcal{L}_2 \in \mathcal{S},$$

implies $\mathcal{L} = \mathcal{L}_1$ or $\mathcal{L} = \mathcal{L}_2$. The notion of a primitive function is fruitful for studying the structure of $\mathcal{S}$. Conrey and Ghosh [59] proved

Theorem 6.2. *Every function in the Selberg class has a factorization into primitive functions.*

Proof. Suppose that $\mathcal{L}$ is not primitive, then there exist functions $\mathcal{L}_1$ and $\mathcal{L}_2$ in $\mathcal{S} \setminus \{1\}$ such that $\mathcal{L} = \mathcal{L}_1 \mathcal{L}_2$. Taking into account (6.3), from

$$N_{\mathcal{L}}(T) = N_{\mathcal{L}_1}(T) + N_{\mathcal{L}_2}(T)$$

we find $d_{\mathcal{L}} = d_{\mathcal{L}_1} + d_{\mathcal{L}_2}$. In view of Theorem 6.1, both $\mathcal{L}_1$ and $\mathcal{L}_2$ have degree at least 1. Thus, each of $d_{\mathcal{L}_1}$ and $d_{\mathcal{L}_2}$ is strictly less than $d_{\mathcal{L}}$. A continuation of this process terminates since the number of factors is $\leq d_{\mathcal{L}}$, which proves the claim.   □

In connection with Theorem 6.1 it follows that Riemann's zeta-function and Dirichlet L-functions are primitive. A more advanced example of primitive elements are L-functions associated with newforms is due to M.R. Murty [269]. On the contrary, Dedekind zeta-functions to cyclotomic fields $\neq \mathbb{Q}$ are not primitive.

Denote by $a_{\mathcal{L}}(n)$ the coefficients of the Dirichlet series representation of $\mathcal{L} \in \mathcal{S}$. The central claim concerning primitive functions is part of

Selberg's Conjectures.

(A) For all $1 \neq \mathcal{L} \in \mathcal{S}$ there exists a positive integer $n_{\mathcal{L}}$ such that

$$\sum_{p \leq x} \frac{|a_{\mathcal{L}}(p)|^2}{p} = n_{\mathcal{L}} \log \log x + \mathrm{O}(1).$$

(B) For any primitive functions $\mathcal{L}_1$ and $\mathcal{L}_2 \in \mathcal{S}$,

$$\sum_{p \leq x} \frac{a_{\mathcal{L}_1}(p)\overline{a_{\mathcal{L}_2}(p)}}{p} = \begin{cases} \log \log x + \mathrm{O}(1) & \text{if } \mathcal{L}_1 = \mathcal{L}_2, \\ \mathrm{O}(1) & \text{otherwise.} \end{cases}$$

In some sense, primitive functions are expected to form an orthonormal system.

In view of the factorization into primitive functions, it is easily seen that Conjecture B implies Conjecture A. In some particular cases, it is not too difficult to verify Selberg's Conjecture A. For instance, $\zeta(s)$ satisfies Selberg's Conjecture A, which is basically due to Euler [76] who already wrote

$$\frac{1}{2} + \frac{1}{3} + \frac{1}{5} + \cdots = \log \log \infty$$

(the first complete proof was given by Mertens [247]). A stronger asymptotic formula is (5.10) which plays an essential role in Bagchi's proof of Voronin's universality theorem. By the prime number theorem for arithmetic progressions (1.29), we get the same asymptotics for Dirichlet L-functions. Taking into account the orthogonality relations for characters, one can also verify

Conjecture B for pairs of Dirichlet L-functions. The Rankin–Selberg convolution method shows that L-functions associated with holomorphic modular forms satisfy some kind of orthogonality (in terms of regularity at $s = 1$) which is related to Selberg's conjectures. Recently, Liu, Wang and Ye [225] proved Conjecture B for automorphic L-functions $L(s, \pi)$ and $L(s, \pi')$, where π and π' are automorphic irreducible cuspidal representations of $\mathsf{GL}_m(\mathbb{Q})$ and $\mathsf{GL}_{m'}(\mathbb{Q})$, respectively; their result holds unconditionally for $m, m' \leq 4$ and in other cases under the assumption of the convergence of

$$\sum_p \frac{|a_\pi(p^k)|^2}{p^k} (\log p)^2$$

for $k \geq 2$, where the $a_\pi(n)$ denote the Dirichlet series coefficients of $L(s, \pi)$. The latter hypothesis is an immediate consequence of the Ramanujan hypothesis. (We shall return to these L-functions in Chap. 13.9.)

Selberg [323] wrote about his conjectures that *"these conjectures, which, by the way, are not unrelated to several other conjectures like the Sato–Tate conjecture, Langlands conjectures, etc., have been verified in a number of cases for Dirichlet series with functional equation and Euler product that occur in number theory, by assuming that the factorizations we can give are actually that a function is really primitive and cannot be factorized further."* For example, M.R. Murty [267] proved that Selberg's Conjecture B implies Artin's conjecture (we return to this topic in Chap. 13.7). Another important consequence is due to Conrey and Ghosh [59].

Theorem 6.3. *Selberg's conjecture B implies that every $\mathcal{L} \in \mathcal{S}$ has a unique factorization into primitive functions.*

Proof. Suppose that $\mathcal{L}$ has two factorizations into primitive functions:

$$\mathcal{L} = \prod_{j=1}^m \mathcal{L}_j = \prod_{k=1}^n \tilde{\mathcal{L}}_k,$$

and assume that no $\tilde{\mathcal{L}}_k$ is equal to $\mathcal{L}_1$. Then it follows from

$$\sum_{j=1}^m a_{\mathcal{L}_j}(p) = \sum_{k=1}^n a_{\tilde{\mathcal{L}}_k}(p)$$

that

$$\sum_{j=1}^m \sum_{p \leq x} \frac{a_{\mathcal{L}_j}(p)\overline{a_{\mathcal{L}_1}(p)}}{p} = \sum_{k=1}^n \sum_{p \leq x} \frac{a_{\tilde{\mathcal{L}}_k}(p)\overline{a_{\mathcal{L}_1}(p)}}{p}.$$

By Selberg's conjecture B, the left-hand side tends to infinity for $x \to \infty$, whereas the right-hand side is bounded, giving the desired contradiction. $\square$

The same argument gives a characterization of primitive functions in terms of the quantity $n_{\mathcal{L}}$ from Selberg's conjecture A:

Theorem 6.4. *If the Selberg conjecture B is true, $\mathcal{L} \in \mathcal{S}$ is primitive if and only if $n_{\mathcal{L}} = 1$.*

Selberg also considered twists with characters. If χ is a primitive Dirichlet character, and $\mathcal{L} \in \mathcal{S}$ is automorphic, then

$$\mathcal{L}_\chi(s) = \sum_{n=1}^\infty \frac{a(n)}{n^s} \chi(n)$$

satisfies the axioms of $\mathcal{S}$ too. It can be shown that then $\mathcal{L}$ and $\mathcal{L}_\chi$ are either both primitive or both not primitive (since they have the same $n_{\mathcal{L}}$). Selberg's twisting conjecture states that *if $\mathcal{L}$ is not primitive, $\mathcal{L}$ and $\mathcal{L}_\chi$ will factor in the same way into primitive factors, in the sense that one gets the factorization of $\mathcal{L}_\chi$ by twisting all factors of $\mathcal{L}$ by χ.*

6.3 Non-Vanishing and Prime Number Theorems

Furthermore, Selberg studied moments and distribution functions for $\mathcal{L} \in \mathcal{S}$ in order to prove an analogue of (1.16). For this aim, he had to assume some unproved hypothesis on the distribution of zeros which hold, or are at least expected to hold, for all known examples of elements in $\mathcal{S}$. Let $N_{\mathcal{L}}(\sigma, T)$ count the number of zeros $\varrho = \beta + i\gamma$ of $\mathcal{L}(s)$ with $\beta > \sigma$ and $|\gamma| < T$ (counting multiplicities). Kaczorowski and Perelli [163] proved

Theorem 6.5. *For every $\mathcal{L} \in \mathcal{S}$,*

$$N_{\mathcal{L}}(\sigma, T) \ll T^{4(d_{\mathcal{L}}+3)(1-\sigma)+\epsilon},$$

uniformly for $\frac{1}{2} \leq \sigma \leq 1$.

Unfortunately, this estimate is only useful for σ close to 1. For smaller σ we note the

Grand Density Hypothesis. *For $\mathcal{L} \in \mathcal{S}$ there is some positive constant c such that for $\sigma > \frac{1}{2}$*

$$N_{\mathcal{L}}(\sigma, T) \ll T^{1-c(\sigma-1/2)+\epsilon}.$$

Selberg proved that if $\mathcal{L} \in \mathcal{S}$ satisfies the Grand density hypothesis and Conjecture A, then the values of

$$\frac{\log \mathcal{L}\left(\frac{1}{2}+it\right)}{\sqrt{\pi n_{\mathcal{L}} \log \log t}}$$

are distributed in the complex plane according to the normal distribution. Furthermore, he investigated the value-distribution of linear combinations of independent elements of $\mathcal{S}$. Selberg's argument was streamlined and extended by Bombieri and Hejhal [39] to independent collections of L-functions having polynomial Euler products with the emphasis just on probabilistic convergence and the goal of applications to the zero-distribution. For this aim they introduced a *strong* version of Selberg's conjecture B.

Strong Selberg's Conjecture B. *For any primitive functions $\mathcal{L}_1$ and $\mathcal{L}_2$ in $\mathcal{S}$ there exist constants $C_{\mathcal{L}_1}, C_{\mathcal{L}_1, \mathcal{L}_2}$ such that*

$$\sum_{p \leq x} \frac{a_{\mathcal{L}_1}(p)\overline{a_{\mathcal{L}_2}(p)}}{p} = \begin{cases} \log\log x + C_{\mathcal{L}_1} + \mathcal{R}(x) & if \quad \mathcal{L}_1 = \mathcal{L}_2, \\ C_{\mathcal{L}_1, \mathcal{L}_2} + \mathcal{R}(x) & otherwise, \end{cases}$$

where $\mathcal{R}(x) \ll \frac{1}{\log x}$.

Their theorem shows the statistical independence of any collection of independent L-functions in any family of elements of $\mathcal{S}$. Furthermore, Bombieri and Hejhal applied their result to the zero-distribution of linear combinations of independent L-functions. Assuming in addition the Grand Riemann hypothesis and a weak conjecture on the well-spacing of the zeros, they proved that almost all zeros of these linear combinations are simple and lie on the critical line. Bombieri and Perelli [40] considered for the same class of functions the distribution of distinct zeros. They proved that, for two different functions $\mathcal{L}_1, \mathcal{L}_2$ of the same degree,

$$\sum_{0 \leq \gamma \leq T} \max\{m_{\mathcal{L}_1}(\varrho) - m_{\mathcal{L}_2}(\varrho), 0\} \gg T \log T, \qquad (6.8)$$

where the sum is taken over the non-trivial zeros $\varrho = \beta + i\gamma$ of $\mathcal{L}_1\mathcal{L}_2(s)$ and $m_{\mathcal{L}_j}(\varrho)$ denotes the multiplicity of the zero ϱ of $\mathcal{L}_j(s)$.

The Selberg conjectures refer to the analytic behaviour at the edge of the critical strip. Conrey and Ghosh [59] proved the non-vanishing on the line $\sigma = 1$ subject to the truth of Selberg's conjecture B:

Theorem 6.6. *Let $\mathcal{L} \in \mathcal{S}$. If Selberg's conjecture B is true, then*

$$\mathcal{L}(s) \neq 0 \quad for \quad \sigma \geq 1.$$

It is conjectured that the Selberg class consists only of automorphic L-functions, and for those Jacquet and Shalika [148] obtained an unconditional non-vanishing theorem.

Proof of Theorem 6.6. In view of the Euler product representation, in the half-plane $\sigma \geq 1$ zeros can only occur on the line $\sigma = 1$. By Theorem 6.2 it suffices to consider primitive functions $\mathcal{L} \in \mathcal{S}$. In case of $\zeta(s)$, it is known that there are no zeros on $\sigma = 1$. It is easily seen that if Selberg's Conjecture B is true and if $\mathcal{L} \in \mathcal{S}$ has a pole at $s = 1$ of order m, then the quotient $\mathcal{L}(s)/\zeta(s)^m$ is an entire function (this can be shown by the argument from the proof of

Theorem 6.3). Hence, we may assume that $\mathcal{L}(s)$ is entire. Then, for any real α, the function $\mathcal{L}(s + i\alpha)$ is a primitive element of $\mathcal{S}$. Selberg's conjecture B applied to $\mathcal{L}(s + i\alpha)$ and $\zeta(s)$ yields

$$\sum_{p \leq x} \frac{a_{\mathcal{L}}(p)}{p^{1+i\alpha}} \ll 1. \tag{6.9}$$

Now suppose that $\mathcal{L}(1 + i\alpha) = 0$. Then

$$\mathcal{L}(s) \sim c(s - (1 + i\alpha))^k$$

as $s = \sigma + i\alpha \to 1 + i\alpha$ for some complex $c \neq 0$ and some positive integer k. It follows that:

$$\log \mathcal{L}(\sigma + i\alpha) \sim k \log(\sigma - 1) \tag{6.10}$$

as $\sigma \to 1+$. Since

$$\log \mathcal{L}(s) = \sum_p \frac{a_{\mathcal{L}}(p)}{p^s} + O(1)$$

for $\sigma > 1$, we get by partial summation

$$\log \mathcal{L}(\sigma + i\alpha) \sim \sum_p \frac{a_{\mathcal{L}}(p)}{p^{\sigma+i\alpha}} = (\sigma - 1) \int_1^\infty \sum_{p \leq x} \frac{a_{\mathcal{L}}(p)}{p^{1+i\alpha}} \frac{\mathrm{d}x}{x^\sigma}.$$

By (6.9) the right-hand side is bounded as $\sigma \to 1+$, which contradicts (6.10). The theorem is proved.   □

Recently, Kaczorowski and Perelli [163] investigated the analogue of the prime number theorem for the Selberg class. It is well known that the non-vanishing of $\zeta(s)$ on the edge of the critical strip is equivalent to the (standard) prime number theorem (without remainder term). For an arbitrary element $\mathcal{L} \in \mathcal{S}$, the corresponding prime number theorem is given in form of the asymptotic formula

$$\psi_{\mathcal{L}}(x) := \sum_{n \leq x} \Lambda_{\mathcal{L}}(n) = k_{\mathcal{L}} x + O(x); \tag{6.11}$$

here $k_{\mathcal{L}} = 0$ if $\mathcal{L}(s)$ is regular at $s = 1$, otherwise $k_{\mathcal{L}}$ is the order of the pole of $\mathcal{L}(s)$ at $s = 1$, and $\Lambda_{\mathcal{L}}(n)$ is the von Mangoldt-function, defined by

$$-\frac{\mathcal{L}'}{\mathcal{L}}(s) = \sum_{n=1}^\infty \frac{\Lambda_{\mathcal{L}}(n)}{n^s},$$

generalizing (1.7), the case of the Riemann zeta-function. Indeed, for polynomial Euler products in the Selberg class one can prove this equivalence by standard arguments involving an appropriate Tauberian theorem. In this case, by Theorem 6.6, Selberg's conjecture B implies (6.11). However, this is not satisfying with respect to $\mathcal{S}$. Kaczorowski and Perelli [163] introduced a weak form of Selberg's conjecture A:

Normality Conjecture. *For all* $1 \neq \mathcal{L} \in \mathcal{S}$ *there exists a non-negative integer* $k_{\mathcal{L}}$ *such that*

$$\sum_{p \leq x} \frac{|a_{\mathcal{L}}(p)|^2}{p} = k_{\mathcal{L}} \log \log x + O(\log \log x).$$

Assuming this hypothesis, they proved the claim of Theorem 6.6, namely the non-vanishing of any $\mathcal{L}(s)$ on the line $\sigma = 1$ and that this statement is equivalent to the prime number theorem (6.11). It should be noted that their proof of $\mathcal{L}(1 + i\mathbb{R}) \neq 0$ for a given $\mathcal{L}$ involves the assumption of the normality conjecture for several elements in $\mathcal{S}$.

6.4 Pair Correlation

Assuming the truth of the Riemann hypothesis, Montgomery [259] studied the distribution of consecutive zeros $\frac{1}{2} + i\gamma, \frac{1}{2} + i\gamma'$ of the Riemann zeta-function. Montgomery's famous pair correlation conjecture states that, *for fixed* α, β *satisfying* $0 < \alpha < \beta$,

$$\lim_{T \to \infty} \frac{1}{N(T)} \sharp \left\{ 0 < \gamma, \gamma' < T : \alpha \leq \frac{(\gamma - \gamma') \log T}{2\pi} \leq \beta \right\}$$
$$= \int_{\alpha}^{\beta} \left(1 - \left(\frac{\sin \pi u}{\pi u} \right)^2 \right) du. \qquad (6.12)$$

Montgomery claims that (6.12) would follow from a sufficiently good estimate for

$$\sum_{n \leq x} \Lambda(n) \Lambda(n + h) - c(h)x$$

in a certain range of h, where $c(h)$ is some quantity depending on h; however, the Hardy–Littlewood twin prime conjecture [117] is too strong of an input into this problem. Heath-Brown [123] obtained for the difficult related problem of the distribution of differences $p_{n+1} - p_n$ of consecutive prime numbers under assumption of the pair correlation conjecture the estimate

$$\sum_{p_n \leq x} (p_{n+1} - p_n)^2 \ll x(\log x)^2.$$

The pair correlation conjecture has many other important consequences. For instance, (6.12) implies that almost all zeros of the zeta-function are simple (actually, this is related to the asymptotic formula (1.12)).

Dyson remarked that the function on the right of (6.12) is the pair correlation function of the eigenvalues of large random Hermitian matrices, or more specifically of the Gaussian Unitary Ensemble. This supports an old idea of Hilbert and Pólya. Their approach towards Riemann's hypothesis was to look for a self-adjoint linear operator of an appropriate Hilbert space whose eigenvalues include the zeros of

$$\xi(t) = \frac{1}{2}s(s-1)\pi^{-s/2}\Gamma\left(\frac{s}{2}\right)\zeta(s)$$

with $s = \frac{1}{2} + \mathrm{i}t$. Since $\xi(t)$ is an entire function which vanishes exactly at the non-trivial zeros ϱ of $\zeta(s)$, the property of self-adjointness would imply that all zeros ϱ lie on the line $\sigma = \frac{1}{2}$. In the last years, remarkable progress in this direction was made. By the work of Odlyzko [283], it turned out that the pair correlation and the nearest neighbour spacing for the zeros of $\zeta(s)$ were amazingly close to those for the Gaussian Unitary Ensemble. There is even more evidence for the pair correlation conjecture than numerical data. In the meantime many results from random matrix theory were found fitting perfectly to certain results on the value-distribution of the Riemann zeta-function (and even other L-functions; see Conrey's survey article [58]). For example, Keating and Snaith [168] showed that certain random matrix ensembles have in some sense the same value-distribution as the zeta-function on the critical line predicted by Selberg's limit law (1.16). Further evidence for the pair correlation conjecture was discovered by Rudnick and Sarnak. Normalize the ordered non-trivial zeros $\varrho_n = \frac{1}{2} + \mathrm{i}\gamma_n$ by setting

$$\tilde{\gamma}_n = \frac{\gamma_n}{2\pi}\log|\gamma_n|,$$

then it follows from the Riemann–von Mangoldt formula (1.5) that the numbers $\tilde{\gamma}_n$ have unit mean spacing. The pair correlation conjecture (6.12) can be rewritten as follows: for any *nice* function f on $(0,\infty)$

$$\lim_{N\to\infty}\sum_{n\leq N} f(\tilde{\gamma}_{n+1} - \tilde{\gamma}_n) = \int_0^\infty f(x)P(x)\,\mathrm{d}x,$$

where P is the distribution of consecutive spacings of the eigenvalues of a large random Hermitean matrix. Rudnick and Sarnak [314] succeeded in showing that the m-dimensional analogue of the latter formula, the m-level correlation, holds for a large class of test functions. Finally, note that Katz and Sarnak [167] proved a function field analogue of Montgomery's pair correlation conjecture without assuming any unproved hypothesis.

Recently, Murty and Perelli [271] extended Montgomery's argument to the Selberg class. For this purpose, they considered two primitive functions $\mathcal{L}_1$ and $\mathcal{L}_2$ from $\mathcal{S}$. To compare the zeros $\frac{1}{2} + \mathrm{i}\gamma_{\mathcal{L}_1}$ of $\mathcal{L}_1$ against the zeros $\frac{1}{2} + \mathrm{i}\gamma_{\mathcal{L}_2}$ of $\mathcal{L}_2$ define

$$\mathcal{F}(\alpha;\mathcal{L}_1,\mathcal{L}_2) = \frac{\pi}{\mathrm{d}_{\mathcal{L}_1} T\log T}\sum_{-T\leq \gamma_{\mathcal{L}_1},\gamma_{\mathcal{L}_2}\leq T} T^{\mathrm{i}\alpha\,\mathrm{d}_{\mathcal{L}_1}(\gamma_{\mathcal{L}_1}-\gamma_{\mathcal{L}_2})}w(\gamma_{\mathcal{L}_1}-\gamma_{\mathcal{L}_2}),$$

where w is a suitable weight function. The pair correlation conjecture for the Selberg class takes then the form:

Pair Correlation Conjecture. *Let $\mathcal{L}_1$ and $\mathcal{L}_2$ be primitive functions in $\mathcal{S}$. Under the assumption of the Grand Riemann hypothesis, uniformly in α, as $T \to \infty$,*

$$\mathcal{F}(\alpha; \mathcal{L}_1, \mathcal{L}_2) \sim \begin{cases} \delta_{\mathcal{L}_1, \mathcal{L}_2} |\alpha| + d_{\mathcal{L}_1} T^{-2|\alpha| \, d_{\mathcal{L}_1}} \log T (1 + O(1)) & \text{if } |\alpha| < 1, \\ \delta_{\mathcal{L}_1, \mathcal{L}_2} |\alpha| & \text{otherwise.} \end{cases}$$

Here

$$\delta_{\mathcal{L}_1, \mathcal{L}_2} := \begin{cases} 1 & \text{if } \mathcal{L}_1 = \mathcal{L}_2, \\ 0 & \text{otherwise,} \end{cases}$$

is the Kronecker-symbol. This conjecture includes Montgomery's pair correlation conjecture. It has plenty of important applications as M.R. Murty and Perelli [271] worked out. For instance, the Artin conjecture follows from the pair correlation conjecture. Further, the pair correlation conjecture implies that almost all zeros of two different primitive functions $\mathcal{L}_1$ and $\mathcal{L}_2$ are simple and distinct. Moreover, if the pair correlation formula holds for at least one value of α, then $\mathcal{S}$ has unique factorization into primitive functions. This shows what a powerful tool pair correlation is. Further, M.R. Murty and Perelli proved that the Grand Riemann hypothesis and the pair correlation conjecture imply the Selberg conjectures. The pair correlation conjecture plays a complementary role to the Riemann hypothesis: vertical vs. horizontal distribution of the non-trivial zeros of $\zeta(s)$. Both together seem to be the key to several unsolved problems in number theory!

In the sequel, we investigate universality for L-functions in the Selberg class. We shall show how our general universality theorem for $\tilde{\mathcal{S}}$ extends to a certain class of polynomial Euler products in $\mathcal{S}$. We expect that it holds for all functions $\mathcal{L} \in \mathcal{S} \setminus \{1\}$. Moreover, we can extend the strip of universality significantly on behalf of the functional equation.

6.5 The Phragmén–Lindelöf Principle

The order of growth of a meromorphic function is of special interest. Recall our observations on the order of growth of Dirichlet series from Sect. 2.1. If the Dirichlet series under investigation satisfies a functional equation, we can obtain more information about the mean-square half-plane. For $\mathcal{L} \in \mathcal{S}$ we define

$$\mu_{\mathcal{L}}(\sigma) = \limsup_{t \to \pm\infty} \frac{\log |\mathcal{L}(\sigma + \mathrm{i}t)|}{\log |t|}.$$

One can show that $\mu_{\mathcal{L}}(\sigma)$ is a convex function of σ. Taking into account the absolute convergence of the defining Dirichlet series we obtain immediately $\mu_{\mathcal{L}}(\sigma) = 0$ for $\sigma > 1$. The order of growth in the half-plane to the left of the critical strip is ruled by the functional equation which we may rewrite as

$$\mathcal{L}(s) = \Delta_{\mathcal{L}}(s)\overline{\mathcal{L}(1 - \overline{s})}, \tag{6.13}$$

where

$$\Delta_{\mathcal{L}}(s) := \omega Q^{1-2s} \prod_{j=1}^{f} \frac{\Gamma(\lambda_j(1-s)+\overline{\mu_j})}{\Gamma(\lambda_j s + \mu_j)}.$$

Applying Stirling's formula (2.17), we get after a short computation

Lemma 6.7. *Let* $\mathcal{L} \in \mathcal{S}$. *For* $t \geq 1$, *uniformly in* σ,

$$\Delta_{\mathcal{L}}(\sigma + \mathrm{i}t) = \left(\lambda Q^2 t^{\,\mathrm{d}_{\mathcal{L}}}\right)^{1/2-\sigma-\mathrm{i}t} \exp\left(\mathrm{i}t\,\mathrm{d}_{\mathcal{L}} + \frac{\mathrm{i}\pi(\mu - \mathrm{d}_{\mathcal{L}})}{4}\right)\left(\omega + \mathrm{O}\left(\frac{1}{t}\right)\right),$$

where

$$\mu := 2\sum_{j=1}^{f}(1-2\mu_j) \quad \text{and} \quad \lambda := \prod_{j=1}^{f}\lambda_j^{2\lambda_j}.$$

Using the so-called Phragmén–Lindelöf principle, we can obtain upper bounds for the order of growth inside the critical strip.

Theorem 6.8. *Let* $\mathcal{L} \in \mathcal{S}$. *Uniformly in* σ, *as* $|t| \to \infty$,

$$\mathcal{L}(\sigma + \mathrm{i}t) \asymp |t|^{(1/2-\sigma)\,\mathrm{d}_{\mathcal{L}}}|\mathcal{L}(1-\sigma+\mathrm{i}t)|.$$

In particular,

$$\mu_{\mathcal{L}}(\sigma) \leq \begin{cases} 0 & \text{if } \sigma > 1, \\ \frac{1}{2}\,\mathrm{d}_{\mathcal{L}}(1-\sigma) & \text{if } 0 \leq \sigma \leq 1, \\ \left(\frac{1}{2}-\sigma\right)\mathrm{d}_{\mathcal{L}} & \text{if } \sigma < 0. \end{cases}$$

Proof. The first assertion follows immediately from the functional equation and Lemma 6.7. This together with the trivial estimate $\mu_{\mathcal{L}}(\sigma) = 0$ for $\sigma > 1$ implies for $\sigma < 0$

$$\mu_{\mathcal{L}}(\sigma) = \left(\frac{1}{2}-\sigma\right)\mathrm{d}_{\mathcal{L}}.$$

The calculation of $\mu_{\mathcal{L}}(\sigma)$ for $0 \leq \sigma \leq 1$ is more difficult. We have to apply a kind of maximum principle for unbounded regions, the theorem of Phragmén–Lindelöf.

Lemma 6.9. *Let* $f(s)$ *be analytic in the strip* $\sigma_1 \leq \sigma \leq \sigma_2$ *with* $f(s) \ll \exp(\epsilon|t|)$. *If*

$$f(\sigma_1 + \mathrm{i}t) \ll |t|^{c_1} \quad \text{and} \quad f(\sigma_2 + \mathrm{i}t) \ll |t|^{c_2},$$

then $f(s) \ll |t|^{c(\sigma)}$ *uniformly in* $\sigma_1 \leq \sigma \leq \sigma_2$, *where* $c(\sigma)$ *is linear with* $c(\sigma_1) = c_1$ *and* $c(\sigma_2) = c_2$.

A proof can be found in the paper of Phragmén and Lindelöf [221] or, for example, in Titchmarsh [352]. Note that there are counterexamples if the growth condition $f(s) \ll \exp(\epsilon|t|)$ is not fulfilled.

We continue with the proof. In view of the axiom concerning the analytic continuation $\mathcal{L}(s)$ is a function of finite order. Thus, Lemma 6.9 shows that $\mu_{\mathcal{L}}(\sigma)$ is non-increasing and convex downwards. By the estimates of $\mu_{\mathcal{L}}(\sigma)$ for σ outside of the critical strip the second assertion of the theorem follows.   □

It should be noticed that we did not use the condition that the μ_j appearing in the gamma factors of the functional equation have positive real part. Thus, if it will turn out that this condition is too restrictive, it does not influence the statement of Theorem 6.8 or what we will deduce from it.

In view of the functional equation, respectively, the convexity of $\mu_{\mathcal{L}}(\sigma)$, the value for $\sigma = \frac{1}{2}$ is essential. In particular, we obtain $\mu_{\mathcal{L}}(\frac{1}{2}) \leq \frac{1}{4}d_{\mathcal{L}}$, or equivalently,

$$\mathcal{L}\left(\frac{1}{2} + \mathrm{i}t\right) \ll |t|^{(d_{\mathcal{L}}/4)+\epsilon} \tag{6.14}$$

for $|t| \geq 1$; this bound is known as the convexity bound. The best known upper bound for the Riemann zeta-function is $\mu_{\zeta}(\frac{1}{2}) \leq \frac{32}{205}$, due to Huxley [138].

Next, we shall apply the following refinement of Carlson's classic theorem on the mean-square of Dirichlet series due to Potter [296].

Theorem 6.10. *Suppose that the functions*

$$A(s) = \sum_{n=1}^{\infty} \frac{a_n}{n^s} \quad and \quad B(s) = \sum_{n=1}^{\infty} \frac{b_n}{n^s}$$

have a half-plane of convergence, are of finite order, and that all singularities lie in a subset of the complex plane of finite area. Further, assume that

$$\sum_{n \leq x} |a_n|^2 \ll x^{b+\epsilon} \quad and \quad \sum_{n \leq x} |b_n|^2 \ll x^{b+\epsilon},$$

as $x \to \infty$, and that $A(s)$ and $B(s)$ satisfy

$$A(s) = h(s)B(1-s),$$

where $h(s) \asymp |t|^{c(a/2-\sigma)}$ uniformly in σ for σ from a finite interval, as $|t| \to \infty$, and a, b, c are some non-negative constants. Then

$$\lim_{T \to \infty} \frac{1}{2T} \int_{-T}^{T} |A(\sigma + \mathrm{i}t)|^2 \, \mathrm{d}t = \sum_{n=1}^{\infty} \frac{|a_n|^2}{n^{2\sigma}}$$

for $\sigma > \max\{\frac{a}{2}, \frac{1}{2}(b+1) - \frac{1}{c}\}$.

Taking into account (2.12) and Theorem 6.8, we may take $a = b = 1$ and $c = d_{\mathcal{L}}$. We obtain

Corollary 6.11. *Let $\mathcal{L} \in \mathcal{S}$. For $\sigma > \max\left\{\frac{1}{2}, 1 - \frac{1}{d_{\mathcal{L}}}\right\}$,*

$$\lim_{T \to \infty} \frac{1}{2T} \int_{-T}^{T} |\mathcal{L}(\sigma + \mathrm{i}t)|^2 \, \mathrm{d}t = \sum_{n=1}^{\infty} \frac{|a(n)|^2}{n^{2\sigma}}.$$

Note that the series on the right-hand side converges on behalf of the Ramanujan hypothesis (1). Further, we remark that the statement of the corollary holds also for L-functions which do not fulfill axiom (4) from the definition of $\mathcal{S}$, since we have used (2.12) and not Lemma 2.3. Corollary 6.11 should be compared with the mean-square estimates of Perelli [289] for his class of L-functions; his argument is different.

Corollary 6.11 improves the range for the existence of the mean-square given by Theorem 2.4, provided $\mathcal{L}(s)$ satisfies a Riemann-type functional equation: the mean-square half-plane of $\mathcal{L}$ then contains the region

$$\sigma > \max\left\{\frac{1}{2}, 1 - \frac{1}{d_{\mathcal{L}}}\right\}.$$

It would be desirable to extend this region further to the half-plane $\sigma > \frac{1}{2}$ in general. However, this is hard. The difficulties arise for large degrees $d_{\mathcal{L}}$. For instance, Chandrasekharan and Narasimhan [52] obtained for Dedekind zeta-functions the estimate

$$\int_0^T |\zeta_{\mathbb{K}}(\sigma + it)|^2 \, dt \ll T^{d(1-\sigma)} (\log T)^d$$

for $\frac{1}{2} \leq \sigma \leq 1 - \frac{1}{d}$, where d is the degree of $\zeta_{\mathbb{K}}(s)$. Potter's theorem only yields an asymptotic formula throughout $\sigma > \frac{1}{2}$ if $d \leq 2$. The difficulties become more obvious by noting that any result on the mean-square of an L-function from the Selberg class of degree d is comparable to the corresponding result for the 2dth moment of the Riemann zeta-function. The fourth moment of $\zeta(s)$ on the critical line is quite well understood (by the work of Hardy and Littlewood [118], Ingham [139], Motohashi [263] and others), however, higher moments are still unsettled; there are only estimates known. Similar problems appear to the right of the critical line. For example, Ivić showed [141] proved for $\frac{5}{8} < \sigma < 1$

$$\int_1^T |\zeta(\sigma + it)|^8 \, dt \ll T^{(11-8\sigma)/6+\epsilon}. \tag{6.15}$$

Recently some new insight was obtained by the analogies to random matrix theory. Extending a conjecture of Conrey and Gonek [61], Keating and Snaith [168] conjectured

$$\frac{1}{T} \int_0^T \left|\zeta\left(\frac{1}{2} + it\right)\right|^{2k} \, dt \sim c(k)(\log T)^{k^2},$$

where

$$c(k) := \frac{G^2(k+1)}{G(2k+1)} \prod_p \left(1 - \frac{1}{p^2}\right)^{k^2} \sum_{m=0}^{\infty} \left(\frac{\Gamma(m+k)}{m!\,\Gamma(k)}\right)^2 p^{-m}$$

with the Barnes double Gamma-function $G(z)$, defined by

$$G(z+1) = (2\pi)^{z/2} \exp\left(-\frac{1}{2}(z(z+1) + \gamma z^2)\right) \prod_{n=1}^{\infty} \left(1 + \frac{z}{n}\right)^n \exp\left(-z + \frac{z^2}{n}\right),$$

and γ is the Euler–Mascheroni constant. For more details on mean-value results and random matrix theory we refer to Ivić's monograph [141], Matsumoto's survey [240], and Conrey's paper [58].

6.6 Universality in the Selberg Class

Now we are going to apply the universality Theorem 5.14 to L-functions in the Selberg class. First of all we compare the classes $\tilde{\mathcal{S}}$ and $\mathcal{S}$. The axioms (i) and (1) are identical. Concerning analytic continuation we note that axiom (2) implies (ii) with any $\sigma_{\mathcal{L}} < 1$. Moreover, any $\mathcal{L}(s)$ satisfying (2) is a function of finite order and together with the functional equation (3) we obtain (iii) with $\mu_{\mathcal{L}} = \mu_{\mathcal{L}}(\sigma)$ (by Theorem 6.8). Only the arithmetic axioms (iv) and (v) cannot be deduced from the axioms of the Selberg class; however, they are expected to hold. As already mentioned, the elements in the Selberg class are automorphic or at least conjecturally automorphic L-functions, and it is conjectured that $\mathcal{S}$ consists of all automorphic L-functions. For all known examples, the Euler product has the form of axiom (iv) in the definition of $\tilde{\mathcal{S}}$: there exists a positive integer m and for each prime p, there are complex numbers $\alpha_j(p)$ such that

$$\mathcal{L}(s) = \prod_p \prod_{j=1}^{m} \left(1 - \frac{\alpha_j(p)}{p^s}\right)^{-1}.$$

The remaining axiom of $\tilde{\mathcal{S}}$ is (v) which states that

$$\lim_{x\to\infty} \frac{1}{\pi(x)} \sum_{p\le x} |a(p)|^2 = \kappa$$

for some positive constant κ. This axiom is closely related to Selberg's conjectures. In fact, it implies a *weak* version of Selberg's conjecture A. By partial summation,

$$\sum_{p\le x} \frac{|a(p)|^2}{p} = \frac{1}{x} \sum_{p\le x} |a(p)|^2 + \int_2^x \sum_{p\le u} |a(p)|^2 \frac{du}{u^2}$$

$$\sim \kappa \int_2^x \frac{du}{\log u} \sim \kappa \log\log x;$$

this is the normality conjecture of Kaczorowski and Perelli [163] for the single function $\mathcal{L}(s)$ (provided $\kappa \in \mathbb{N}$ which we do not require); indeed, the full normality conjecture is equivalent to the validity of axiom (v) for the whole of $\mathcal{S}$.[1] Conversely, the *strong* Selberg Conjecture B with a remainder term

$$\mathcal{R}(x) = \mathrm{O}\left(\frac{1}{\log x}\right)$$

implies axiom (v) on the prime mean square with $\kappa = n_{\mathcal{L}}$. Moreover, from this strong version it would follow that $\mathcal{S} \cap \tilde{\mathcal{S}}$ is multiplicatively closed.

Now we can formulate a rather general universality theorem for L-functions in the Selberg class.

Theorem 6.12. *Let $\mathcal{L} \in \mathcal{S} \cap \tilde{\mathcal{S}}$ (i.e., $\mathcal{L}(s)$ satisfies the axioms (2), (3), (iv), and (v)), let $\mathcal{K}$ be a compact subset of the strip $\sigma_m < \sigma < 1$ (where $\sigma_m \leq \max\{\frac{1}{2}, 1 - \frac{1}{d_{\mathcal{L}}}\} < 1$) with connected complement, and let $g(s)$ be a non-vanishing continuous function on $\mathcal{K}$ which is analytic in the interior of $\mathcal{K}$. Then, for any $\epsilon > 0$,*

$$\liminf_{T \to \infty} \frac{1}{T} \, \mathrm{meas} \, \left\{ \tau \in [0, T] : \max_{s \in \mathcal{K}} |\mathcal{L}(s + i\tau) - g(s)| < \epsilon \right\} > 0.$$

Obvious examples for elements in $\mathcal{S} \cap \tilde{\mathcal{S}}$ are the Riemann zeta-function and Dirichlet L-functions $L(s, \chi)$ to primitive characters χ. Further examples of elements are normalized L-functions associated with holomorphic newforms, Dedekind zeta-functions, Rankin–Selberg L-functions. This can be shown by using analogues of the prime number theorem (e.g., the prime ideal theorem). If one is willing to accept some widely believed conjectures, then many L-functions belong to this class. However, later we shall prove universality for Dirichlet L-functions to imprimitive characters which are not elements of the Selberg class (by lack of having the appropriate functional equation) and further L-functions which are conjectured to lie in $\mathcal{S}$. This marks an important advantage of dealing with the class $\tilde{\mathcal{S}}$ rather than the Selberg class.

We shall briefly discuss some refinements of the universality Theorem 6.12 with respect to the strip of universality. In some particular cases of functions $\mathcal{L} \in \tilde{\mathcal{S}}$ with degree $d_{\mathcal{L}} > 2$ the existence of the mean-square

$$\lim_{T \to \infty} \frac{1}{T} \int_1^T |\mathcal{L}(\sigma + \mathrm{i}t)|^2 \, \mathrm{d}t$$

is known for some $\sigma \leq 1 - \frac{1}{d_{\mathcal{L}}}$. For instance, let $L(s, \chi)$ be an arbitrary Dirichlet L-function to a primitive character χ. Then $\zeta(s)^2 L(s, \chi)$ is an element of $\tilde{\mathcal{S}}$ of degree 3, so Theorem 6.12 gives universality for $\frac{2}{3} < \sigma < 1$. Montgomery [258] proved that

$$\sum_{\substack{\chi \bmod q \\ \text{primitive}}} \int_{-T}^{T} \left| L\left(\frac{1}{2} + \mathrm{i}t, \chi\right) \right|^4 \mathrm{d}t \ll \varphi(q) T (\log(qT))^4.$$

---

[1] This remark is due to Giuseppe Molteni; see MathReviews MR1981179 (2004c:11165). Selberg's conjecture A for a single L-function seems to be insufficient for a proof of universality (by lack of a sufficiently good error term).

It is easily seen that the same bound holds for the fourth moment on vertical lines to the right as well. In combination with Ivić's eigth-moment estimate (6.15), the Cauchy–Schwarz inequality yields

$$\int_1^T |\zeta(\sigma+it)^2 L(\sigma+it,\chi)|^2 \, dt$$

$$\ll \left(\int_1^T |\zeta(\sigma+it)|^8 \, dt \int_1^T |L(\sigma+it,\chi)|^4 \, dt \right)^{1/2} .$$

$$\ll T^{(11-8\sigma)/12+\epsilon} T^{1/2+\epsilon} \ll T$$

for any $\sigma > \frac{5}{8}$ with sufficiently small ϵ. Thus $\zeta(s)^2 L(s,\chi)$ is universal in the strip $\frac{5}{8} < \sigma < 1$. In this example, we can even do better by applying the joint universality Theorem 1.10 which improves the range to $\frac{1}{2} < \sigma < 1$.

In Sect. 6.7, we show that the strip of universality for $\mathcal{L} \in \mathcal{S} \cap \tilde{\mathcal{S}}$ can be extended subject to a growth condition for $\mathcal{L}(s)$ on the critical line.

6.7 Lindelöf's Hypothesis

For many applications in number theory, it is useful to assume Riemann's hypothesis but quite often it suffices to work with weaker conjectures.

Lindelöf [220] conjectured that the order of growth of the zeta-function is much smaller than the one the Phragmén–Lindelöf principle gives. More precisely, he expressed his belief that $\zeta(s)$ is bounded if $\sigma \geq \frac{1}{2} + \epsilon$ for any fixed positive ϵ. In terms of the μ-function from Sect. 6.5 this would imply $\mu_\zeta(\frac{1}{2}) = 0$ or equivalently

$$\zeta\left(\frac{1}{2}+it\right) \ll t^\epsilon$$

as $t \to \infty$. The last statement is now known as Lindelöf's hypothesis and it is yet unproved. However, the boundedness conjecture is false as one can easily deduce, for example, from Voronin's universality theorem. It was proved by Littlewood [223] that the Lindelöf hypothesis follows from the truth of the Riemann hypothesis. On the contrary, Backlund [8] proved that the Lindelöf hypothesis is equivalent to the much less drastic but yet unproved hypothesis that for every $\sigma > \frac{1}{2}$

$$N(\sigma, T+1) - N(\sigma, T) = O(\log T). \tag{6.16}$$

Furthermore, the Lindelöf hypothesis implies the classic density hypothesis which claims

$$N(\sigma, T) \ll T^{2(1-\sigma)+\epsilon}. \tag{6.17}$$

This should be compared with the density Theorems 1.2 and 6.5.

In the case of the Lerch zeta-function (1.41) the analogue of the density hypothesis is not true in general. For instance, the function given by (6.7) can be rewritten as $L\left(\frac{1}{2}, 1, s\right)$, and obviously it has infinitely many zeros off the critical line $\sigma = \frac{1}{2}$, at least the zeros $s = 1 - \frac{2\pi i}{\log 2} k$ with $k \in \mathbb{Z}\backslash\{0\}$, coming from the factor $1 - 2^{1-s}$. This violates the analogue of the Riemann hypothesis and, moreover, it violates the analogue of (6.17) as well. On the other side $L(\frac{1}{2}, 1, s)$ satisfies the analogue of the Lindelöf hypothesis provided $\zeta(s)$ satisfies the Lindelöf hypothesis. This example does not seem to be special. It is known that a generic Lerch zeta-function has many zeros off the critical line. Denote by $\varrho = \beta + i\gamma$ the non-trivial zeros of $L(\lambda, \alpha, s)$ (where the notion *non-trivial* is defined quite similarly as for $\zeta(s)$). Garunkštis and Steuding [91] (see also [89]) proved, for $0 < \lambda, \alpha \leq 1$,

$$\sum_{|\gamma| \leq T} \left(\beta - \frac{1}{2}\right) = \frac{T}{2\pi} \log \frac{\alpha}{\sqrt{\lambda^+ \lambda^-}} + O(\log T)$$

(for the definition of the quantities λ^+ and λ^- see (1.43)). Consequently, any $L(\lambda, \alpha, s)$ with $\alpha^2 > \lambda^+ \lambda^-$ has $\gg T$ many zeros in the region $\sigma > \frac{1}{2}$, $0 < t \leq T$. The proportion of the parameters $\lambda, \alpha \in (0, 1]$ which satisfy $\alpha^2 > \lambda^+ \lambda^-$ is equal to

$$1 - \int_0^1 \sqrt{\lambda(1 - \lambda)} \, d\lambda = 1 - \frac{\pi}{8}.$$

Since *most* of these zeros off the critical line could lie arbitrarily close to $\sigma = \frac{1}{2}$, this alone does not violate the analogue of the density hypothesis for the Lerch zeta-function. However, Garunkštis [85] proved that for transcendental α the number of zeros $\varrho = \beta + i\gamma$ of $L(\lambda, \alpha, s)$ with $\beta > \sigma$, $|\gamma| \leq T$ is $\gg T$ for any fixed $\sigma \in (\frac{1}{2}, 1]$; the proof relies on the strong universality property of $L(\lambda, \alpha, s)$ (see also Garunkštis and Laurinčikas [89]). This indicates that if $L(\lambda, \alpha, s)$ has no Euler product representation, the density hypothesis does not follow from the Lindelöf hypothesis. On the other hand, Garunkštis and Steuding [92] showed that the analogue of the Lindelöf hypothesis for Lerch zeta-functions seems reasonable.

In [60] Conrey and Ghosh generalized the Lindelöf hypothesis to the Selberg class. They proved that if $\mathcal{L} \in \mathcal{S}$ is an entire function which satisfies the Riemann hypothesis, the Euler product is of the form (iv), and that all λ_j appearing in the functional equation are equal to $\frac{1}{2}$, then there exists a constant C, depending only on $\epsilon > 0$ and the degree $d_{\mathcal{L}}$, such that

$$\left| \mathcal{L}\left(\frac{1}{2} + it\right) \right| \leq C \left(Q(1 + |t|)^{d_{\mathcal{L}}/2} \prod_{j=1}^{f} (1 + |\mu_j|) \right)^{\epsilon}.$$

The proof follows the lines of Littlewood's proof that the Riemann hypothesis implies the Lindelöf hypothesis (in fact, it relies on a combination of the Phragmén–Lindelöf principle, the Borel–Carathéodory theorem and

Hadamard's three circle theorem). In view of the Phragmén–Lindelöf principle one can prove unconditionally

$$\left| \mathcal{L}\left(\frac{1}{2} + it\right) \right| \ll (cQ^2(1 + |t|)^{d_{\mathcal{L}}})^{(1/4) + \epsilon}$$

for some positive constant $c = c(\lambda, \mu)$ depending only on the data of the functional equation. We already obtained with (6.14) a similar bound in the t-aspect. The so-called subconvexity problem is to find a $\delta > 0$ such that the exponent on the right-hand can be replaced by $\frac{1}{4} - \delta$. For certain L-functions subconvexity bounds are known but in general, this seems to be a difficult problem. However, solutions of the subconvexity problem (in the Q-aspect) lead to several important applications in number theory, e.g., Hilbert's eleventh problem on the representation of integers by a given quadratic form; we refer for details and further examples to the survey of Iwaniec and Sarnak [145]. Here, we are only interested in the t-aspect.

Grand Lindelöf Hypothesis. *For $\mathcal{L} \in \tilde{\mathcal{S}}$ and any $\epsilon > 0$,*

$$\mathcal{L}\left(\frac{1}{2} + it\right) \ll |t|^\epsilon \quad for \quad |t| \geq 1. \tag{6.18}$$

The implicit constant may depend on ϵ and $\mathcal{L}$. This hypothetical estimate coincides with Perelli's Lindelöf hypothesis for his class of L-functions [289]. Among other interesting results, Perelli showed for his L-functions that Riemann's hypothesis implies the Lindelöf hypothesis, and that Backlund's reformulation (6.16) of the Lindelöf hypothesis in terms of their zero-distribution off the critical line holds. We observe that the Lindelöf hypothesis (6.18) for $\mathcal{L}(s)$ is equivalent to

$$\mu_{\mathcal{L}}(\sigma) = d_{\mathcal{L}} \max\left\{0, \frac{1}{2} - \sigma\right\}.$$

There are several further interesting reformulations of the Lindelöf hypothesis in the particular case of the Riemann zeta-function. One, given in terms of moments on the critical line, was found by Hardy and Littlewood [119]. They proved that the Lindelöf hypothesis is true if and only if for any $k \in \mathbb{N}$

$$\frac{1}{T} \int_1^T \left| \zeta\left(\frac{1}{2} + it\right) \right|^{2k} dt \ll T^\epsilon. \tag{6.19}$$

Another equivalent form is due to Laurinčikas [195]. He proved that the Lindelöf hypothesis is equivalent to the asymptotic formula

$$\frac{1}{T} \operatorname{meas}\left\{ t \in [0, T] : \left| \zeta\left(\frac{1}{2} + it\right) \right| < xT^\epsilon \right\} = 1 - \mathrm{O}\left(\frac{\delta(T)}{1 + x^A}\right)$$

with positive ϵ, A and sufficiently large x, where $\delta(T)$ is an arbitrary positive function which tends to zero as $T \to \infty$.

Now we consider the mean-square on vertical lines to the right of the critical line subject to the Lindelöf hypothesis. For this purpose we return to Sect. 2.4. Assuming the Lindelöf hypothesis, we may take $\sigma_{\mathcal{L}} = \frac{1}{2}$ and $\mu_{\mathcal{L}} = \mu_{\mathcal{L}}(\frac{1}{2}) = 0$ in Theorem 2.4. Thus, for $\mathcal{L} \in \mathcal{S} \cap \tilde{\mathcal{S}}$, the Lindelöf hypothesis implies

$$\lim_{T \to \infty} \frac{1}{T} \int_1^T |\mathcal{L}(\sigma + it)|^2 \, dt = \sum_{n=1}^{\infty} \frac{|a(n)|^2}{n^{2\sigma}}$$

for any $\sigma > \frac{1}{2}$. In fact, there are no big difficulties to consider also higher moments, and also to derive an equivalent statement for Lindelöf's hypothesis analogous to (6.19).

If the Lindelöf hypothesis is true, then the strip of universality can be extended to the open right half of the critical strip.

Corollary 6.13. *Assume that the Lindelöf hypothesis is true for $\mathcal{L} \in \mathcal{S} \cap \tilde{\mathcal{S}}$. Let $\mathcal{K}$ be a compact subset of the strip $\frac{1}{2} < \sigma < 1$ with connected complement, and let $g(s)$ be a non-vanishing continuous function on $\mathcal{K}$ which is analytic in the interior of $\mathcal{K}$. Then, for any $\epsilon > 0$,*

$$\liminf_{T \to \infty} \frac{1}{T} \operatorname{meas} \left\{ \tau \in [0, T] \ : \ \max_{s \in \mathcal{K}} |\mathcal{L}(s + i\tau) - g(s)| < \epsilon \right\} > 0.$$

In view of results on the frequency of the values taken in neighbourhood of the critical line (see Theorem 7.6) it seems to be impossible to have universality (in the sense of Voronin's theorem) in any region covering the critical line.

6.8 Symmetric Power L-Functions

We conclude this chapter with an example. Symmetric power L-functions became important by Serre's reformulation of the Sato–Tate conjecture. For the sake of simplicity, here we shall only consider the case where f is a normalized eigenform of weight k to the full modular group. According to the normalization from Sect. 2.2, we suppose that its Fourier expansion is given by

$$f(z) = \sum_{n=1}^{\infty} a(n) n^{(k-1)/2} \exp(2\pi i n z)$$

and the attached L-function has an Euler product representation of the form

$$L(s, f) = \prod_p \left(1 - \frac{a(p)}{p^s} + \frac{1}{p^{2s}} \right)^{-1}.$$

In view of Deligne's estimate (1.34), for each prime p, we may define an angle $\theta_p \in [0, \pi]$ by setting

$$a(p) = 2 \cos \theta_p; \tag{6.20}$$

this should be compared with the angles defined in equation (5.20). Then, for any non-negative integer m the symmetric mth power L-function attached to f is for $\sigma > 1$ given by

$$L_m(s,f) = \prod_p \prod_{j=0}^{m} \left(1 - \frac{\exp(\mathrm{i}\theta_p(m-2j))}{p^s} \right)^{-1}. \qquad (6.21)$$

Then

$$L_0(s,f) = \zeta(s), \quad L_1(s,f) = L(s,f),$$

and

$$L_2(s,f) = \frac{\zeta(2s)}{\zeta(s)} L(s, f \otimes f),$$

where $L(s, f \otimes f)$ is the Rankin–Selberg L-function (introduced in Sect. 1.6). Shimura [328] obtained the analytic continuation and a functional equation of Riemann-type for the case $m = 2$. By the powerful methods of the Langlands program, Shahidi [326] obtained the analytic continuation of $L_m(s,f)$ to $\sigma \geq 1$ for $m \leq 4$. In particular cases more is known; for example, Kim and Shahidi [170] showed that certain third symmetric power L-function attached to non-monomial cusp forms of GL_2 over any number field are entire, and Shahidi [327] proved that the third and the fourth symmetric power L-function are cuspidal unless the underlying cusp form is either of dihedral or tetrahedral type (in the terminology of Klein's classification of finite subgroups of $\mathrm{PGL}_2(\mathbb{C})$). However, the case of $m > 4$ is open. Serre [324] conjectured that if f does not have complex multiplication, the angles θ_p are uniformly distributed with respect to the Sato–Tate measure

$$\frac{2}{\pi}(\sin\theta)^2 \, \mathrm{d}\theta, \qquad (6.22)$$

if p ranges over the set of prime numbers (in analogy to a similar conjecture for elliptic curves E due to Sato and Tate which was recently solved by Taylor [351] assuming a mild condition on E). Furthermore, Serre proved that the non-vanishing of $L_m(s,f)$ on the abscissa of convergence $\sigma = 1$ for all $m \in \mathbb{N}$ implies the Sato–Tate conjecture for newforms, namely

$$\lim_{x\to\infty} \frac{1}{\pi(x)} \sharp\{p \leq x : \alpha < \theta_p < \beta\} = \frac{2}{\pi} \int_{\alpha}^{\beta} (\sin\theta)^2 \, \mathrm{d}\theta. \qquad (6.23)$$

Ogg [285] has shown that if for each $m \leq 2M$, the function $L_m(s,f)$ has an analytic continuation to a half-plane $\sigma > \frac{1}{2} - \epsilon$, then $L_m(s,f)$ does not vanish on the line $\sigma = 1$. V.K. Murty [272] proved that it suffices to have an analytic continuation of any $L_m(s,f)$ to $\sigma \geq 1$ for proving the Sato–Tate conjecture. In this context, M.R. Murty [265] (see also [270]) proved asymptotic formulae for the $2m$th power moments of $\cos\theta_p$.

Theorem 6.14. *Let f be a normalized newform. If $L_m(s, f)$ has an analytic continuation up to $\sigma \geq \frac{1}{2}$ for all $m \leq 2(M + 1)$, then, for $m \leq M + 1$,*

$$\lim_{x \to \infty} \frac{1}{\pi(x)} \sum_{p \leq x} (2 \cos \theta_p)^{2m} = \frac{1}{m+1} \binom{2m}{m},$$

and, for $m \leq M$,

$$\lim_{x \to \infty} \frac{1}{\pi(x)} \sum_{p \leq x} (2 \cos \theta_p)^{2m+1} = 0.$$

In view of Shimura's results, the asymptotic formulae of the theorem hold unconditionally for $M \leq 1$. This includes Rankin's asymptotic formula (1.37). If we expand the Euler product (6.21) into a Dirichlet series, the pth coefficient is given by

$$a_p := \sum_{j=0}^{m} \exp(i\theta_p(m - 2j)) = \frac{\sin((m+1)\theta_p)}{\sin \theta_p} = U_m(\cos \theta_p),$$

where $U_m(x)$ is the Chebyshev polynomial of the second kind, defined by $U_0(x) = 1$, $U_1(x) = x$ and $U_m(x) = xU_{m-1}(x) - U_{m-2}(x)$ for $m \geq 2$. In view of Theorem 6.14 it follows that:

$$\sum_{p \leq x} |a_p|^2 = \sum_{p \leq x} U_m(\cos \theta_p)^2 \sim \kappa \pi(x)$$

as $x \to \infty$, where κ is a positive constant. Taking into account the functional equation for $L_m(s, f)$ (see [219], resp. Cogdell and Michel [56]), it follows that $L_m(s, f)$ is an element of degree $m + 1$ in the Selberg class $\mathcal{S}$, unconditionally for $m \leq 4$, and conditionally for $m > 4$ (depending on the analytic continuation which actually is a consequence of the Langlands conjectures). It is not difficult to see that we also have $L_m(s, f) \in \tilde{\mathcal{S}}$ under the same conditions. Thus Theorem 5.14 yields

Corollary 6.15. *Let f be a normalized newform, let $\mathcal{K}$ be a compact subset of the strip $\max\{\frac{1}{2}, 1 - \frac{1}{m+1}\} < \sigma < 1$ with connected complement, and let $g(s)$ be a non-vanishing continuous function on $\mathcal{K}$ which is analytic in the interior of $\mathcal{K}$. Then, for any $\epsilon > 0$,*

$$\liminf_{T \to \infty} \frac{1}{T} \operatorname{meas} \left\{ \tau \in [0, T] : \max_{s \in \mathcal{K}} |L_m(s + i\tau, f) - g(s)| < \epsilon \right\} > 0,$$

unconditionally if $m \leq 4$, and for $m \geq 5$ under the assumption of the holomorphy of $L_m(s, f)$ throughout $\mathbb{C}$ and the existence of a functional equation.

With slightly more effort, one can show that if f is a holomorphic newform of some level $N \geq 1$ and integer weight $k \geq 1$ which is not of complex multiplication type, then for $m = 2, 3, 4$ the mth symmetric power L-function $L_m(s, f)$ is universal. Corollary 6.15 was independently obtained by Li and Wu [219].

7

Value-Distribution in the Complex Plane

Une fonction entière, qui ne devient jamais ni à a ni à b est nécessaire-
ment une constante.   Emile Picard

Many beautiful results on the value-distribution of L-functions follow from
the general theory of Dirichlet series like the Big Picard theorem (see Boas [26]
and Mandelbrojt [234]), but more advanced statements can only be proved
by exploiting the characterizing properties (the functional equation and the
Euler product). In this chapter, we study the distribution of values of Dirichlet
series satisfying a Riemann-type functional equation. These results are due to
Steuding [346, 347] and their proofs follow in the main part the methods of
Levinson [217], Levinson and Montgomery [218], and Nevanlinna theory.

7.1 Sums Over c-Values

Let c be any complex number. Levinson [217] proved that all but
$\ll N(T)(\log \log T)^{-1}$ of the roots of $\zeta(s) = c$ in $T < t < 2T$ lie in

$$\left| \sigma - \frac{1}{2} \right| < \frac{(\log \log T)^2}{\log T}.$$

Thus, the c-values of the zeta-function are clustered around the critical line.
In particular, we see that the density estimate (1.13) alone does not indicate
the truth of the Riemann hypothesis. As we shall show in this chapter, this
distribution of c-values is typical for Dirichlet series satisfying a Riemann-type
functional equation.

Throughout this chapter, we shall assume that

$$\mathcal{L}(s) = \sum_{n=1}^{\infty} \frac{a(n)}{n^s}$$

satisfies the axioms (1)–(3) from the definition of the Selberg class $\mathcal{S}$, and so we may define the degree $d_{\mathcal{L}}$ of $\mathcal{L}$ by (6.2); we shall not make use of axiom (4) (neither do we use the condition on the real parts of the complex numbers μ_j in the Gamma-factors of the functional equation), however, for simplicity we suppose that $a(1) = 1$. In some places we shall assume the Lindelöf hypothesis for $\mathcal{L}(s)$; by that we mean the estimate (6.18).

We give an example of a function satisfying these axioms which does not have an Euler product. The Davenport–Heilbronn zeta-function is given by

$$L(s) = \frac{1 - i\kappa}{2} L(s, \chi) + \frac{1 + i\kappa}{2} L(s, \overline{\chi}), \tag{7.1}$$

where

$$\kappa := \frac{\sqrt{10 - 2\sqrt{5}} - 2}{\sqrt{5} - 1}$$

and χ is the character mod 5 with $\chi(2) = i$. It is easily seen that the Davenport–Heilbronn zeta-function satisfies the functional equation

$$\left(\frac{5}{\pi}\right)^{s/2} \Gamma\left(\frac{s+1}{2}\right) L(s) = \left(\frac{5}{\pi}\right)^{(1-s)/2} \Gamma\left(1 - \frac{s}{2}\right) L(1 - s).$$

Davenport and Heilbronn [65] introduced this function as an example for a Dirichlet series having zeros in the half-plane $\sigma > 1$ although $L(s)$ satisfies a Riemann-type functional equation; see Balanzario [14] for more examples of a similar type.

The c-values of $\mathcal{L}(s)$ are the roots of the equation

$$\mathcal{L}(s) = c, \tag{7.2}$$

which we denote by $\varrho_c = \beta_c + i\gamma_c$. Our first aim is to prove estimates for sums taken over c-values, weighted with respect to their real parts.

Theorem 7.1. *Assume that $\mathcal{L}(s)$ satisfies the axioms (1)–(3) with $a(1) = 1$ and let $c \neq 1$. Then, for any $b > \max\{\frac{1}{2}, 1 - \frac{1}{d_{\mathcal{L}}}\}$,*

$$\sum_{\substack{\beta_c > b \\ T < \gamma_c \leq 2T}} (\beta_c - b) \ll T.$$

Assuming the truth of Lindelöf's hypothesis for $\mathcal{L}(s)$,

$$\sum_{\substack{\beta_c > \frac{1}{2} \\ T < \gamma_c \leq 2T}} \left(\beta_c - \frac{1}{2}\right) = O(T \log T).$$

The case $c = 1$ is exceptional since $1 = a(1)$ is the limit of $\mathcal{L}(s)$ as $\sigma \to \infty$:

$$\mathcal{L}(s) = 1 + \mathrm{O}(2^{-\sigma}). \tag{7.3}$$

We will briefly discuss this case at the end of Sect. 7.2.

Proof. In view of (7.3) there exists a positive real number A depending on c such that all real parts β_c of c-values satisfy $\beta_c < A$. Put

$$\ell(s) = \frac{\mathcal{L}(s) - c}{1 - c}.$$

Obviously, the zeros of $\ell(s)$ correspond exactly to the c-values of $\mathcal{L}(s)$. Next we will apply Littlewood's lemma which relates the zeros of an analytic function $f(s)$ with a contour integral over $\log f(s)$.

Lemma 7.2 (Littlewood). *Let $b < a$ and let $f(s)$ be analytic on $\mathcal{R} := \{s \in \mathbb{C} : b \le \sigma \le a, |t| \le T\}$. Suppose that $f(s)$ does not vanish on the right edge $\sigma = a$ of $\mathcal{R}$. Let $\mathcal{R}'$ be $\mathcal{R}$ minus the union of the horizontal cuts from the zeros of f in $\mathcal{R}$ to the left edge of $\mathcal{R}$, and choose a single-valued branch of $\log f(s)$ in the interior of $\mathcal{R}'$. Denote by $\nu(\sigma, T)$ the number of zeros $\varrho = \beta + \mathrm{i}\gamma$ of $f(s)$ inside the rectangle with $\beta > \sigma$ including zeros with $\gamma = T$ but not those with $\gamma = -T$. Then*

$$\int_{\partial \mathcal{R}} \log f(s)\, \mathrm{d}s = -2\pi\mathrm{i} \int_b^a \nu(\sigma, T)\, \mathrm{d}\sigma.$$

This is an integrated version of the principle of the argument. We give a sketch of the simple proof. Cauchy's theorem implies $\int_{\partial \mathcal{R}'} \log f(s)\, \mathrm{d}s = 0$, and so the left-hand side of the formula of the lemma, $\int_{\partial \mathcal{R}}$, is minus the sum of the integrals around the paths hugging the cuts. Since the function $\log f(s)$ jumps by $2\pi\mathrm{i}$ across each cut (assuming for simplicity that the zeros of f in $\mathcal{R}$ are simple and have different height; the general case is no harder), $\int_{\partial \mathcal{R}}$ is $-2\pi\mathrm{i}$ times the total length of the cuts, which is the right-hand side of the formula in the lemma. For more details we refer to Titchmarsh [353, Sect. 9.9], or Littlewood's original paper [224].

Let $\nu(\sigma, T)$ denote the number of zeros ϱ_c of $\ell(s)$ with $\beta_c > \sigma$ and $T < \gamma_c \le 2T$ (counting multiplicities). Now let a be a parameter with $a > \max\{A + 1, b\}$. Then Littlewood's Lemma 7.2, applied to the rectangle $\mathcal{R}$ with vertices $a + \mathrm{i}T, a + 2\mathrm{i}T, b + \mathrm{i}T, b + 2\mathrm{i}T$, gives

$$\int_{\mathcal{R}} \log \ell(s)\, \mathrm{d}s = -2\pi\mathrm{i} \int_b^a \nu(\sigma, T)\, \mathrm{d}\sigma.$$

Since

$$\int_b^a \nu(\sigma, T)\, \mathrm{d}\sigma = \sum_{\substack{\beta_c > b \\ T < \gamma \le 2T}} \int_b^{\beta_c} \mathrm{d}\sigma = \sum_{\substack{\beta_c > b \\ T < \gamma_c \le 2T}} (\beta_c - b) \tag{7.4}$$

and this quantity is real-valued, we get

$$
2\pi \sum_{\substack{\beta_c > b \\ T < \gamma_c \le 2T}} (\beta_c - b) = \int_T^{2T} \log |\ell(b + it)| \, dt - \int_T^{2T} \log |\ell(a + it)| \, dt +
$$

$$
- \int_b^a \arg \ell(\sigma + iT) \, d\sigma + \int_b^a \arg \ell(\sigma + 2iT) \, d\sigma
$$

$$
= \sum_{j=1}^4 I_j, \tag{7.5}
$$

say. To define $\log \ell(s)$ and $\log \mathcal{L}(s)$ we may choose the principal branch of the logarithm on the real axis, as $\sigma \to \infty$; for other points s the value of the logarithm is obtained by continuous variation along line segments (this is in agreement with Lemma 7.2).

We start with the vertical integrals. Obviously,

$$
I_1(T, b) := I_1 = \int_T^{2T} \log |\mathcal{L}(b + it) - c| \, dt - T \log |1 - c|. \tag{7.6}
$$

By Jensen's inequality the integral is

$$
\le \frac{T}{2} \log \left(\frac{1}{T} \int_T^{2T} |\mathcal{L}(b + it)|^2 \, dt \right) + O(T).
$$

By Corollary 6.11 (which also applies to L-functions satisfying just axioms (1)–(3) as already remarked) this is $\ll T$ for $b > \max\{\frac{1}{2}, 1 - \frac{1}{d_\mathcal{L}}\}$. Thus we get $I_1(T, b) \ll T$ unconditionally. An immediate consequence of Lindelöf's hypothesis is

$$
\int_T^{2T} \left| \mathcal{L}\left(\frac{1}{2} + it \right) \right|^2 \, dt \ll T^{1+\epsilon}
$$

for any positive ϵ. Thus, assuming the truth of Lindelöf's hypothesis we get

$$
I_1 \left(T, \frac{1}{2} \right) \ll \epsilon T \log T.
$$

Next we consider I_2. Since $a > 1$ we have

$$
\ell(a + it) = 1 + \frac{1}{1 - c} \sum_{n=2}^\infty \frac{a(n)}{n^{a+it}}, \tag{7.7}
$$

and in view of (7.3) the absolute value of the series is less than 1 for sufficiently large a. Therefore, we find by the Taylor expansion of the logarithm

$$
\log |\ell(a + it)| = \operatorname{Re} \sum_{k=1}^\infty \frac{(-1)^k}{k(1 - c)^k} \sum_{n_1=2}^\infty \cdots \sum_{n_k=2}^\infty \frac{a(n_1) \cdots a(n_k)}{(n_1 \cdots n_k)^{a+it}}.
$$

This leads to the estimate

$$I_2 = \operatorname{Re} \sum_{k=1}^{\infty} \frac{(-1)^k}{k(1-c)^k} \sum_{n_1=2}^{\infty} \cdots \sum_{n_k=2}^{\infty} \frac{a(n_1)\cdots a(n_k)}{(n_1\cdots n_k)^a} \int_T^{2T} \frac{dt}{(n_1\cdots n_k)^{it}}$$

$$\ll \sum_{k=1}^{\infty} \frac{1}{k} \left(\sum_{n=2}^{\infty} \frac{1}{n^{a-\epsilon}} \right)^k \ll 1 \qquad (7.8)$$

for sufficiently large a. It remains to estimate the horizontal integrals I_3, I_4.

Suppose that $\operatorname{Re} \ell(\sigma + iT)$ has N zeros for $b \leq \sigma \leq a$. Divide the interval $[b,a]$ into at most $N+1$ subintervals in each of which $\operatorname{Re} \ell(\sigma+iT)$ is of constant sign. Then

$$|\arg \ell(\sigma + iT)| \leq (N+1)\pi. \qquad (7.9)$$

To estimate N let

$$g(z) = \frac{1}{2}\left(\ell(z+iT) + \overline{\ell(\overline{z}+iT)} \right).$$

Then we have $g(\sigma) = \operatorname{Re} \ell(\sigma + iT)$. Let $R = a - b$ and choose T so large that $T > 2R$. Now, $\operatorname{Im}(z + iT) > 0$ for $|z - a| < T$. Thus $\ell(z+iT)$, and hence $g(z)$ is analytic for $|z - a| < T$. Let $n(r)$ denote the number of zeros of $g(z)$ in $|z - a| \leq r$. Obviously, we have

$$\int_0^{2R} \frac{n(r)}{r}\, dr \geq n(R) \int_R^{2R} \frac{dr}{r} = n(R) \log 2.$$

With Jensen's formula (see for example, Titchmarsh [353, Sect. 3.61]),

$$\int_0^{2R} \frac{n(r)}{r}\, dr = \frac{1}{2\pi} \int_0^{2\pi} \log \left| g\left(a + 2Re^{i\theta}\right) \right|\, d\theta - \log|g(a)|, \qquad (7.10)$$

we deduce

$$n(R) \leq \frac{1}{2\pi \log 2} \int_0^{2\pi} \log \left| g\left(a + 2Re^{i\theta}\right) \right|\, d\theta - \frac{\log|g(a)|}{\log 2}.$$

By (7.7) it follows that: $\log|g(a)|$ is bounded. By Theorem 6.8, in any vertical strip of bounded width,

$$\mathcal{L}(s) \ll |t|^B$$

as $|t| \to \infty$ with a certain positive constant B. Obviously, the same estimate holds for $g(z)$. Thus, the integral above is $\ll \log T$, and $n(R) \ll \log T$. Since the interval (b, a) is contained in the disc $|z - a| \leq R$, the number N is less than or equal to $n(R)$. Therefore, with (7.9), we get

$$|I_4| \leq \int_b^a |\arg \ell(\sigma + iT)|\, d\sigma \ll \log T.$$

Obviously, I_3 can be bounded in the same way.

Collecting all estimates, the assertions of the theorem follow. $\square$

Now we want to include *most* of the c-values into our observations. In view of Lemma 6.7 and Theorem 6.8 there exist positive constants C', T' such that there are no c-values in the region $\sigma < -C', t \geq T'$. Therefore, assume that $b < -C' - 1$ and $T \geq T' + 1$. By the functional equation in the form (6.13),

$$\log |\mathcal{L}(s) - c| = \log |\Delta_{\mathcal{L}}(s)| + \log |\overline{\mathcal{L}}(1 - \overline{s})| + O\left(\frac{1}{|\Delta_{\mathcal{L}}(s)\overline{\mathcal{L}}(1 - \overline{s})|}\right).$$

In view of Lemma 6.7

$$\log |\Delta_{\mathcal{L}}(s)| = \left(\frac{1}{2} - \sigma\right)(d_{\mathcal{L}} \log t + \log(\lambda Q^2)) + O\left(\frac{1}{t}\right).$$

Thus

$$\int_T^{2T} \log |\mathcal{L}(b + it) - c| \, dt$$

$$= \left(\frac{1}{2} - b\right) \int_T^{2T} (d_{\mathcal{L}} \log t + \log(\lambda Q^2)) \, dt$$

$$+ \int_T^{2T} \log |\mathcal{L}(1 - b - it)| \, dt + O(\log T).$$

Now suppose that $c \neq 1$. The first integral on the right-hand side is easily calculated by elementary methods. The second integral is small if $-b$ is chosen sufficiently large (see (7.8)). Together with (7.6) we get

$$I_1 = \left(\frac{1}{2} - b\right)\left(d_{\mathcal{L}} T \log \frac{4T}{e} + T \log(\lambda Q^2)\right) - T \log |1 - c| + O(\log T).$$

By (7.5) and with the estimates for the I_j's from the proof of Theorem 7.1, we obtain

Theorem 7.3. *Assume that $\mathcal{L}(s)$ satisfies the axioms (1)–(3) with $a(1) = 1$ and let $c \neq 1$. Then, for sufficiently large negative b,*

$$2\pi \sum_{T < \gamma_c \leq 2T} (\beta_c - b) = \left(\frac{1}{2} - b\right)\left(d_{\mathcal{L}} T \log \frac{4T}{e} + T \log(\lambda Q^2)\right)$$

$$- T \log |1 - c| + O(\log T).$$

7.2 Riemann–von Mangoldt-Type Formulae

We can rewrite the sum over c-values from Sect. 7.1 as follows:

$$\sum_{\beta_c} (\beta_c - b) = \left(\frac{1}{2} - b\right) \sum_{\beta_c} 1 + \sum_{\beta_c} \left(\beta_c - \frac{1}{2}\right).$$

The first sum on the right counts the number of c-values and the second sum measures the distances of the c-values from the critical line. Let $\mathcal{N}^c(T)$ count the number of c-values of $\mathcal{L}(s)$ with $T < \gamma_c \leq 2T$. Then, subtracting the formula of Theorem 7.3 with $b + 1$ instead of b from the formula with b, we obtain

Corollary 7.4. *Assume that $\mathcal{L}(s)$ satisfies the axioms (1)–(3) with $a(1) = 1$. Then, for $c \neq 1$,*

$$\mathcal{N}^c(T) = \frac{\mathrm{d}_{\mathcal{L}}}{2\pi} T \log \frac{4T}{e} + \frac{T}{2\pi} \log(\lambda Q^2) + \mathrm{O}(\log T).$$

Furthermore,

Corollary 7.5. *Assume that $\mathcal{L}(s)$ satisfies the axioms (1)–(3) with $a(1) = 1$. Then, for $c \neq 1$,*

$$\sum_{T < \gamma_c \leq 2T} \left(\beta_c - \frac{1}{2} \right) = -\frac{T}{2\pi} \log |1 - c| + \mathrm{O}(\log T).$$

Thus, for $c \neq 1$ satisfying $|1 - c| \neq 1$, the c-values, weighted with respect to their distance to the critical line, lie asymmetrically distributed. Nevertheless, our next aim is to show that *most* of the c-values lie close to the critical line. Unfortunately, for this purpose we have to assume the Lindelöf hypothesis. Define the counting functions (according multiplicities)

$$\mathcal{N}_+^c(\sigma, T) = \sharp\{\varrho_c : T < \gamma_c \leq 2T, \beta_c > \sigma\},$$

and

$$\mathcal{N}_-^c(\sigma, T) = \sharp\{\varrho_c : T < \gamma_c \leq 2T, \beta_c < \sigma\}.$$

Then

Theorem 7.6. *Assume that $\mathcal{L}(s)$ satisfies the axioms (1)–(3) with $a(1) = 1$ and let $c \neq 1$. Then, for any $\sigma > \max\{\frac{1}{2}, 1 - \frac{1}{d_{\mathcal{L}}}\}$,*

$$\mathcal{N}_+^c(\sigma, T) \ll T, \tag{7.11}$$

and assuming the truth of the Lindelöf hypothesis, for any $\delta > 0$,

$$\mathcal{N}_-^c \left(\frac{1}{2} - \delta, T \right) + \mathcal{N}_+^c \left(\frac{1}{2} + \delta, T \right) \ll \delta T \log T.$$

Proof. First of all, let $\sigma > \max\{\frac{1}{2}, 1 - \frac{1}{d_{\mathcal{L}}}\}$ and fix $\sigma_1 \in (\max\{\frac{1}{2}, 1 - \frac{1}{d_{\mathcal{L}}}\}, \sigma)$. Then

$$\mathcal{N}_+^c(\sigma, T) \leq \frac{1}{\sigma - \sigma_1} \sum_{\substack{\beta_c > \sigma \\ T < \gamma_c \leq 2T}} (\beta_c - \sigma_1).$$

The sum on the right hand-side is less than or equal to

$$\sum_{\substack{\beta_c > \sigma_1 \\ T < \gamma_c \leq 2T}} (\beta_c - \sigma_1) \ll \int_T^{2T} \log |\ell(\sigma_1 + \mathrm{i}t)| \, \mathrm{d}t + \mathrm{O}(\log T),$$

where we used Littlewood's Lemma 7.2 and the techniques from Sect. 7.1 for the latter inequality. In view of the unconditional estimate for (7.6) in the proof of Theorem 7.1 we obtain (7.11). Assuming the truth of the Lindelöf hypothesis, we get analogously

$$\mathcal{N}_+^c\left(\frac{1}{2}+\delta,T\right) \ll \frac{\epsilon}{\delta}T\log T \qquad (7.12)$$

for any positive ϵ.

Next we consider $\mathcal{N}_-^c$. Let b be a sufficiently large constant. We have

$$\sum_{\substack{\beta_c\geq\frac{1}{2}-\delta \\ T<\gamma_c\leq 2T}}(\beta_c-b) \leq \left(\frac{1}{2}-b\right)\sum_{\substack{\beta_c\geq\frac{1}{2}-\delta \\ T<\gamma_c\leq 2T}}1 + \sum_{\substack{\beta_c\geq\frac{1}{2} \\ T<\gamma_c\leq 2T}}\left(\beta_c-\frac{1}{2}\right).$$

Hence

$$\sum_{T<\gamma_c\leq 2T}(\beta_c-b) = \sum_{\substack{\beta_c<\frac{1}{2}-\delta \\ T<\gamma_c\leq 2T}}\left(\frac{1}{2}-b+\beta_c-\frac{1}{2}\right) + \sum_{\substack{\beta_c\geq\frac{1}{2}-\delta \\ T<\gamma_c\leq 2T}}(\beta_c-b)$$

$$\leq \left(\frac{1}{2}-b\right)\mathcal{N}^c(T) + \sum_{\substack{\beta_c<\frac{1}{2}-\delta \\ T<\gamma_c\leq 2T}}\left(\beta_c-\frac{1}{2}\right)$$

$$+ \sum_{\substack{\beta_c>\frac{1}{2} \\ T<\gamma_c\leq 2T}}\left(\beta_c-\frac{1}{2}\right).$$

By Theorem 7.1, the second sum on the right is bounded by $\epsilon T\log T$. Since any term in the first sum on the right is $<-\delta$, we obtain

$$-\delta\mathcal{N}_-^c\left(\frac{1}{2}-\delta,T\right) \geq \sum_{T<\gamma_c\leq 2T}(\beta_c-b) - \left(\frac{1}{2}-b\right)\mathcal{N}^c(T) + \mathrm{O}(\epsilon T\log T).$$

In view of Theorem 7.3 and Corollary 7.4 we get

$$\mathcal{N}_-^c\left(\frac{1}{2}-\delta,T\right) \ll \frac{\epsilon}{\delta}T\log T.$$

This is the same bound as for $\mathcal{N}_+^c$ in (7.12). Putting $\epsilon=\delta^2$ we obtain the assertion of the theorem.   □

Thus, subject to the truth of the Lindelöf hypothesis, we get by comparing Corollary 7.4 and Theorem 7.6, for any positive ϵ,

$$\mathcal{N}_-^c\left(\frac{1}{2}-\epsilon,T\right) + \mathcal{N}_+^c\left(\frac{1}{2}+\epsilon,T\right) \ll \epsilon\mathcal{N}^c(T),$$

so the c-values are clustered around the critical line for any c. This extra-ordinary value distribution shows that if the Lindelöf hypothesis for $\mathcal{L}(s)$ is true, the critical line is a so-called Julia line from the classical theory of functions. Julia [151] improved the Big Picard theorem by showing: if the analytic function f has an essential singularity at a, then there exist a real θ_0 and at most one complex number z such that for every sufficiently small $\epsilon > 0$

$$\mathbb{C} \setminus \{z\} \subset f(\{a + r\exp(i\theta) : |\theta - \theta_0| < \epsilon, 0 < r < \epsilon\});$$

the ray $\{a + r\exp(i\theta_0) : r > 0\}$ is called a Julia line. For more details on Julia's theorem we refer to Burckel [48, Sect. XII.4].

The distribution of the c-values close to the real axis is quite regular. It can be shown that there is always a c-value in some neighbourhood of any trivial zero of $\mathcal{L}(s)$ with sufficiently large negative real part, and with finitely many exceptions there are no other in the left half-plane. The main ingredients for the proof are Rouché's theorem (Theorem 8.1) and Stirling's formula (2.17). With regard to (6.6), thus the number of these c-values having real part in $[-R, 0]$ is asymptotically $\frac{1}{2} \, d_{\mathcal{L}} R$. On the other side, by (7.3) the behaviour nearby the positive real axis is very regular. Note that all results from above hold as well with respect to c-values from the lower half-plane.

Now let $N_{\mathcal{L}}^c(\sigma, T)$ count the number of c-values $\varrho_c = \beta_c + i\gamma_c$ of $\mathcal{L}(s)$ satisfying $\beta_c > \sigma, |\gamma_c| \leq T$. Using Corollary 7.4 with $2^{-n}T$ for $n \in \mathbb{N}$ instead of T and adding up, we get, for fixed $\sigma \leq 0$,

$$N_{\mathcal{L}}^c(\sigma, T) = 2 \sum_{n=1}^{\infty} \mathcal{N}^c(\sigma, 2^{-n}T)$$

$$= \left(\frac{d_{\mathcal{L}}}{\pi} T \log \frac{T}{e} + \frac{T}{\pi} \log(\lambda Q^2) \right) \sum_{n=1}^{\infty} \frac{1}{2^n}$$

$$+ \frac{d_{\mathcal{L}}}{\pi} T \sum_{n=1}^{\infty} \frac{\log 4 - n \log 2}{2^n} + O(\log T).$$

The appearing infinite series are equal to 1 and 0, respectively. Hence, this summation removes the factor 4 in the logarithmic term, and we have proved

Theorem 7.7. *Assume that $\mathcal{L}(s)$ satisfies the axioms (1)–(3) with $a(1) = 1$. For any fixed $\sigma \leq 0$ and any complex $c \neq 1$,*

$$N_{\mathcal{L}}^c(\sigma, T) = \frac{d_{\mathcal{L}}}{\pi} T \log \frac{T}{e} + \frac{T}{\pi} \log(\lambda Q^2) + O(\log T).$$

The case $c = \sigma = 0$ (the non-trivial zeros of $\mathcal{L}(s)$) is a precise Riemann–von Mangoldt formula (1.5). Similar results were obtained by Perelli [289] and Lekkerkerker [214] for other classes of Dirichlet series. It should be noticed that Tsang [355] investigated the number of c-values of $\zeta(s)$ with respect to short intervals for the imaginary parts. Let $\sigma < \frac{1}{2}$, $T^{(1/2)+\epsilon} \leq H \leq T$, and

c be a complex number satisfying $\epsilon \leq |1 - c| \leq \frac{1}{\epsilon}$ with sufficiently small ϵ. Assuming the truth of the Riemann hypothesis, Tsang proved

$$N_\zeta^c(\sigma, T + H) - N_\zeta^c(\sigma, T) \sim \frac{H}{\pi} \log \frac{T}{2\pi}$$

with an explicit error term depending on ϵ, H and T; his result holds unconditionally provided $\sigma \leq 0$.

An immediate consequence of Theorem 7.7 is that the multiplicity of nontrivial zeros ϱ of $\mathcal{L}(s)$ is bounded by $1 + \log |\gamma|$. More advanced results on the multiplicities of the zeros were obtained by Ivić [142] in the case of the Riemann zeta-function.

We conclude with another result from Selberg [323] for L-functions from $\mathcal{S}$. Assuming the truth of the Riemann hypothesis and of conjecture A, he obtained for $c \neq 1$ the asymptotic formula

$$\sum_{\substack{\beta_c > \frac{1}{2} \\ 0 < \gamma_c < T}} \left(\beta_c - \frac{1}{2} \right) = \frac{\sqrt{n_\mathcal{L}}}{4\pi^{3/2}} T \sqrt{\log \log T} + \frac{T}{4\pi} \log \frac{|c|}{1 - |c|^2} +$$

$$+ O \left(T \frac{(\log \log \log T)^3}{\sqrt{\log \log T}} \right).$$

Furthermore, for

$$\sigma(T) := \frac{1}{2} - \nu \frac{\sqrt{\log \log T}}{\log T} \qquad \text{and} \qquad \xi := \frac{d_\mathcal{L} \nu}{2\sqrt{\pi n_\mathcal{L}}}$$

with positive ν, he proved

$$\sum_{\substack{\beta_c > \sigma(T) \\ 0 < \gamma_c < T}} (\beta_c - \sigma(T))$$

$$= \frac{1}{2} \sqrt{\frac{n_\mathcal{L}}{\pi}} \left(\frac{\exp(-\pi \xi^2)}{2\pi} + \xi - \xi \int_\xi^\infty \exp(-\pi x^2) \, dx \right) T \sqrt{\log \log T}$$

$$+ \left(\log |c| \int_\xi^\infty \exp(-\pi x^2) \, dx - \log |1 - c| \right) \frac{T}{2\pi}$$

$$+ O \left(T \frac{(\log \log \log T)^3}{\sqrt{\log \log T}} \right).$$

From these results Selberg deduced that about half of the c-values lie to the left of the critical line, statistically well distributed at distances of order

$$\frac{\sqrt{\log \log T}}{\log T}$$

off $\sigma = \frac{1}{2}$, and that

$$N_\mathcal{L}^c(\sigma(T), T) \sim N_\mathcal{L}^c(T) \int_{-\xi}^\infty \exp(-\pi x^2) \, dx.$$

Most of the remaining c-values lie rather close to the critical line at distances of order not exceeding

$$\frac{(\log\log\log T)^3}{\log T\sqrt{\log\log T}}.$$

This improves some previous results of Selberg (unpublished) and Joyner [150] and gives a much more detailed description of the clustering of the c-values around the critical line.

In the exceptional case $c = 1$ one has to consider the function

$$\ell(s) = \frac{q^s}{a(q)}(\mathcal{L}(s) - 1),$$

where $q \geq 1$ is the least integer such that $a(q) \neq 0$. Then, by a similar reasoning as in the proof of Theorem 7.7, one gets analogous results. For the special case of the zeta-function this is carried out in Steuding [348, 349] where Levinson's method is applied to Epstein zeta-functions. These methods also allow to drop the condition $a(1) = 1$.

7.3 Nevanlinna Theory

Nevanlinna theory was created by Nevanlinna [281] in the 1920's to tackle the value-distribution of meromorphic functions in general. We recall some basic facts which, for example, can be found in Nevanlinna's monograph [281, Chaps. VI and IX].

Let f be a meromorphic function and denote the number of poles of $f(s)$ in $|s| \leq r$ by $\mathbf{n}(f, \infty, r)$ (counting multiplicities). The number of c-values of f is given by

$$\mathbf{n}(f, c, r) = \mathbf{n}\left(\frac{1}{f - c}, \infty, r\right).$$

The integrated counting function is

$$\mathbf{N}(f, c, r) = \int_0^r (\mathbf{n}(f, c, \varrho) - \mathbf{n}(f, c, 0))\frac{d\varrho}{\varrho} + \mathbf{n}(f, c, 0)\log r.$$

The proximity function is defined by

$$\mathbf{m}(f, r) = \frac{1}{2\pi}\int_0^{2\pi} \log^+ |f(r\exp(i\theta))|\, d\theta,$$

and, for $c \in \mathbb{C}$, by

$$\mathbf{m}(f, c, r) = \mathbf{m}\left(\frac{1}{f - c}, r\right),$$

where $\log^+ x := \max\{0, \log x\}$. The function $\mathbf{m}(f, c, r)$ indicates how close $f(s)$ is to the value c on the circle $|s| = r$. The characteristic function of f is defined by

$$\mathbf{T}(f, r) = \mathbf{N}(f, \infty, r) + \mathbf{m}(f, r).$$

Furthermore, let
$$\mathbf{T}(f,c,r) = \mathbf{N}(f,c,r) + \mathbf{m}(f,c,r)$$

for $c \in \mathbb{C}$. The first main theorem in Nevanlinna theory states states that $\mathbf{T}(f,c,r)$ differs from the characteristic function by a bounded quantity:

Theorem 7.8. *Let f be a meromorphic function and let c be any complex number. Then*
$$\mathbf{T}(f,c,r) = \mathbf{T}(f,r) + \mathrm{O}(1),$$
where the error term depends on f and c.

The proof relies on Jensen's formula (7.10).

Thus, $\mathbf{T}(f,c,r)$ for different values of c is invariant up to additive terms that are bounded. The invariant, the characteristic function $\mathbf{T}(f,r)$, encodes information about the analytic behaviour of f.

The quantity
$$\delta(f,c) := 1 - \limsup_{r \to \infty} \frac{\mathbf{N}(f,c,r)}{\mathbf{T}(f,r)}$$

is called the deficiency of the value c of f. This deficiency is positive only if there are *relatively few* c-values. The second main theorem in Nevanlinna theory implies the so-called deficiency relation which states that
$$\sum_{c \in \mathbb{C} \cup \{\infty\}} \delta(f,c) \leq 2 \, ;$$

note that only for countably many values of c the deficiency can differ from zero. Another consequence is the Big Picard theorem.

Only recently Ye [372] computed the Nevanlinna functions for the Riemann zeta-function. Without big effort we can extend his results to the class of Dirichlet series under investigation. The Nevanlinna functions for those $\mathcal{L}(s)$ are determined by the Gamma-factors in the functional equation.

First, let $\sigma_0 > 1$ be fixed. We write $s = r \exp(i\theta)$, so $\sigma = r \cos\theta$. It is easily seen that
$$\frac{1}{2\pi} \int_{\{\theta : r \cos\theta > \sigma_0\}} \log^+ |\mathcal{L}(r \exp(i\theta))| \, \mathrm{d}\theta \ll 1.$$

Further, in view of Theorem 6.8,
$$\frac{1}{2\pi} \int_{\{\theta : 1 - \sigma_0 \leq r \cos\theta \leq \sigma_0\}} \log^+ |\mathcal{L}(r \exp(i\theta))| \, \mathrm{d}\theta \ll \log r ;$$

note that the Lebesgue measure of the set
$$\{\theta \in [0, 2\pi] : \sigma = r \cos\theta \in [1 - \sigma_0, \sigma_0]\}$$

is bounded by $\frac{1}{r}$. Finally, for $\sigma \leq 1 - \sigma_0$ we deduce from the functional equation in the form (6.13) that

$$\log^+ |\mathcal{L}(r\exp(i\theta))| \le \sum_{j=1}^{f} \Big\{ \log^+ |\Gamma(\lambda_j(1 - r\exp(i\theta)) + \overline{\mu_j})|$$

$$+ \log^+ |\Gamma(\lambda_j r\exp(i\theta) + \mu_j)| \Big\} + \mathrm{O}(r).$$

Now we shall use Ye's decomposition of the Gamma-function. For any $z = r\exp(i\theta)$, there is an integer n_0 with $n_0 < r \le n_0 + 1$ such that

$$\frac{1}{\Gamma(z)} = F_1(z)\,F_2(z) \quad \text{with} \quad F_1(z) := z\left(\gamma z - \sum_{n=1}^{2n_0} \frac{z}{n}\right),$$

where γ is the Euler–Mascheroni constant, and $F_2(z)$ is an entire function with $\mathbf{m}(F_2, r) \ll r$. The order of growth of $\Gamma(z)$ is ruled by the order of growth of $F_1(z)$. Ye computed

$$\log |F_1(z)| = -r\log r \cos\theta + \mathrm{O}(r).$$

If λ is a positive real number and μ an arbitrary complex number, Ye's estimate leads to

$$\frac{1}{2\pi} \int_{\{\theta:r\cos\theta<1-\sigma_0\}} \log^+ |\Gamma(\lambda(1 - r\exp(i\theta)) + \mu)|\,d\theta$$

$$\le \frac{\lambda}{2\pi} \int_{-\pi/2}^{\pi/2} r\log r \cos\theta\,d\theta + \mathrm{O}(r) = \frac{\lambda}{\pi} r\log r + \mathrm{O}(r),$$

and, similarly,

$$\frac{1}{2\pi} \int_{\{\theta:r\cos\theta<1-\sigma_0\}} \log^+ |\Gamma(\lambda r\exp(i\theta)) + \mu)|\,d\theta \le \frac{\lambda}{\pi} r\log r + \mathrm{O}(r).$$

Thus, we get

$$\frac{1}{2\pi} \int_{\{\theta:r\cos\theta<1-\sigma_0\}} \log^+ |\mathcal{L}(r\exp(i\theta))|\,d\theta \le \frac{d_{\mathcal{L}}}{\pi} r\log r + \mathrm{O}(r).$$

Adding the estimates for the other cases, we obtain for the proximity function of $\mathcal{L}(s)$

$$\mathbf{m}(\mathcal{L}, r) \le \frac{d_{\mathcal{L}}}{\pi} r\log r + \mathrm{O}(r).$$

Since $\mathcal{L}(s)$ is regular except for at most a pole at $s = 1$,

$$\mathbf{N}(\mathcal{L}, \infty, r) \ll \int_1^r \frac{d\varrho}{\varrho} = \log r. \tag{7.13}$$

Thus, we get

$$\mathbf{T}(\mathcal{L}, r) \le \frac{d_{\mathcal{L}}}{\pi} r\log r + \mathrm{O}(r). \tag{7.14}$$

It follows from Theorem 7.7 that:

$$\mathbf{N}(\mathcal{L}, 0, r) = \frac{\mathrm{d}_\mathcal{L}}{\pi} r \log r + \mathrm{O}(r). \tag{7.15}$$

The first main Theorem 7.8 implies

$$\mathbf{N}(\mathcal{L}, 0, r) \leq \mathbf{T}(\mathcal{L}, 0, r) = \mathbf{T}(\mathcal{L}, r) + \mathrm{O}(1).$$

In view of (7.14) and (7.15) we get an asymptotic formula for the characteristic function:

Theorem 7.9. *For $\mathcal{L}$ satisfying axioms (1)–(3) with $a(1) = 1$,*

$$\mathbf{T}(\mathcal{L}, r) = \frac{\mathrm{d}_\mathcal{L}}{\pi} r \log r + \mathrm{O}(r).$$

We deduce from this and (7.13) for the deficiency value of infinity:

$$\delta(\mathcal{L}, \infty) = 1 - \limsup_{r \to \infty} \frac{\mathbf{N}(\mathcal{L}, \infty, r)}{\mathbf{T}(\mathcal{L}, r)} = 1.$$

In view of Theorem 7.7 the deficiency values for $c \neq 1, \infty$ are equal to zero.

In combination with Theorem 7.7 the asymptotic formula of the theorem shows that the counting function $\mathbf{N}(\mathcal{L}, c, r)$ *dominates* the proximity function $\mathbf{m}(\mathcal{L}, c, r)$, at least for any complex value $c \neq 1$. In the exceptional case $c = 1$, by the first main Theorem 7.8, we may deduce from Theorem 7.9 that

$$N_\mathcal{L}^1(T) \leq \frac{\mathrm{d}_\mathcal{L}}{\pi} T \log T + \mathrm{O}(T).$$

A more sophisticated analysis would show that this is actually an equality. However, we do not go into the details. In Sect. 9.7 we return to the distribution of c-values in the half-plane of absolute convergence.

We conclude with a description of the analytic behaviour of the Dirichlet series $\mathcal{L}$ under investigation in terms of the notion of finite order. A positive function $t(r)$ is said to be of finite order λ if

$$\limsup_{r \to \infty} \frac{\log t(r)}{\log r} = \lambda;$$

$t(r)$ is of maximum, mean or minimum type of order λ if the upper limit

$$\limsup_{r \to \infty} \frac{t(r)}{r^\lambda}$$

is infinite, finite and positive, or zero. A meromorphic function is defined to be of the same order and the same type as its characteristic function $\mathbf{T}(r, f)$. Thus, by Theorem 7.9, we get

Corollary 7.10. *Every $\mathcal{L}$ satisfying the axioms (1)–(3) with $a(1) = 1$ is of order one and of maximum type.*

7.4 Uniqueness Theorems

We say that two meromorphic functions f and g share a value $c \in \mathbb{C} \cup \{\infty\}$ if the sets of pre-images of the value c under f and under g are equal, for short

$$f^{-1}(c) := \{s \in \mathbb{C} : f(s) = c\} = g^{-1}(c). \tag{7.16}$$

We say that f and g share the value c counting multiplicities (CM) if (7.16) holds and if the roots of the equations

$$f(s) = c \quad \text{and} \quad g(s) = c$$

have the same multiplicities; if there is no restriction on the multiplicities, f and g are said to share the value c ignoring multiplicities (IM). Nevanlinna [280] proved two fundamental results on shared values. His remarkable five-point theorem states that any two non-constant meromorphic functions which share five distinct values are equal. Since $f(s) = \exp(s)$ and $g(s) = \exp(-s)$ share the four values $0, \pm 1, \infty$, the number 5 in Nevanlinna's statement is best possible. If multiplicities are taken into account, Nevanlinna proved that if two meromorphic functions f and g share four distinct values $c_1, \ldots, c_4$ CM, then either $f \equiv g$ or there exists a linear fractional transformation M such that $g \equiv M \circ f$ and

$$M(c_1) = c_1, \quad M(c_2) = c_2, \quad M(c_3) = c_4, \quad \text{and} \quad M(c_4) = c_3;$$

in the latter case f and g do not assume the values c_3 and c_4. Also the number 4 for the upper bound of shared values CM is best possible. The result can be sharpened if two of the four values are allowed to be shared IM (see [111]). In [347], Steuding, investigated how many values L-functions can share. In this special case better estimates are possible than those which Nevanlinna's theorems provide. It is expected that *independent* L-functions cannot share any complex value which is actually taken.

First of all, we trivially note that two L-functions from the Selberg class share the value ∞ CM if and only if both are entire or if they both have a pole at $s = 1$ of the same order (other poles cannot occur), e.g., the Riemann zeta-function $\zeta(s)$ and a Dedekind zeta-function to a quadratic number field. If the orders of the poles differ, they share the value ∞ IM. Further, we observe that two different L-functions in the Selberg class cannot share the value zero CM. This follows immediately from a theorem of M.R. Murty and V.K. Murty [268]. To see that denote the non-trivial zeros of $\mathcal{L} \in \mathcal{S}$ by ϱ and let $m_{\mathcal{L}}(\varrho)$ be the multiplicity of ϱ. Further, define for $\mathcal{L}_1, \mathcal{L}_2 \in \mathcal{S}$ the function

$$D_{\mathcal{L}_1, \mathcal{L}_2}(T) = \sum_{\varrho} |m_{\mathcal{L}_1}(\varrho) - m_{\mathcal{L}_2}(\varrho)|,$$

where the summation is taken over all non-trivial zeros ϱ of $\mathcal{L}_1$ and $\mathcal{L}_2$ (counting multiplicities). Then M.R. Murty and V.K. Murty proved that $\mathcal{L}_1, \mathcal{L}_2 \in \mathcal{S}$ are either equal or

$$\liminf_{T \to \infty} \frac{1}{T} D_{\mathcal{L}_1, \mathcal{L}_2}(T) > 0$$

(see also the related result of Bombieri and Perelli (6.8)). However, the trivial example $\zeta(s)$ and $\zeta(s)^2$ shows that different elements of $\mathcal{S}$ can share the value zero IM.

Concerning CM-shared values we shall prove that two different Dirichlet series satisfying our axioms do not share any complex value CM. For sharing values IM we shall only obtain an improvement of the five-point theorem under an additional assumption on the number of distinct c-values. For this purpose let $\tilde{N}_{\mathcal{L}}^c(T)$ count the number of distinct roots ϱ_c of the equation $\mathcal{L}(s) = c$ lying in the rectangle $0 \leq \sigma \leq 1, |t| \leq T$.

Theorem 7.11. *Assume that $\mathcal{L}_1, \mathcal{L}_2$ satisfy the axioms (1)–(3) with $a(1) = 1$.*

(i) If $\mathcal{L}_1, \mathcal{L}_2$ share a value $c \neq \infty$ CM, then $\mathcal{L}_1 \equiv \mathcal{L}_2$.
(ii) If $\mathcal{L}_1, \mathcal{L}_2$ satisfy the same functional equation and share two distinct values $c_1, c_2 \neq \infty$ IM such that

$$\liminf_{T \to \infty} \frac{\tilde{N}_{\mathcal{L}_j}^{c_1}(T) + \tilde{N}_{\mathcal{L}_j}^{c_2}(T)}{N_{\mathcal{L}_j}^{c_1}(T) + N_{\mathcal{L}_j}^{c_2}(T)} > \frac{1}{2} + \epsilon \tag{7.17}$$

for some positive ϵ with either $j = 1$ or 2, then $\mathcal{L}_1 \equiv \mathcal{L}_2$.

We briefly discuss the second assertion of the theorem. Condition (7.17) reflects that more than 50% of the c_1- and c_2-values of $\mathcal{L}_j(s)$ are supposed to be distinct. It should be noted that such conditions are very difficult to verify. For instance, Farmer [78] proved that more than 63% of the zeros of $\zeta(s)$ are distinct; however, any extension to L-functions of larger degree seems to be hard to realize.

Proof. We start with the first assertion. In view of Theorem 7.7 two L-functions satisfying the axioms (1)–(3) can only share a value $c \neq \infty$ CM if they have the same degree, d say. First of all assume that $\mathcal{L}_1, \mathcal{L}_2$ are both entire functions and share the value $c \neq \infty$ CM. Define the function

$$\ell(s) = \frac{\mathcal{L}_1(s) - c}{\mathcal{L}_2(s) - c}.$$

Since $\mathcal{L}_1(s)$ assumes the value c if and only if $\mathcal{L}_2(s) = c$ and since for any such root the multiplicities coincide, $\ell(s)$ is a non-vanishing entire function. In view of the first main Theorem 7.8 and Theorem 7.9

$$\mathbf{T}(\mathcal{L}_2, c, r) = \mathbf{T}(\mathcal{L}_2, r) + O(1) = \frac{d}{\pi} r \log r + O(r)$$
$$= \mathbf{T}(\mathcal{L}_1 - c, r) + O(r).$$

For a meromorphic function f denote its order by $\lambda(f)$. Then it follows that:

$$\lambda\left(\frac{1}{\mathcal{L}_1 - c}\right) = \lambda(\mathcal{L}_2 - c) = \lambda(\mathcal{L}_2) = 1.$$

It is easily seen that the order of a finite product of functions of finite order is less than or equal to the maximum of the order of the factors. Thus $\lambda(\ell) \leq 1$. By Hadamard's factorization theorem (see [281, Sect. VIII.2]) this implies that $\ell(s)$ is of the form

$$\ell(s) = \exp(P(s)),$$

where P is a polynomial of degree at most $\lambda(\ell) \leq 1$. Since $\mathcal{L}_j(s)$ tends to one as $s \to \infty$ for $j = 1, 2$, we have

$$\lim_{s \to \infty} \ell(s) = \frac{1-c}{1-c} = 1.$$

This implies that the polynomial P is vanishing identically, which implies $\mathcal{L}_1 \equiv \mathcal{L}_2$.

If $\mathcal{L}_1(s)$ or $\mathcal{L}_2(s)$ has a pole at $s = 1$ of order k, we may replace $\mathcal{L}_j(s)$ by $(s-1)^k \mathcal{L}_j(s)$ and repeat the argument from above. This proves the first assertion.

Now we shall prove the second statement. If $\mathcal{L}_1$ and $\mathcal{L}_2$ satisfy the same functional equation, they both have the same degree, d say. Now consider the function

$$\ell(s) := \mathcal{L}_1(s) - \mathcal{L}_2(s).$$

Obviously, also $\ell(s)$ satisfies the common functional equation for the $\mathcal{L}_j$'s. Then the number $N_\ell(T)$ of zeros of $\ell(s)$ in the rectangle $0 \leq \sigma \leq 1, |t| \leq T$ (counting multiplicities) is asymptotically given by

$$N_\ell(T) \sim \frac{\mathrm{d}}{\pi} T \log T. \tag{7.18}$$

Now suppose that $\mathcal{L}_1$ and $\mathcal{L}_2$ share two distinct complex values c_1, c_2 IM. Then $\ell(s)$ vanishes also for the pre-images of the c_k's. Hence, we obtain a lower bound for the number of zeros of $\ell(s)$ in terms of the c_1- and c_2-value counting functions, namely

$$N_\ell(T) \geq \tilde{N}_{\mathcal{L}_j}^{c_1}(T) + \tilde{N}_{\mathcal{L}_j}^{c_2}(T), \tag{7.19}$$

where we can take $j = 1$ or $j = 2$. Taking into account Theorem 7.7 and (7.18) we can replace $N_\ell(T)$ by

$$\frac{1}{2}\left(N_{\mathcal{L}_j}^{c_1}(T) + N_{\mathcal{L}_j}^{c_2}(T)\right) + \mathrm{O}(T).$$

Thus we can rewrite (7.19) as

$$\frac{\tilde{N}_{\mathcal{L}_j}^{c_1}(T) + \tilde{N}_{\mathcal{L}_j}^{c_2}(T)}{N_{\mathcal{L}_j}^{c_1}(T) + N_{\mathcal{L}_j}^{c_2}(T)} \leq \frac{1}{2} + o(1).$$

This contradicts (7.17). Hence $\mathcal{L}_1$ and $\mathcal{L}_2$ can share at most one value $c \in \mathbb{C}$. Theorem 7.11 is proved.   □

The functions $\mathcal{L}(s)$ and $\mathcal{L}(s)^2$ share the value zero IM. This is a special example for two reasons. First, these functions are *not* independent in the sense that they have the same primitive functions in their factorizations. Second, they share the zeros. Bombieri and Hejhal [40] proved, assuming some widely believed but yet unproved hypotheses, that almost all zeros of pairwise independent L-functions are distinct. Of course, we expect the same to hold for other c-values too. With respect to condition (7.17) this leads us to conjecture that zero is the only possible shared value and that this happens only in cases of *dependent* L-functions.

We conclude with a few words about the significance of such studies. Some problems in arithmetic (see Chap. 13.7) could be solved if one could show that, given distinct primitive L-functions $\mathcal{L}_1(s), \ldots, \mathcal{L}_m(s)$ (in the sense of the Selberg class), then $\mathcal{L}_j(\varrho_k) = 0$ holds only for $j = k$, where the ϱ_k denote the non-trivial zeros of $\mathcal{L}_k(s)$. Clearly, this would also imply the unique factorization into primitive elements.

8

The Riemann Hypothesis

> ...und es ist sehr wahrscheinlich, dass alle Wurzeln reell sind. Hiervon
> wäre allerdings ein strenger Beweis zu wünschen; ich habe indess die
> Aufsuchung desselben nach einigen flüchtigen vergeblichen Versuchen
> vorläufig bei Seite gelassen...   Bernhard Riemann

There is an interesting link between universality and the zero-distribution.
As we will show in this chapter, the question whether the zeta-function can
approximate itself in the right half of the critical strip turns out to be equivalent to the Riemann hypothesis. This reformulation dates back to Bohr [30]
who proved its analogue for Dirichlet L-functions to non-principal characters.
Bagchi [9] extended this result to the Riemann zeta-function. We shall also
consider further generalizations.

8.1 Uniform Approximation and Zeros

In view of the phenomenon of universality a natural question arises: is the
condition on the non-vanishing of $g(s)$ in the universality Theorem 5.14 necessary or *is it possible to approximate uniformly functions having zeros by a
universal L-function?* The answer is negative. For the sake of simplicity we
consider only the case of the Riemann zeta-function. Assume that there is
an analytic function $g(s)$ defined on an admissible set $\mathcal{K}$ with a zero in the
interior and

$$\lim_{j \to \infty} \zeta(s + i\tau_j) = g(s)$$

on $\mathcal{K}$ for some infinite sequence τ_j which tends to infinity as $j \to \infty$. Then,
by Hurwitz's Theorem 5.13, the zero of $g(s)$ is limit point of zeros of the
shifts $\zeta(s + i\tau_j)$ and, in particular, there would exist nontrivial zeros off the
critical line – a violation of the Riemann hypothesis. However, the Riemann
hypothesis is still unproved and we shall sketch an unconditional argument
which can be fixed with a bit more effort by the techniques of Sect. 8.2.

In order to see that $\zeta(s)$ cannot approximate uniformly a function with a zero, recall Rouché's theorem:

Theorem 8.1. *Let $f(s)$ and $g(s)$ be analytic for $|s| \le r$. If*

$$|f(s) - g(s)| < |g(s)|$$

on $|s| = r$, then $f(s)$ and $g(s)$ have the same number of zeros in $|s| < r$.

This classical result follows from a simple application of the argument principle; for details see Burckel [48, Sect. VIII.3] or Titchmarsh [352, Sect. 3.42].

Now assume that $g(s)$ is an analytic function on $|s| \le r$, where $0 < r < \frac{1}{4}$, which has a zero ξ with $|\xi| < r$, but which is non-vanishing on the boundary. An application of Rouché's theorem shows that whenever the inequality

$$\max_{|s|=r} \left| \zeta \left(s + \frac{3}{4} + i\tau \right) - g(s) \right| < \min_{|s|=r} |g(s)| \tag{8.1}$$

holds, $\zeta \left(s + \frac{3}{4} + i\tau \right)$ has to have a zero inside $|s| < r$. The zeros of an analytic function lie either discretely distributed or the function vanishes identically, and thus the inequality (8.1) holds if the left-hand side is sufficiently small. If now for any $\epsilon > 0$

$$\liminf_{T \to \infty} \frac{1}{T} \operatorname{meas} \left\{ \tau \in [0, T] : \max_{|s| \le r} \left| \zeta \left(s + \frac{3}{4} + i\tau \right) - g(s) \right| < \epsilon \right\} > 0,$$

then we expect $\gg T$ many τ in the interval $[0, T]$ each of which corresponds via (8.1) to a complex zeros of $\zeta(s)$ in the strip $\frac{3}{4} - r < \sigma < \frac{3}{4} + r$ up to T (for a rigorous proof one has to consider the densities of values τ satisfying (8.1); this can be done along the lines of the proof of Theorem 8.3). This contradicts the density Theorem 1.2, which gives

$$N \left(\frac{3}{4} - r, T \right) = \mathrm{O}(T).$$

Thus, a given function having zeros cannot be approximated uniformly by the zeta-function (in the sense of Voronin's theorem)! $\zeta(s)$ is not strongly universal.

The above reasoning shows that the location of the complex zeros of Riemann's zeta-function is closely connected with the universality property.

8.2 Bagchi's Theorem

Bohr introduced the fruitful notion of almost periodicity to analysis. An analytic function $f(s)$, defined on some vertical strip $a < \sigma < b$, is called almost periodic if, for any positive ε, and any α, β with $a < \alpha < \beta < b$, there exists a

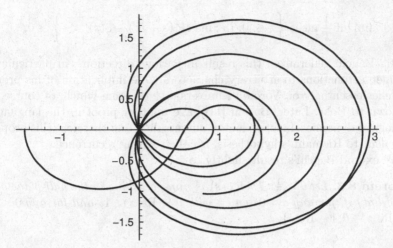

Fig. 8.1. $\zeta(\frac{1}{2} + it)$ for $t \in [0, 40]$. In this range, there are six non-trivial zeros lying on the critical line

length $\ell = \ell(f, \alpha, \beta, \varepsilon) > 0$ such that every interval (t_1, t_2) of length ℓ contains an almost period of f relatively to ε in the closed strip $\alpha \leq \sigma \leq \beta$, i.e., there exists a number $\tau \in (t_1, t_2)$ such that

$$|f(\sigma + it + i\tau) - f(\sigma + it)| < \varepsilon \quad \text{for any} \quad \alpha \leq \sigma \leq \beta, t \in \mathbb{R}. \qquad (8.2)$$

Bohr [30] proved.

Theorem 8.2. *Every Dirichlet series is almost-periodic in its half-plane of absolute convergence.*

The most important open problem in the theory of the Riemann zeta-function is the Riemann hypothesis on the location of the nontrivial zeros (see Fig. 8.1 for the values taken by the zeta-function on the critical line). Bohr discovered an interesting relation between the Riemann hypothesis and almost periodicity; indeed, his aim in introducing the concept of almost periodicity might have been Riemann's hypothesis. Bohr showed that if χ is a non-principal character, then the Riemann hypothesis for the Dirichlet L-function $L(s, \chi)$ is equivalent to the almost periodicity of $L(s, \chi)$ in the half-plane $\sigma > \frac{1}{2}$. The condition on the character looks artificial but it is necessary for Bohr's reasoning. His argument relies in the main part on diophantine approximation applied to the coefficients of the Dirichlet series representation. The Dirichlet series for $L(s, \chi)$ with a non-principal character χ converges throughout the critical strip, but the Dirichlet series for the zeta-function does not.

More than half a century later Bagchi [9] proved that the Riemann hypothesis is true if and only if for any compact subset $\mathcal{K}$ of the strip $\frac{1}{2} < \sigma < 1$ with connected complement and for any $\epsilon > 0$

$$\liminf_{T\to\infty} \frac{1}{T} \, \text{meas} \left\{ \tau \in [0,T] : \max_{s\in\mathcal{K}} |\zeta(s+i\tau) - \zeta(s)| < \epsilon \right\} > 0.$$

In [10], Bagchi generalized this result in various directions; in particular for Dirichlet L-functions to arbitrary characters. One implication of his proof in [10] relies essentially on Voronin's universality theorem which, of course, was unknown to Bohr. Later, Bagchi [11] gave another proof in the language of topological dynamics, independent of universality, and therefore this property, equivalent to Riemann's hypothesis, is called strong recurrence.

We extend Bagchi's result slightly to

Theorem 8.3. *Let $\theta \geq \frac{1}{2}$. Then $\zeta(s)$ is non-vanishing in the half-plane $\sigma > \theta$ if and only if, for any $\epsilon > 0$, any z with $\theta < \text{Re } z < 1$, and for any $0 < r < \min\{\text{Re } z - \theta, 1 - \text{Re } z\}$,*

$$\liminf_{T\to\infty} \frac{1}{T} \, \text{meas} \left\{ \tau \in [0,T] : \max_{|s-z|\leq r} |\zeta(s+i\tau) - \zeta(s)| < \epsilon \right\} > 0.$$

Proof. If Riemann's hypothesis is true we can apply Voronin's universality Theorem 1.7 with $g(s) = \zeta(s)$, which implies the strong recurrence. More generally, the non-vanishing of $\zeta(s)$ for $\sigma > \theta$ would allow to approximate $\zeta(s)$ by shifts $\zeta(s+i\tau)$ uniformly on appropriate subsets of the strip $\theta < \sigma < 1$. The idea for the proof of the other implication is that if there is at least one zero to the right of the line $\sigma = \theta$, then the strong recurrence property implies the existence of *many* zeros, in fact *too many* with regard to the classic density Theorem 1.2.

Suppose that there exists a zero ξ of $\zeta(s)$ with $\text{Re } \xi > \theta$. Without loss of generality we may assume that $\text{Im } \xi > 0$. We have to show that there exists a disk with center z and radius r, satisfying the conditions of the theorem, and a positive ϵ such that

$$\liminf_{T\to\infty} \frac{1}{T} \, \text{meas} \left\{ \tau \in [0,T] : \max_{|s-z|\leq r} |\zeta(s+i\tau) - \zeta(s)| < \epsilon \right\} = 0. \qquad (8.3)$$

Locally, the zeta-function has the expansion

$$\zeta(s) = c(s-\xi)^m + \text{O}\left(|s-\xi|^{m+1}\right) \qquad (8.4)$$

with some non-zero $c \in \mathbb{C}$ and $m \in \mathbb{N}$. Now assume that for a neighbourhood $\mathcal{K}_\delta := \{s \in \mathbb{C} : |s-\xi| \leq \delta\}$ of ξ the relation

$$\max_{s\in\mathcal{K}_\delta} |\zeta(s+i\tau) - \zeta(s)| < \epsilon \leq \min_{|s|=\delta} |\zeta(s)| \qquad (8.5)$$

holds; the second inequality is fulfilled for sufficiently small ϵ (by an argument already discussed in Sect. 8.1). Then Rouché's Theorem 8.1 implies the existence of a zero ϱ of $\zeta(s)$ in

$$\mathcal{K}_\delta + i\tau := \{s \in \mathbb{C} : |s - i\tau - \xi| \leq \delta\}.$$

We say that the zero ϱ of $\zeta(s)$ is *generated* by the zero ξ. With regard to (8.4) and (8.5) the zeros ξ and $\varrho = \beta + i\gamma$ are intimately related; more precisely,

$$\epsilon > |\zeta(\varrho) - \zeta(\varrho - i\tau)| = |\zeta(\varrho - i\tau)| \geq |c| \cdot |\varrho - i\tau - \xi|^m + O(\delta^{m+1}).$$

Hence,

$$|\varrho - i\tau - \xi| \leq \left(\frac{\epsilon}{|c|}\right)^{1/m} + O\left(\delta^{1+(1/m)}\right).$$

In particular,

$$\frac{1}{2} < \operatorname{Re}\xi - 2\left(\frac{\epsilon}{|c|}\right)^{1/m} < \beta < 1,$$

and

$$|\gamma - (\tau + \operatorname{Im}\xi)| < 2\left(\frac{\epsilon}{|c|}\right)^{1/m},$$

for sufficiently small ϵ and $\delta = O(\epsilon^{m+1})$. Next we have to count the generated zeros in terms of τ. Two different shifts τ_1 and τ_2 can lead to the same zero ϱ, but their distance is bounded by

$$|\tau_1 - \tau_2| < 4\left(\frac{\epsilon}{|c|}\right)^{1/m}.$$

If we now write

$$\mathcal{I}(T) := \bigcup_j \mathcal{I}_j(T) := \left\{\tau \in [0, T] : \max_{s \in \mathcal{K}_\delta} |\zeta(s + i\tau) - \zeta(s)| < \epsilon\right\},$$

where the $\mathcal{I}_j(T)$ are disjoint intervals, it follows that there are

$$\geq \left[\frac{1}{4}\left(\frac{|c|}{\epsilon}\right)^{1/m} \operatorname{meas}\mathcal{I}_j(T)\right] + 1 > \frac{1}{4}\left(\frac{|c|}{\epsilon}\right)^{1/m} \operatorname{meas}\mathcal{I}_j(T)$$

many distinct zeros according to $\tau \in \mathcal{I}_j(T)$, generated by ξ. The number of generated zeros is a lower bound for the number of all zeros. For the number of all zeros having real part $> \operatorname{Re}\xi - 2(\frac{\epsilon}{|c|})^{1/m}$ up to level T it follows that

$$\sharp\left\{\varrho = \beta + i\gamma : \beta > \operatorname{Re}\xi - 2\left(\frac{\epsilon}{|c|}\right)^{1/m}, 0 < \gamma < T + \operatorname{Im}\xi + 2\left(\frac{\epsilon}{|c|}\right)^{1/m}\right\}$$

$$\geq 1/4\left(\frac{|c|}{\epsilon}\right)^{1/m} \operatorname{meas}\mathcal{I}(T).$$

This and the density Theorem 1.2 lead to

$$\operatorname{meas}\mathcal{I}(T) = O(T),$$

which implies (8.3). The theorem is proved. □

Using the same reasoning, Reich's discrete version of Voronin's universality theorem, Theorem 5.16, yields a discrete version of Theorem 8.3: Riemann's hypothesis is true if and only if for any $\epsilon > 0$, any real number $\Delta \neq 0$, any z with $\theta < \mathrm{Re}\ z < 1$, and any $0 < r < \min\{\mathrm{Re}\ z - \theta, 1 - \mathrm{Re}\ z\}$,

$$\liminf_{N \to \infty} \frac{1}{N} \sharp \left\{ 1 \le n \le N : \max_{|s-z| \le r} |\zeta(s + i\Delta n) - \zeta(s)| < \epsilon \right\} > 0.$$

The expected strong recurrence of $\zeta(s)$ may be regarded as a kind of *self-similarity*. Assuming the truth of Riemann's hypothesis, this has a nice interpretation. Consider the amplitude of light which is a physical bound for the size of objects which human eyes can see, or the Planck length $l_P \approx 1.616\times 10^{-35}$ meter which is the smallest size of objects in quantum mechanics. If we assume that ϵ is less than one of these quantities, then we cannot physically distinguish between $\zeta(s)$ and $\zeta(s + i\tau)$ for s from a compact subset $\mathcal{K}$ of the right half of the critical strip, whenever

$$\max_{s \in \mathcal{K}} |\zeta(s + i\tau) - \zeta(s)| < \epsilon.$$

This shows that we cannot decide where in the analytic landscape of $\zeta(s)$ we actually are without moving to the boundary. The zeta-function is an *amazing maze!*

8.3 A Generalization

It is obvious how Bagchi's Theorem 8.3 can be generalized to other universal L-functions, e.g., Dirichlet L-functions, for which appropriate density estimates are known. Assume that $\mathcal{L}(s)$ is strongly recurrent in the strip $\sigma_m < \sigma < 1$ in the sense of Theorem 8.3. In view of the positive lower density for the set of shifts τ with which the target $\mathcal{L}(s)$ can be uniformly approximated by $\mathcal{L}(s + i\tau)$ we need that the number of zeros with real part greater than σ_m up to level T is bounded by $o(T)$ as $T \to \infty$ to deduce the non-vanishing of $\mathcal{L}(s)$ in the strip $\sigma_m < \sigma < 1$. Thus, strong recurrence allows the conclusion that if there are not too many zeros, then there are none!

For the set of L-functions in $\mathcal{S} \cap \tilde{\mathcal{S}}$ we can replace the density Theorem 1.2 by Theorem 6.5 from Kaczorowski and Perelli [163]. We observe for the number of zeros in question that

$$N_{\mathcal{L}}(\sigma, T) \ll T^{4(d_{\mathcal{L}}+3)(1-\sigma)+\epsilon} = o(T)$$

if

$$\sigma > \sigma_* = \sigma_*(\mathcal{L}) := 1 - \frac{1}{4(d_{\mathcal{L}} + 3)},$$

which is larger than the left abscissa of the strip of universality in Theorem 6.12. This yields the non-vanishing of $\mathcal{L}(s)$ in some strip inside the critical strip provided $\mathcal{L}(s)$ is strongly recurrent. For the line $\sigma = 1$, we notice that axiom (v) on the mean-square for the Dirichlet series coefficients of $\mathcal{L} \in \tilde{\mathcal{S}}$ on the primes implies the normality conjecture (as remarked in Sect. 6.6) and Kaczorowski and Perelli [163] proved that the latter conjecture implies $\mathcal{L}(1 + i\mathbb{R}) \neq 0$ (see Sect. 6.2). Hence,

Theorem 8.4. *Let $\mathcal{L} \in \mathcal{S} \cap \tilde{\mathcal{S}}$. Then $\mathcal{L}(s)$ is non-vanishing in the half-plane $\sigma > \sigma_*$ if and only if, for any $\epsilon > 0$, any z with $\sigma_* < \mathrm{Re}\, z < 1$ and any $0 < r < \min\{\mathrm{Re}\, z - \sigma_*, 1 - \mathrm{Re}\, z\}$,*

$$\liminf_{T \to \infty} \frac{1}{T} \, \mathrm{meas} \left\{ \tau \in [0,T] : \max_{|s-z| \leq r} |\mathcal{L}(s + i\tau) - \mathcal{L}(s)| < \epsilon \right\} > 0.$$

Of course, assuming the validity of the density hypothesis or the Lindelöf hypothesis, one can also prove a conditional result valid in the strip $\frac{1}{2} < \sigma < 1$.

In many cases one may obtain better results with a general density estimate from Perelli [289]. For this purpose we introduce Perelli's class of L-functions [289] which has many similarities with the Selberg class.

Perelli's class $\mathcal{P}$ of L-functions consists of Dirichlet series

$$L(s) := \sum_{n=1}^{\infty} \frac{a(n)}{n^s},$$

which satisfy the following axioms.

- *Analytic continuation.* $L(s)$ can be analytically continued to a meromorphic function with at most one simple pole at $s = 1$.
- *Polynomial Euler product.* There exists a positive integer m such that for any prime p there exists an $m \times m$ matrix A_p with complex entries and eigenvalues $\alpha_j(p)$ with $|\alpha_j(p)| = 1$ for all but finitely many p, and $|\alpha_j(p)| < 1$ otherwise, such that

$$L(s) = \prod_p \prod_{j=1}^{m} \left(1 - \frac{\alpha_j(p)}{p^s}\right)^{-1}.$$

- *Finite order.* There exist positive constants c and δ for which

$$L(s) \ll \exp(\exp(c|t|))$$

uniformly in $-\delta \leq \sigma \leq 1 + \delta$.
- *Functional equation.* There is a sequence of matrices A_p^*, all satisfying the axiom on the polynomial Euler product, such that for $L^*(s)$, defined according to the axiom on the polynomial Euler product and satisfying all previous axioms,

$$\Lambda_L(s) = \omega \Lambda_{L^*}(1 - s),$$

where

$$\Lambda_L(s) := L(s)Q^s \prod_{j=1}^{f} \Gamma(\lambda_j s + \mu_j),$$

and where Q, λ_j are real numbers and μ_j and ω with $|\omega| = 1$ are complex numbers.

For $L \in \mathcal{P}$ define the quantities d_L and μ, as for the elements of the Selberg class $\mathcal{S}$, by

$$d_L = 2 \sum_{j=1}^{f} \lambda_j \quad \text{and} \quad \mu = 2 \sum_{j=1}^{f} (1 - 2\mu_j).$$

Perelli's class of L-functions is rather similar to $\tilde{\mathcal{S}}$ but it does not contain $\tilde{\mathcal{S}}$ since, for example, $\zeta(s)^2 \notin \mathcal{P}$ by the axiom on the analytic continuation (there is only a simple pole at $s = 1$ allowed). All axioms for $\mathcal{P}$, apart from the one on the analytic continuation, are covered by the related axioms for $\tilde{\mathcal{S}}$. This follows more or less directly from the axioms in addition with the Phragmén–Lindelöf principle. The growth restriction in $\mathcal{P}$ follows via Theorem 6.8 from the estimate

$$\mathcal{L}(s) \ll t^{\mu_{\mathcal{L}}(\sigma)+\epsilon} \ll t^{d_{\mathcal{L}}(\frac{1}{2}+\delta)+\epsilon},$$

which is valid for $-\delta \leq \sigma \leq 1 + \delta$ with any positive δ. We do not know any example of an element of $\mathcal{P}$ which does not lie in $\tilde{\mathcal{S}}$.

Besides other interesting results (which we partially mentioned in Sect. 6.7), Perelli proved zero-free regions subject to a natural condition on the matrices A_p, a mean-square estimate, and an approximate functional equation. Moreover, using Montgomery's zero detection method [258], he obtained a density theorem comparable with Theorem 1.2, respectively, (6.5). In many instances this yields, for any $\sigma > \max\{\frac{1}{2}, 1 - \frac{1}{d_{\mathcal{L}}}\}$,

$$N_{\mathcal{L}}(\sigma, T) = \mathrm{O}(T).$$

Actually, Perelli proved stronger estimates for the whole right half of the critical strip which are even uniform in Q, but this is not of interest for our investigations.

8.4 An Approach Towards Riemann's Hypothesis?

Bagchi's theorem offers an interesting but up to now insufficient approach toward the Riemann hypothesis.

In view of Voronin's proof we may have the naive idea to start with a truncated Euler product

$$\zeta_M(s) = \prod_{p \leq M} \left(1 - \frac{1}{p^s}\right)^{-1},$$

which converges throughout the half-plane $\sigma > 0$ (clearly, we are not allowed to work with the logarithm of the zeta-function since we are interested in its zeros). Obviously,

$$|\zeta(s+\mathrm{i}\tau) - \zeta(s)| \leq |\zeta(s+\mathrm{i}\tau) - \zeta_M(s)| + |\zeta_M(s) - \zeta(s)|.$$

Since $\zeta_M(s)$ is a non-vanishing analytic function in the half-plane $\sigma > 0$, by Voronin's universality theorem, the first quantity on the right-hand side can be made as small as we please for all s in some disk $|s - z| \leq r$ for a set $\mathcal{T}$ of values τ with positive lower density. For a proof of Riemann's hypothesis it would be sufficient to show that also the second quantity on the right-hand side can be made arbitrarily small on a subset of $\mathcal{T}$ with positive lower density. The second term is independent of τ but unfortunately, we cannot simply approximate $\zeta(s)$ by the truncated Euler product $\zeta_M(s)$. It is not difficult to show that

$$\lim_{M\to\infty}\lim_{T\to\infty} \frac{1}{T}\int_0^T |\zeta_M(\sigma+\mathrm{i}t) - \zeta(\sigma+\mathrm{i}t)|^2 \, \mathrm{d}t = 0$$

for any $\sigma > \frac{1}{2}$ (see [353, Sect. 7.11]; see also Lemma 4.8). This implies that $\zeta_M(s)$ approximates $\zeta(s)$ almost everywhere: given $\epsilon > 0$, there exist M_0, T_0 such that for any $M \geq M_0$, $T \geq T_0$ we have

$$\frac{1}{T}\int_0^T |\zeta_M(\sigma+\mathrm{i}t) - \zeta(\sigma+\mathrm{i}t)|^2 \, \mathrm{d}t < \epsilon.$$

Hence the set

$$\{t \in [0,T] : |\zeta_M(\sigma+\mathrm{i}t) - \zeta(\sigma+\mathrm{i}t)| \geq \epsilon\}$$

has Lebesgue measure $o_\epsilon(T)$ and so $\zeta_M(\sigma+\mathrm{i}t) - \zeta(\sigma+\mathrm{i}t)$ is small on average. However, this is not sufficient for $|\zeta_M(s) - \zeta(s)|$ being small in general.

8.5 Further Equivalents of the Riemann Hypothesis

Mishou and Nagoshi [253] obtained necessary and sufficient conditions for the truth of the Riemann hypothesis in terms of the functional distribution of quadratic L-functions $L(s, \chi_d)$ in the right half of the critical strip; here $L(s, \chi_d)$ denotes the Dirichlet L-functions to the character χ_d mod $|d|$ given by the Kronecker symbol $(\frac{d}{\cdot})$ to a fundamental discriminant d (see Sect. 1.4). They proved that the Riemann hypothesis for $\zeta(s)$ is true if and only if, for any compact subset K of the open strip $\frac{1}{2} < \sigma < 1$ and any positive ϵ,

$$\liminf_{X\to\infty} \frac{\#\{1 \leq d \leq X : \max_{s\in K} |L(s,\chi_d) + \zeta(s)| < \epsilon\}}{\#\{1 \leq d \leq X\}} > 0;$$

an analogous result is stated and proved with respect to prime discriminants d. Their proof makes use of the universality theorems from Mishou and Nagoshi's paper [251, 252] (see (1.31)) and a zero density estimate for Dirichlet L-functions of Jutila in combination with Rouché's theorem.

Another equivalent condition for the Riemann hypothesis was given by Šleževičienė–Steuding [334]. Remarkably, the Riemann zeta-function does not appear in its formulation. In order to state this equivalent we first recall Beurling's generalized Euler products. In 1937, Beurling [20] investigated prime number theorems for generalized multiplicative structures. For this aim he introduced the so-called Beurling primes. We assume that we are given a set $\mathcal{P}$ of positive real numbers p_n which can be arranged as

$$\mathsf{p}_1 < \mathsf{p}_2 < \cdots < \mathsf{p}_n < \mathsf{p}_{n+1} < \cdots$$

and satisfy $\lim_{n \to \infty} \mathsf{p}_n = \infty$. Then the set $\mathcal{G}$ consisting of all possible products of powers of elements p_n including the positive integer 1 (the empty product) forms a semigroup. The attached Euler product is defined by

$$\zeta_{\mathcal{P}}(s) = \prod_{\mathsf{p}_n \in \mathcal{P}} \left(1 - \frac{1}{\mathsf{p}_n^s} \right)^{-1}.$$

Assuming a certain distribution of the elements of $\mathcal{G}$, Beurling [20] proved that

$$\pi_{\mathcal{P}}(x) := \sum_{\mathcal{P} \ni \mathsf{p}_n \leq x} 1 \sim \frac{x}{\log x};$$

later Nyman improved this result by giving an error term. It follows from Reich's universality theorem [305] that Beurling zeta-functions $\zeta_{\mathcal{P}}(s)$ are universal. In [110], Grosswald and Schnitzer suggested an interesting approach toward the Riemann hypothesis. They considered a prime system $\mathcal{Q} = \{q_n\}$ formed with arbitrary but fixed real numbers q_n satisfying

$$p_n \leq q_n \leq p_{n+1},$$

where p_n denotes the nth rational prime number. Then the associated Beurling zeta-function $\zeta_{\mathcal{Q}}(s)$ has an analytic continuation to the half-plane $\sigma > 0$ (but in general the line $\sigma = 0$ is a border over which we cannot expect analytic continuation) except for a simple pole at $s = 1$ and, most remarkably, for $\sigma > 0$ it has the same zeros with the same multiplicity as the Riemann zeta-function $\zeta(s)$. Müller [264] showed that the same statement holds if the condition of Grosswald and Schnitzer is replaced by the weaker assumption

$$\pi_{\mathcal{Q}}(x) = \pi(x) + \mathrm{O}(x^\epsilon)$$

for any $\epsilon > 0$. Combining Reich's universality theorem, the zero-distribution of the latter Beurling zeta-functions, and the ideas behind Bagchi's Theorem 8.3, Šleževičienė–Steuding [334] showed that the Riemann hypothesis for $\zeta(s)$ is true if and only if $\zeta_\mathcal{Q}(s)$ can approximate itself uniformly, i.e., for certain disks $\mathcal{K}$ and any $\epsilon > 0$,

$$\liminf_{T \to \infty} \frac{1}{T} \text{ meas} \left\{ \tau \in [0, T] : \max_{s \in \mathcal{K}} |\zeta_\mathcal{Q}(s + i\tau) - \zeta_\mathcal{Q}(s)| < \epsilon \right\} > 0;$$

here $\zeta_\mathcal{Q}(s)$ is assumed to satisfy certain natural side-conditions. Furthermore, in [334] with

$$q_n = p_n \exp\left(\frac{1}{p_n}\right),$$

an example of such a Beurling zeta-function $\zeta_\mathcal{Q}(s)$ was given.

Effective Results

Kronecker's theorem is one of those mathematical theorems which assert, roughly, that what is not impossible will happen sometimes however improbable it may be.   G.F. Hardy and E.M. Wright

In this chapter, we shall use ideas from the previous chapter in order to obtain certain effective results on the value-distribution of L-functions. The first sections deal with the density of the approximating τ in universality theorems. The derived upper bounds are due to Steuding [342, 350]. In the following sections explicit estimates for c-values in the half-plane of absolute convergence are obtained. These results are due to Girondo and Steuding [100] and rely on a theorem of Rieger [310], resp. a quantified version of Steuding [344], on effective inhomogeneous diophantine approximation.

9.1 The Problem of Effectivity

The known proofs of universality theorems are ineffective, giving neither an estimate for the first approximating shift τ nor bounds for the positive lower density with the exception of some attempts by Good, Laurinčikas, and Garunkštis which we shall now discuss shortly.

If the Riemann hypothesis is true, then

$$\log\left|\zeta\left(\frac{1}{2}+\mathrm{i}t\right)\right| = \mathrm{O}\left(\frac{\log t}{\log\log t}\right) \tag{9.1}$$

as $t \to \infty$ (see [353]). This is a significant improvement of the bound for $\zeta(s)$ on the critical line predicted by the Lindelöf hypothesis, but we may ask whether it is the correct order? On the contrary, Montgomery [260] proved, for fixed $\frac{1}{2} < \sigma < 1$ and any real θ,

$$\mathrm{Re}\left\{\exp(\mathrm{i}\theta)\log\zeta(\sigma+\mathrm{i}t)\right\} = \Omega\left(\frac{(\log t)^{1-\sigma}}{(\log\log t)^{\sigma}}\right), \tag{9.2}$$

and the same estimate is valid for $\sigma = \frac{1}{2}$ under assumption of the truth of the Riemann hypothesis. By a different method, Balasubramanian and Ramachandra [15] obtained the same estimate for $\sigma = \frac{1}{2}$ unconditionally. These Ω-results were only slight improvements of earlier results and some probabilistic heuristics suggest these estimates to be best possible, i.e., the quantity in (9.2) describes the exact order of growth of $\zeta(s)$. However, the random matrix model predicts significantly larger values: in analogy to large deviations for characteristic polynomials one may expect that the estimate in (9.1) gives the true order (see Hughes [136] for details). Steuding [338] showed that Montgomery's estimate holds on any rectifiable curve inside the right half of the critical strip as well. In particular, assuming the truth of Riemann's hypothesis, for any rectifiable curve $t \mapsto \eta(t) + \mathrm{i}t, t \in \mathbb{R}$, in the right half of the critical strip,

$$\zeta(\eta(t) + \mathrm{i}t) = \Omega \left(\exp\left(\frac{1}{20} \frac{(\log t)^{1-\eta(t)}}{(\log\log t)^{\eta(t)}} \right) \right).$$

Thus, the analytic landscape of the Riemann zeta-function over the critical strip does not contain *long* valleys.

The proofs of Voronin's theorem do not give any information about the question how soon a given target function is approximated by $\zeta(s + \mathrm{i}\tau)$ within a given range of accuracy, and Montgomery's approach does not give us any idea of the shape of the set of values of $\zeta(s)$ on vertical lines. Good [105] combined Voronin's universality theorem with the work of Montgomery on extreme values of the zeta-function. This enabled him to complement Voronin's qualitative picture with Montgomery's quantitative estimates. For sufficiently large T and ν with $\nu \leq \log T$ and r sufficiently small (but *not too small*), he showed the existence of a certain constant c and more than $T \exp(-\frac{c\nu}{\log \nu})$ positive numbers $t_n \leq T$ with $t_{n+1} > t_n + 2r$ for $n = 1, 2, \ldots$, such that the annulus

$$\left\{ z \in \mathbb{C} : \exp\left(-\frac{c\nu^{1-\sigma+r}}{\log \nu} \right) \leq |z| \leq \exp\left(\frac{c\nu^{1-\sigma+r}}{\log \nu} \right) \right\}$$

is contained in $\zeta(\sigma + \mathrm{i}t_n + B_r)$, where $B_r := \{z : |z| \leq r\}$. This extends Bohr's classic result on the denseness of the values $\zeta(s)$ taken on vertical lines (see Sect. 1.2). The other results of Good are too complicated and too lengthy to be given here. He proved some kind of quantitative version of Voronin's universality Theorem 1.7 but his dissection of the target function leads to certain function spaces which are not very well described yet. Recently, Garunkštis [86] proved another, more satisfying effective universality theorem along the lines of Voronin's proof in addition with some of Good's ideas. In particular, his remarkable result shows that if $f(s)$ is analytic in $|s| \leq 0.05$ with $\max_{|s|\leq0.05} |f(s)| \leq 1$, then for any $0 < \epsilon < \frac{1}{2}$ there exists a

$$0 \leq \tau \leq \exp\left(\exp\left(10\epsilon^{-13} \right) \right) \tag{9.3}$$

such that

$$\max_{|s|\leq 0.0001} \left| \log \zeta \left(s + \frac{3}{4} + i\tau \right) - f(s) \right| < \epsilon,$$

and further

$$\liminf_{T\to\infty} \frac{1}{T} \text{ meas } \left\{ \tau \in [0,T] : \max_{|s|\leq 0.0001} \left| \log \zeta \left(s + \frac{3}{4} + i\tau \right) - f(s) \right| < \epsilon \right\}$$
$$\geq \exp\left(-\epsilon^{-13} \right). \tag{9.4}$$

The original theorem is too complicated to be given here. Laurinčikas [193] (see also [89]) found another approach which gives conditional effective results subject to certain assumptions on the speed of convergence of the related limit distribution. However, the rate of convergence of weakly convergent probability measures related to the space of analytic functions is not understood very well.

All these attempts to quantify universality are remarkable but the given quantitative universality results obtained so far have led to more open problems than they actually solved.

9.2 Upper Bounds for the Density of Universality

We shall prove effective upper bounds for the upper density of universality.

Denote by B_r the closed disc of radius $r > 0$ with center in the origin. For a meromorphic function $L(s)$, an analytic function $f : B_r \to \mathbb{C}$ with fixed $r \in \left(0, \frac{1}{4}\right)$, and a positive ϵ, we define the densities

$$\underline{d}(\epsilon, f, L) = \liminf_{T\to\infty} \frac{1}{T} \text{ meas } \left\{ \tau \in [0,T] : \max_{|s|\leq r} \left| L \left(s + \frac{3}{4} + i\tau \right) - f(s) \right| < \epsilon \right\},$$

and

$$\overline{d}(\epsilon, f, L) = \limsup_{T\to\infty} \frac{1}{T} \text{ meas } \left\{ \tau \in [0,T] : \max_{|s|\leq r} \left| L \left(s + \frac{3}{4} + i\tau \right) - f(s) \right| < \epsilon \right\}.$$

For continuity sets we do not have to distinguish between the lower and the upper density of universality, since in this case, by the Portmanteau Theorem 3.1, the limit exists.

For sufficiently large classes of functions $L(s)$ and of functions $f(s)$ we shall prove effective upper bounds for the upper density $\overline{d}(\epsilon, f, L)$ which tend to zero as $\epsilon \to 0$. For this purpose we consider analytic isomorphisms $f : B_r \to B_1$, i.e., the inverse f^{-1} exists and is analytic. Obviously, such a function f has exactly one simple zero ξ in the interior of B_r. By the Schwarz lemma (from the classical theory of functions, see [352, Sect. 5.2]) it turns out that such a function f has the representation

$$f(s) = r \exp(\mathrm{i}\phi)\frac{\xi - s}{r^2 - \overline{\xi}s} \quad \text{with} \quad \phi \in \mathbb{R} \quad \text{and} \quad |\xi| < r. \tag{9.5}$$

Denote by $\mathcal{A}_r$ the class of analytic isomorphisms from the closed disk B_r (with fixed $0 < r < \frac{1}{4}$) to the unit disk. Further, let $N_L(\sigma_1, \sigma_2, T)$ count the number of zeros of $L(s)$ in $\frac{1}{2} < \sigma_1 < \sigma < \sigma_2 < 1$, $0 \le t < T$ (counting multiplicities). The main result of this section is

Theorem 9.1. *Let $f \in \mathcal{A}_r$. Assume that $L(s)$ is analytic in $\sigma \ge \frac{3}{4} - r$ except for at most $\mathrm{O}(T)$ many singularities inside $\sigma \ge \frac{3}{4} - r$, $0 \le t \le T$, as $T \to \infty$, and that $\underline{\mathrm{d}}(\epsilon, f, L) > 0$ for all $\epsilon > 0$. Then, for any $\epsilon \in \left(0, \frac{1}{2r}\left(\frac{1}{4} + \operatorname{Re}|\xi|\right)\right)$,*

$$\overline{\mathrm{d}}(\epsilon, f, L) \le \frac{8r^3 \epsilon}{r^2 - |\xi|^2} \tag{9.6}$$

$$\times \limsup_{T \to \infty} \frac{1}{T} N_L\left(\frac{3}{4} + \operatorname{Re}\xi - 2r\epsilon, \frac{3}{4} + \operatorname{Re}\xi + 2r\epsilon, T\right).$$

This theorem relates the density of universality to the value-distribution of L. Note that one can obtain similar estimates for other c-values instead of $c = 0$ whenever c lies in the interior of B_r. Since *too many well distributed* zeros of $L(s)$ in $\frac{1}{2} < \sigma < 1$ *violate* the universality property, very likely the limit in (9.6) exists in general.

Proof. The idea of proof is that the zero ξ of f is related to some zeros of $L(s)$ in $\frac{1}{2} < \sigma < 1$. Since f maps the boundary of B_r onto the unit circle, Rouché's Theorem 8.1 implies the existence of one simple zero λ of $L(z)$ in

$$K_\tau := \left\{z = s + \frac{3}{4} + \mathrm{i}\tau : s \in B_r\right\}$$

whenever

$$\max_{s \in B_r}\left|L\left(s + \frac{3}{4} + \mathrm{i}\tau\right) - f(s)\right| < \epsilon < 1 = \min_{|s|=r}|f(s)|. \tag{9.7}$$

Recall that, in the language of Sect. 8.2, the zero λ of $L(s)$ is generated by the zero ξ of $f(s)$.

Universality is a phenomenon that happens in intervals. We prove an upper bound for the distance of different shifts generating the same zero λ of $L(s)$:

Lemma 9.2. *Suppose that a zero λ of $L(s)$, generated by ξ, lies in two different sets K_{τ_1} and K_{τ_2}. Then*

$$|\tau_1 - \tau_2| < \frac{8r^3 \epsilon}{r^2 - |\xi|^2}.$$

Proof. Suppose that there exist complex numbers

$$s_j = \operatorname{Re}\lambda - \frac{3}{4} + \mathrm{i}t_j \in B_r,$$

and $\tau_j \in \mathbb{R}$ for which

$$L\left(s_j + \frac{3}{4} + \mathrm{i}\tau_j\right) = 0 \qquad \text{for} \quad j = 1, 2,$$

such that

$$\lambda = s_1 + \frac{3}{4} + \mathrm{i}\tau_1 = s_2 + \frac{3}{4} + \mathrm{i}\tau_2.$$

In view of (9.5),

$$|f(s_2) - f(s_1)| = r\frac{r^2 - |\xi|^2}{|r^2 - \overline{\xi}s_1||r^2 - \overline{\xi}s_2|}|s_2 - s_1|.$$

We deduce from (9.7) that $|f(s_j)| < \epsilon$ for $j = 1, 2$, and therefore

$$|\tau_1 - \tau_2| = |t_2 - t_1| \le \frac{4r^3}{r^2 - |\xi|^2}|f(s_2) - f(s_1)| < \frac{8r^3\epsilon}{r^2 - |\xi|^2},$$

which proves the lemma.   □

We continue with the proof of the theorem. Denote by $\mathcal{I}_j(T)$ the disjoint intervals in $[0, T]$ such that (9.7) is valid exactly for

$$\tau \in \bigcup_j \mathcal{I}_j(T) =: \mathcal{I}(T).$$

Using Lemma 9.2, in every interval $\mathcal{I}_j(T)$, there lie at least

$$1 + \left[\frac{r^2 - |\xi|^2}{8r^3\epsilon}\operatorname{meas}\mathcal{I}_j(T)\right] \ge \frac{r^2 - |\xi|^2}{8r^3\epsilon}\operatorname{meas}\mathcal{I}_j(T)$$

zeros λ of $L(s)$ in the strip $\frac{1}{2} < \sigma < 1$. Therefore, the number $\mathcal{N}(T)$ of such zeros λ satisfies the estimate

$$\frac{8r^3\epsilon}{r^2 - |\xi|^2}\mathcal{N}(T) \ge \operatorname{meas}\mathcal{I}(T). \tag{9.8}$$

The next step is to replace $\mathcal{N}(T)$ by the zero counting function appearing in the theorem.

The value distribution of $L(z)$ in K_τ is ruled by that of $f(s)$ in B_r. This gives a restriction on the real parts of the zeros λ.

Lemma 9.3. *Let λ be a zero of $L(s)$ generated by ξ. Then*

$$\left|\operatorname{Re}\lambda - \frac{3}{4} - \operatorname{Re}\xi\right| < 2r\epsilon.$$

Proof. Let $s \in B_r$. First of all, we consider the case $|f(s)| \geq \epsilon$. Then, in view of (9.7),

$$\left| L\left(s + \frac{3}{4} + i\tau\right) \right| \geq |f(s)| - \left| f(s) - L\left(s + \frac{3}{4} + i\tau\right) \right| > 0.$$

Since (9.5) implies

$$|f(s)| \geq \frac{|\xi - s|}{2r}, \tag{9.9}$$

we obtain for $s = \lambda - \frac{3}{4} - i\tau$ by (9.7) that

$$\left| \xi - \left(\lambda - \frac{3}{4} - i\tau\right) \right| \leq 2r \left| f\left(\lambda - \frac{3}{4} - i\tau\right) \right|$$

$$\leq 2r|L(\lambda)| + \max_{s \in B_r} \left| f(s) - L\left(s + \frac{3}{4} + i\tau\right) \right|$$

$$< 2r\epsilon,$$

which yields the estimate of the lemma by taking the real parts.

If $|f(s)| < \epsilon$, then we may deduce the desired estimate directly from (9.9). The lemma is proved.   □

Now we are in the position to finish the proof of the theorem. In view of Lemma 9.3 we find

$$\mathcal{N}(T) \leq N_L\left(\frac{3}{4} + \operatorname{Re}\xi - 2r\epsilon, \frac{3}{4} + \operatorname{Re}\xi + 2r\epsilon, T\right), \tag{9.10}$$

where N_L is the zero-counting function of L. On the other side, since $\underline{d}(\epsilon, f, L) > 0$, for any $\delta > 0$, there exists an increasing sequence $\{T_k\}$ with $\lim_{k \to \infty} T_k = \infty$ such that

$$\operatorname{meas}\left(\mathcal{I}(T_k)\right) \geq (\overline{d}(\epsilon, f, L) - \delta)T_k.$$

Consequently, this together with (9.10) leads in (9.8) to

$$\frac{8r^3\epsilon}{r^2 - |\xi|^2} N_L\left(\frac{3}{4} + \operatorname{Re}\xi - 2r\epsilon, \frac{3}{4} + \operatorname{Re}\xi + 2r\epsilon, T_k\right) \geq (\overline{d}(\epsilon, f, L) - \delta)T_k.$$

Sending $\delta \to 0$ yields the estimate (9.6) of the theorem. Since the set of singularities of $L(s)$ in $\sigma \geq \frac{3}{4} - r$ has zero density but $\underline{d}(\epsilon, f, L) > 0$, the singularities do not affect the above reasoning. The theorem is proved.   □

In the case of the Riemann zeta-function we can get a slightly stronger result. For this purpose we apply Theorem 9.1 to $L(s) = \log \zeta(s)$. In view of the density Theorem 1.2 the set of singularities of $\log \zeta(s)$ has density zero. As shown by Bohr and Jessen [32] (Hilfssatz 6 to Theorem 1.5), the limit

$$\lim_{T \to \infty} \frac{1}{T} N_{\log \zeta} \left(\frac{3}{4} + \operatorname{Re} \xi - 2r\epsilon, \frac{3}{4} + \operatorname{Re} \xi + 2r\epsilon, T \right) \qquad (9.11)$$

exists and tends to zero as $\epsilon \to 0$. Now we proceed as above and obtain, for $f \in \mathcal{A}_r$,

$$\overline{d}(\epsilon, \exp f, \zeta(s)) = O(\epsilon).$$

Steuding [350] extended the above argument in order to obtain upper bounds for the density of universality with respect to approximation of a rather general class of functions $g(s) = \exp f(s)$; however, the implicit constants cannot be given explicitly. Here we sketch the new idea in the case of the zeta-function.

Assume that $g(s)$ is a non-constant, non-vanishing analytic function defined on B_r. Then there exists a complex number c in the interior of $g(B_r)$ (which is not empty since $g(s)$ is not constant) such that

$$g(s) = c + \gamma(s - \lambda_c) + O\left(|s - \lambda_c|^2 \right) \qquad (9.12)$$

for some λ_c of modulus less than r and some $\gamma \neq 0$; this means that λ_c is a c-value of $g(s)$ of multiplicity one. To see this suppose that for all c in the interior of $g(B_r)$ the local expansion is different than (9.12), i.e., $g'(s)$ vanishes identically in the interior. Then g is a constant function, a contradiction to the assumption of the theorem.

Now suppose that

$$\max_{|s|=r} \left| \left\{ \zeta \left(s + \frac{3}{4} + i\tau \right) - c \right\} - \{ g(s) - c \} \right| < \min_{|s|=r} |g(s) - c|.$$

Then, by Rouché's theorem, $\zeta(z)$ has at least one c-value ϱ_c in $\{ z = s + \frac{3}{4} + i\tau : |s| < r \}$. We rewrite the latter inequality as

$$\max_{|s| \leq r} \left| \zeta \left(s + \frac{3}{4} + i\tau \right) - g(s) \right| < \epsilon \leq \min_{|s|=r} |g(s) - c|. \qquad (9.13)$$

Since the zeta-function is universal, the first inequality holds for a set of τ with positive lower density. The second one follows for sufficiently small ϵ from the fact that $c = g(\lambda_c)$ has positive distance to the boundary of $g(B_r)$. Thus, a c-value of $g(s)$ generates many c-values of $\zeta(z)$.

Assume that $\varrho_c = s_j + \frac{3}{4} + i\tau_j$ with $|s_j| < r$ for $j = 1, 2$. It follows from (9.13) that

$$|g(\lambda_c) - g(s_j)| = |c - g(s_j)| < \epsilon. \qquad (9.14)$$

Since $g'(\lambda_c) = \gamma \neq 0$, there exists a neighborhood of c where the inverse function g^{-1} exists and is a one-valued continuous function. By continuity, (9.14) implies

$$|s_j - \lambda_c| < \varepsilon = \varepsilon(\epsilon), \qquad (9.15)$$

where $\varepsilon(\epsilon)$ tends with ϵ to zero; since $g(s)$ behaves locally as a linear function by (9.12), we have $\varepsilon(\epsilon) \asymp \epsilon$. Now (9.15) implies

$$|\tau_2 - \tau_1| = |s_1 - s_2| \le |s_1 - \lambda_c| + |s_2 - \lambda_c| < 2\varepsilon. \tag{9.16}$$

Now we can proceed as in the proof of Theorem 9.1 in addition with (9.11). This yields:

Theorem 9.4. *Suppose that $g(s)$ is a non-constant, non-vanishing analytic function defined on $|s| \le r$, where $r \in (0, \frac{1}{4})$. Then, for any sufficiently small $\epsilon > 0$,*

$$\overline{\mathrm{d}}(\epsilon, r, g, \zeta) = \mathrm{O}(\epsilon).$$

Thus, the decay of $\overline{\mathrm{d}}(\epsilon, g, \zeta)$ with $\epsilon \to 0$ is more than linear in ϵ for any non-constant, non-vanishing analytic function g.

9.3 Value-Distribution on Arithmetic Progressions

Now we consider the special case of *discrete* universality. The argument in the proof of Theorem 9.4 which gave us a factor ϵ for the upper bound does not apply if we consider discrete shifts and so, in general, we do not get an upper bound which tends with ϵ to zero (as in Lemma 9.2 or (9.16)). Anyway, for the zeta-function we obtain via Reich's discrete universality Theorem 5.16 and (9.11)

$$\limsup_{N \to \infty} \frac{1}{N} \sharp \left\{ 1 \le n \le N : \max_{|s| \le r} \left| \zeta \left(s + \frac{3}{4} + in\Delta \right) - g(s) \right| < \epsilon \right\} = \mathrm{O}(1) \tag{9.17}$$

for any real $\Delta \ne 0$, as $\epsilon \to 0$. This is of interest with respect to an estimate of Reich concerning small values of Dirichlet series on arithmetic progressions. Let $F(s)$ be a Dirichlet series, not identically zero, which has a half-plane of absolute convergence $\sigma > \sigma_a$, an analytic continuation to $\sigma > \sigma_m$ ($\sigma_m < \sigma_a$) except for at most a finite number of poles on the line $\sigma = \sigma_a$, such that its mean square exists and $F(s)$ is of finite order of growth in any closed strip in $\sigma_m < \sigma < \sigma_a$. In [307], Reich proved under these assumptions (which are rather similar to our axioms (ii) and (iii)) for any $\sigma > \sigma_m, \sigma \ne \sigma_a$, any sufficiently small $\epsilon > 0$, and any real Δ, neither being equal to zero nor of the form $2\pi\ell \cos(\frac{q}{r})$ with positive integers ℓ, q, r and $q \ne r$, that the relation

$$\limsup_{N \to \infty} \frac{1}{N} \sharp \{ 1 \le n \le N : |F(\sigma + in\Delta)| < \epsilon \} < 1$$

holds. In particular, it follows that $F(\sigma + i\Delta n)$ cannot converge to zero as $n \to \infty$, and hence $s_n = \sigma + i\Delta n$ cannot be a sequence of zeros of $F(s)$. It should be noticed that Reich's theorem also includes estimates for c-values on arithmetic progressions (since with $F(s)$ also $F(s) - c$ satisfies the conditions).

In the special case of the Riemann zeta-function we note the following improvement of Reich's theorem:

Theorem 9.5. *Let c be any constant and $\sigma \in (\frac{1}{2}, 1)$, and $\Delta \neq 0$ be real. Then*

$$\lim_{\epsilon \to 0} \limsup_{N \to \infty} \frac{1}{N} \sharp \{1 \leq n \leq N : |\zeta(\sigma + in\Delta) - c| < \epsilon\} = 0.$$

In particular, there does not exist an arithmetic progression $s_n = \sigma + i\Delta n$ (with σ and Δ as in the theorem) on which $\zeta(s)$ converges to any complex number c.

We sketch the easy proof. Let $g(s)$ be a non-constant, non-vanishing, analytic function defined on a small disk centered at $\sigma \in (\frac{1}{2}, 1)$ such that its closure lies inside the strip of universality for the zeta-function. Further assume that

$$|g(s) - c| < \epsilon;$$

this choice for $g(s)$ is certainly possible for any given complex number c. By the triangle inequality,

$$|\zeta(\sigma + in\Delta) - c| \leq |\zeta(\sigma + in\Delta) - g(s)| + |g(s) - c|$$

for any s. Hence, applying (9.17) yields

$$\limsup_{N \to \infty} \frac{1}{N} \sharp \{1 \leq n \leq N : |\zeta(\sigma + in\Delta) - c| < 2\epsilon\} \ll \epsilon.$$

This is the assertion of the theorem.

An alternative proof can be given by using the deep *Hauptsatz I* of Bohr and Jessen [32]; this approach does not depend on the universality property of $\zeta(s)$. As a matter of fact, this theorem may also be used to obtain the estimate $\overline{d}(\epsilon, r, g, \zeta) \ll \epsilon^2$ for constant functions $g \not\equiv 0$.

There are remarkable results for a related problem. Putnam [297, 298] showed that $\zeta(s)$ does not have an infinite vertical arithmetic progressions of zeros (or even approximate zeros). Lapidus and van Frankenhuijsen [179, Chap. 9] gave a different proof of Putnam's theorem. Watkins (cf. [80]) was the first to give upper bounds for the length of such arithmetic progressions (valid for any Dirichlet L-functions). Recently, van Frankenhuijsen [80] improved these bounds by showing that

$$\zeta(\sigma + in\Delta) = 0 \quad \text{for} \quad 0 < |n| < N$$

with $\sigma, \Delta > 0$ and $N \geq 2$ cannot hold for

$$N \geq 60 \left(\frac{\Delta}{2\pi}\right)^{(1/\sigma)-1} \log \Delta$$

(his method also applies to Dirichlet L-functions). It is conjectured that there are no arithmetic progressions at all; there are even no zeros known of the form $\frac{1}{2} + i\gamma$ and $\frac{1}{2} + i2\gamma$. It is conjectured that the ordinates of the nontrivial zeros of $\zeta(s)$ are linearly independent over $\mathbb{Q}$. Ingham [140] showed that this conjecture implies large values for the sum of the values of the Möbius μ-function:

$$\liminf_{x\to\infty} \frac{M(x)}{x^{1/2}} = -\infty \quad \text{and} \quad \limsup_{x\to\infty} \frac{M(x)}{x^{1/2}} = +\infty,$$

which would improve the unconditional bounds (3.1). The methods of Putnam, Lapidus and van Frankenhuijsen do not apply to c-values.

9.4 Making Universality Visible

We return to the problem of effectivity in the universality theorem for $\zeta(s)$. Comparing the lower bound (9.4) of Garunkštis [86] from Sect. 9.1 with the upper bounds of Steuding [342, 350] from Theorem 9.4, it makes sense to ask which estimate is more close to the truth. If a given function $g(s)$ is sufficiently *nice*, i.e., if its logarithm $f(s)$ satisfies the condition of Garunkštis' theorem, then

$$\exp\left(-\epsilon^{-13}\right) \ll \underline{d}(\epsilon, g, \zeta) \leq \overline{d}(\epsilon, g, \zeta) = O(\epsilon).$$

Given a positive ϵ and a sufficiently small disk $\mathcal{K}$ located in the right half of the critical strip, in principle, the estimate (9.3) allows to find algorithmically an approximating τ such that

$$\max_{s\in\mathcal{K}} |\mathcal{L}(s+i\tau) - g(s)| < \epsilon;$$

however, we cannot expect a reasonable running time for such an algorithm when ϵ is small. This idea was indeed considered in a project by Garunkštis, Šleževičienė–Steuding and Steuding. For certain *smooth* functions $g(s)$ and rather *large* values for ϵ approximating shifts τ were computed. Quite many of these τ were found. However, it is impossible to deduce any information about the density of universality as long as the running time of the underlying algorithm cannot be significantly improved. Nevertheless, we shall illustrate this attempt toward effective universality by some data.

Our first example (Fig. 9.1) is the exponential function on a small disk centered at the origin. For example, we have

$$\max_{|s|\leq 0.006} \left|\zeta\left(s + \frac{3}{4} + 12\,963\,i\right) - \exp(s)\right| < 0.05.$$

The shift τ is a positive integer since the discrete variant of universality was used in order to simplify the algorithm.

In view of the assumptions on the set $\mathcal{K}$ in the universality Theorem 1.9, we may choose $\mathcal{K}$ to be a line segment in the right half of the critical strip; in this case the interior of $\mathcal{K}$ is empty and thus the target function needs only to be continuous. Here is an example for such a function (Fig. 9.2):

$$g(s) = \frac{1}{2} + \left|s - \frac{3}{4}\right| \quad \text{for} \quad s \in I = \left[\frac{3}{4} - \frac{1}{10}, \frac{3}{4} + \frac{1}{10}\right].$$

It was computed that

$$\max_{s\in I} |\zeta(s + 411\,744i) - g(s)| < 0.05.$$

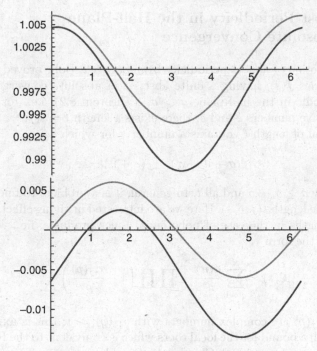

Fig. 9.1. $\zeta(s + \frac{3}{4} + 12\,963\,i) \approx \exp(s)$ for $s = 0.006\exp(\mathrm{i}\phi)$ with $0 \le \phi \le 2\pi$. On the left the real parts, on the right the imaginary parts are plotted; the zeta-function is given in black, exp in *grey*

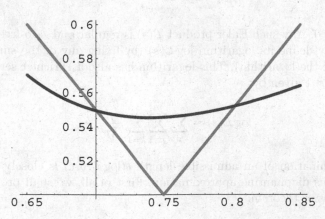

Fig. 9.2. $\zeta(s + \frac{3}{4} + \mathrm{i}\,411\,744) \approx g(s) := \frac{1}{2} + |s - \frac{3}{4}|$; the zeta-function in black, $g(s)$ in *grey*

9.5 Almost Periodicity in the Half-Plane of Absolute Convergence

Recall the notion of almost periodicity from Sect. 8.2. Bohr proved that every Dirichlet series $f(s)$, having a finite abscissa of absolute convergence σ_a, is almost periodic in the half-plane $\sigma > \sigma_a$ (Theorem 8.2); i.e., for any given pair of positive numbers ε and δ, there exists a length $\ell = \ell(f, \delta, \varepsilon)$ such that every interval of length ℓ contains a number τ for which

$$|f(\sigma + \mathrm{i}t + \mathrm{i}\tau) - f(\sigma + \mathrm{i}t)| < \varepsilon$$

holds for any $\sigma \geq \sigma_a + \delta$ and all t. In general, it is a problem to find explicitly an admissible length $\ell(f, \delta, \varepsilon)$. Here we are interested in giving effective bounds for these lengths in the case of polynomial Euler products; i.e., we consider functions of the form

$$\mathcal{L}(s) = \sum_{n=1}^{\infty} \frac{a(n)}{n^s} = \prod_p \prod_{j=1}^{m} \left(1 - \frac{\alpha_j(p)}{p^s}\right)^{-1}, \tag{9.18}$$

where the $\alpha_j(p)$ are complex numbers with $|\alpha_j(p)| \leq 1$; this is axiom (iv) in addition with a bound for the local roots which is equivalent to the Ramanujan hypothesis (i) by Lemma 2.2. Clearly, the Dirichlet series coefficients $a(n)$ are multiplicative and satisfy

$$a(n) = \prod_{p|n} \sum_{\substack{k_1,\ldots,k_m \geq 0 \\ k_1+\cdots+k_m=\nu_p(n)}} \alpha_j(p)^{k_j} \ll n^\epsilon$$

for any $\epsilon > 0$. Any such Euler product $\mathcal{L}(s)$ is regular and zero-free for $\sigma > 1$ and we may define its logarithm $\log \mathcal{L}(s)$ (by fixing any of the single-valued branches of the logarithm). This logarithm has also a Dirichlet series expansion, for $\sigma > 1$ given by

$$\log \mathcal{L}(s) = \sum_p \sum_{j=1}^{m} \sum_{k=1}^{\infty} \frac{\alpha_j(p^k)}{kp^{ks}}. \tag{9.19}$$

The determination of an admissible length $\ell(\log \mathcal{L}, \delta, \epsilon)$ is closely related to simultaneous diophantine approximation. First of all, we shall prove the following transfer theorem:

Theorem 9.6. *Let* $M(x, \epsilon)$ *be a positive function, defined for* $0 < \epsilon \leq \frac{1}{2}$ *and* $x \geq 2$, *such that every interval* $(t_1, t_2) \subset \mathbb{R}$ *of length* $M(x, \epsilon)$ *contains a number* τ *for which*

$$\left\|\tau \frac{\log p}{2\pi}\right\| < \epsilon \quad \text{for all} \quad p \leq x, \tag{9.20}$$

where $\|z\|$ denotes the minimal distance of a real number z to the nearest integer and p is prime. Suppose that $\mathcal{L}$ is given by (9.18) and let $\delta > 0$. Then, for sufficiently small ε,

$$\ell(\log \mathcal{L}, \delta, \varepsilon) = M\left(\left(\frac{\delta\varepsilon}{\sqrt{5}\pi m}\right)^{-1/\delta}, \frac{\delta}{\sqrt{5}\pi m}(\varepsilon + \mathrm{O}(1))\right),$$

$$\ell(\mathcal{L}, \delta, \varepsilon) = \ell\left(\log \mathcal{L}, \delta, \left(\frac{\delta}{1+\delta}\right)^m \varepsilon\right)$$

$$= M\left(\left(\frac{1+\delta}{\delta}\right)^{m/\delta}\left(\frac{\delta\varepsilon}{\sqrt{5}\pi m}\right)^{-1/\delta}, \left(\frac{\delta}{1+\delta}\right)^m \frac{\delta}{\sqrt{5}\pi m}(\varepsilon + \mathrm{O}(1))\right).$$

This holds uniformly in $\delta \geq \delta_0 > 0$.

We argue quite similar as Bohr [30] in his proof for almost periodicity of Dirichlet series in their half-plane of absolute convergence.

Proof. It follows from (9.19) that

$$\log \mathcal{L}(s) - \log \mathcal{L}(s + i\tau) = \sum_p \sum_{j=1}^m \sum_{k=1}^\infty \frac{\alpha_j(p)^k}{kp^{ks}}\left(1 - \frac{1}{p^{ik\tau}}\right).$$

Since the $\alpha_j(p)$ all have absolute value less than or equal to one, we obtain

$$|\log \mathcal{L}(s) - \log \mathcal{L}(s + i\tau)| \tag{9.21}$$

$$\leq m\left\{\sum_{p \leq x}\sum_{k \leq y} + \sum_{p \leq x}\sum_{k > y} + \sum_{p > x}\sum_{k=1}^\infty\right\}\frac{1}{kp^{k\sigma}}\left|1 - \frac{1}{p^{ik\tau}}\right|,$$

where x and y are parameters ≥ 2, which will be chosen later. First of all, we shall bound the first sum on the right-hand side for some values of τ.

In order to find real numbers τ such that $p^{-ik\tau}$ lies for $p \leq x$ and $k \leq y$ sufficiently close to 1 we apply (9.20). It follows that, for any $\epsilon \in (0, \frac{1}{2}]$ and any real number t_1, there exist integers x_p and a real number $\tau \in (t_1, t_1 + M(x, \epsilon))$ such that

$$|\tau \log p - 2\pi x_p| < 2\pi\epsilon \quad \text{for all} \quad p \leq x.$$

In particular,

$$\cos(k\tau \log p) > \cos(2\pi k\epsilon) \geq 1 - \frac{(2\pi k\epsilon)^2}{2},$$

and

$$\sin(k\tau \log p) < \sin(2\pi k\epsilon) \leq 2\pi k\epsilon$$

for all primes $p \leq x$, provided that $k \leq y < \frac{1}{4\epsilon}$. Then we obtain

$$|1 - p^{-ik\tau}|^2 = (1 - \cos(k\tau \log p))^2 + \sin^2(k\tau \log p) < \frac{5}{4}(2\pi k\epsilon)^2 \tag{9.22}$$

for $p \leq x$ and $k \leq y$. This yields

$$\sum_{p \leq x} \sum_{k \leq y} \frac{1}{kp^{k\sigma}} \left| 1 - \frac{1}{p^{i\tau}} \right| < \sqrt{5}\pi\epsilon \sum_{p \leq x} \sum_{k \leq y} \frac{1}{p^{k\sigma}}. \tag{9.23}$$

In order to estimate the double sum on the right, note that by the uniqueness of the prime factorization of the integers

$$\sum_{p \leq x} \sum_{k \leq y} \frac{1}{p^{k\sigma}} \leq \sum_{n \leq x^y} \frac{1}{n^{\sigma}}.$$

Since

$$\sum_{n=2}^{z} \frac{1}{n^{\sigma}} < \sum_{n=2}^{z} \int_{n-1}^{n} \frac{du}{u^{\sigma}} = \int_{1}^{z} \frac{du}{u^{\sigma}} < \frac{1}{\sigma - 1},$$

we may replace (9.23) by

$$\sum_{p \leq x} \sum_{k \leq y} \frac{1}{kp^{k\sigma}} \left| 1 - \frac{1}{p^{i\tau}} \right| < \frac{\sqrt{5}\pi\epsilon}{\sigma - 1}. \tag{9.24}$$

Next we have to estimate the second and the third term on the right-hand side of (9.21). For the sake of simplicity we shall not give the optimal and rather complicated bounds. Later, it will turn out that these rough estimates do not effect our result.

First of all,

$$\sum_{p \leq x} \sum_{k > y} \frac{1}{kp^{k\sigma}} \left| 1 - \frac{1}{p^{ik\tau}} \right| \leq \frac{2}{y} \sum_{p \leq x} \sum_{k > y} \frac{1}{p^{k\sigma}} \leq \frac{2}{y} \sum_{p \leq x} \sum_{k > y} \left(\frac{1}{2^{\sigma}} \right)^{k}.$$

It is easily seen that the sum over k is less than

$$\sum_{k > y} 2^{-k} \leq 2^{-y} \sum_{\ell=0}^{\infty} 2^{-\ell} = 2^{1-y}.$$

Thus, we have

$$\sum_{p \leq x} \sum_{k > y} \frac{1}{kp^{k\sigma}} \left| 1 - \frac{1}{p^{ik\tau}} \right| < \frac{4\pi(x)}{y2^{y}}. \tag{9.25}$$

It remains to bound

$$\sum_{p > x} \sum_{k=1}^{\infty} \frac{1}{kp^{k\sigma}} \left| 1 - \frac{1}{p^{ik\tau}} \right| \leq 2 \left\{ \sum_{p > x} \frac{1}{p^{\sigma}} + \sum_{p > x} \sum_{k=2}^{\infty} \frac{1}{kp^{k\sigma}} \right\}. \tag{9.26}$$

We start with the second term on the right-hand side. Obviously,

$$\sum_{p > x} \sum_{k=2}^{\infty} \frac{1}{kp^{k\sigma}} < \frac{1}{2} \sum_{p > x} \sum_{k=2}^{\infty} \left(\frac{1}{p^{\sigma}} \right)^{k} = \frac{1}{2} \sum_{p > x} \frac{1}{p^{\sigma} - 1}.$$

Since

$$\frac{1}{p^\sigma - 1} = \frac{1}{p^\sigma}\frac{p^\sigma}{p^\sigma - 1} < \frac{1}{p^\sigma}\frac{x^\sigma}{x^\sigma - 1} < \frac{1}{p^\sigma}\frac{x}{x - 1} \leq \frac{2}{p^\sigma}$$

for $p > x$, it follows that

$$\sum_{p>x}\sum_{k=2}^{\infty}\frac{1}{kp^{k\sigma}} < \sum_{p>x}\frac{1}{p^\sigma}.$$

This gives in (9.26)

$$\sum_{p>x}\sum_{k=1}^{\infty}\frac{1}{kp^{k\sigma}}\left|1 - \frac{1}{p^{ik\tau}}\right| < 4\sum_{p>x}\frac{1}{p^\sigma}. \tag{9.27}$$

To estimate the sum on the right we make use of the prime number Theorem 1.1. By partial integration we obtain

$$\sum_{p>x}\frac{1}{p^\sigma} = \int_x^\infty u^{-\sigma}\,\mathrm{d}\pi(u) = (1 + O(1))\int_x^\infty u^{-\sigma}\left(\frac{u}{\log u}\right)'\,\mathrm{d}u.$$

The integral is equal to

$$\int_x^\infty u^{-\sigma}\frac{\log u - 1}{(\log u)^2}\,\mathrm{d}u < \frac{1}{\log x}\int_x^\infty u^{-\sigma}\,\mathrm{d}u = \frac{x^{1-\sigma}}{(\sigma - 1)\log x}.$$

Substituting this estimate in (9.27) yields

$$\sum_{p\leq x}\sum_{k=1}^{\infty}\frac{1}{kp^{k\sigma}}\left|1 - \frac{1}{p^{ik\tau}}\right| < \frac{(4 + O(1))x^{1-\sigma}}{(\sigma - 1)\log x}. \tag{9.28}$$

In view of (9.24), (9.25), and (9.28) we obtain in (9.21)

$$|\log\mathcal{L}(s) - \log\mathcal{L}(s + i\tau)|$$
$$< m\left\{\frac{\sqrt{5}\pi\epsilon}{\sigma - 1} + \frac{4\pi(x)}{y2^y} + \frac{(4 + O(1))x^{1-\sigma}}{(\sigma - 1)\log x}\right\}$$

as x tends to infinity. Taking the maximum over $\sigma \geq 1 + \delta$ and incorporating (1.1) we find

$$\max_{\sigma\geq 1+\delta}|\log\mathcal{L}(s) - \log\mathcal{L}(s + i\tau)|$$
$$< m\left\{\frac{\sqrt{5}\pi\epsilon}{\delta} + \frac{(4 + O(1))x}{y2^y\log x} + \frac{(4 + O(1))}{\delta x^\delta\log x}\right\} \tag{9.29}$$

for any τ which satisfies (9.20). For a suitable choice of our parameters x, y we can make the right-hand side of (9.29) as small as we please. Let

$$x = \epsilon^{-1/\delta} \quad \text{and} \quad y = -\left(1 + \frac{1}{\delta}\right) \log \epsilon; \tag{9.30}$$

then the condition $4\epsilon y < 1$ for (9.22) is fulfilled for sufficiently small ϵ. It follows that:

$$\max_{\sigma \geq 1+\delta} |\log \mathcal{L}(s) - \log \mathcal{L}(s + i\tau)| < m \frac{\sqrt{5}\pi}{\delta}(\epsilon + O(1)) =: \tilde{\epsilon}.$$

This proves the assertion of the theorem for $\log \mathcal{L}$.

Since $|\alpha_j(p)| \leq 1$, we have for $\sigma > 1$

$$|\mathcal{L}(s)| \leq \prod_{j=1}^{m} \prod_{p} \left(1 - \frac{|\alpha_j|}{p^\sigma}\right)^{-1} \leq \prod_{p} \left(1 - \frac{1}{p^\sigma}\right)^{-m} = \zeta(\sigma)^m.$$

Hence

$$\max_{\sigma \geq 1+\delta} |\mathcal{L}(s)| \leq \zeta(1 + \delta)^m.$$

Furthermore,

$$\zeta(1 + \delta) = 1 + \sum_{n=2}^{\infty} \frac{1}{n^{1+\delta}} < 1 + \int_{1}^{\infty} u^{-(1+\delta)} \, du = \frac{1 + \delta}{\delta}.$$

Obviously,

$$\mathcal{L}(s) - \mathcal{L}(s + i\tau) = \mathcal{L}(s)(1 - \exp\{\log \mathcal{L}(s + i\tau) - \log \mathcal{L}(s)\}).$$

It thus follows that

$$\ell(\mathcal{L}, \delta, \varepsilon) = \ell\left(\log \mathcal{L}, \delta, \left(\frac{\delta}{1 + \delta}\right)^m \varepsilon\right).$$

This implies the assertion of the theorem for $\mathcal{L}$.   □

We are interested in giving effective estimates. Application of Weyl's approximation theorem, Lemma 1.8, would lead to an explicit lower bound for the density of shifts $\mathcal{L}(s + i\tau)$ approximating $\mathcal{L}(s)$. However, this approach does not give an upper bound for the first approximating τ. We can solve this problem by applying effective results from the theory of inhomogeneous diophantine approximation which we derive in the following section.

9.6 Effective Inhomogeneous Diophantine Approximation

Bohr and Landau [33, 36] (see also [353, Sect. 8.8]) proved for given N the existence of a real number τ with $0 \leq \tau \leq \exp(N^6)$ such that

$$\cos(\tau \log p_\nu) < -1 + \frac{1}{N} \quad \text{for} \quad 1 \le \nu \le N,$$

where p_ν denotes the νth prime number. This can be regarded as a first effective version of Kronecker's approximation theorem, with a bound for τ similar to the bound in Dirichlet's approximation theorem. Using the idea of Bohr and Landau in connection with Baker's estimate for linear forms, Rieger [309] proved the following remarkable discrete approximation theorem:

Theorem 9.7. *Let* $v, N \in \mathbb{N}$, $b \in \mathbb{Z}$, $1 \le \omega, U \in \mathbb{R}$. *Let* $p_1 < \ldots < p_N$ *be prime numbers (not necessarily consecutive) and*

$$u_\nu \in \mathbb{Z}, \quad 0 < |u_\nu| \le U, \quad \beta_\nu \in \mathbb{R} \quad \text{for} \quad 1 \le \nu \le N.$$

Then there exist $h_\nu \in \mathbb{Z}$, $0 \le \nu \le N$, *and an effectively computable number* $C = C(N, p_N) > 0$, *depending on* N *and* p_N *only, so that*

$$\left| h_0 \frac{u_\nu}{v} \log p_\nu - \beta_\nu - h_\nu \right| < \frac{1}{\omega} \quad \text{for} \quad 1 \le \nu \le N, \tag{9.31}$$

and $b \le h_0 \le b + (2Uv\omega)^C$.

However, we are interested in an explicit bound for C. Therefore, we repeat Rieger's proof and add in the crucial step a result on an explicit lower bound for linear forms in logarithms due to Waldschmidt.

Let $\mathbb{K}$ be a number field of degree d over $\mathbb{Q}$ and denote by $L_\mathbb{K}$ the set of logarithms of the elements of $\mathbb{K} \setminus \{0\}$, more precisely,

$$L_\mathbb{K} = \{\ell \in \mathbb{C} : \exp(\ell) \in \mathbb{K}\}.$$

If α is an algebraic number with minimal polynomial $P(X)$ over $\mathbb{Z}$, then define the absolute logarithmic height of α by

$$h(\alpha) = \frac{1}{d} \int_0^1 \log |P(\exp(2\pi i\phi))| \, d\phi;$$

note that $h(\alpha) = \log |\alpha|$ for every integer $\alpha \ne 0$, and $h(0) = 0$. Waldschmidt [365] proved

Theorem 9.8. *Let* $\ell_\nu \in L_\mathbb{K}$ *and* $\beta_\nu \in \mathbb{Q}$ *for* $1 \le \nu \le N$, *not all equal zero. Define* $a_\nu = \exp(\ell_\nu)$ *for* $1 \le \nu \le N$, *and*

$$\Lambda = \beta_0 + \beta_1 \log a_1 + \cdots + \beta_N \log a_N.$$

Let E, W *and* V_ν, $1 \le \nu \le N$, *be positive real numbers, satisfying*

$$W \ge \max_{1 \le \nu \le N} \{h(\beta_\nu)\}, \quad \frac{1}{d} \le V_1 \le \ldots \le V_N,$$

$$V_\nu \geq \max \left\{ h(a_\nu), \frac{|\log a_\nu|}{d} \right\} \quad \text{for} \quad 1 \leq \nu \leq N,$$

and

$$1 < E \leq \min \left\{ \exp(V_1), \min_{1 \leq \nu \leq N} \left\{ \frac{4 \, d V_\nu}{|\log a_\nu|} \right\} \right\}.$$

Finally, define $V_\nu^+ = \max\{V_\nu, 1\}$ *for* $\nu = N$ *and* $\nu = N - 1$, *with* $V_1^+ = 1$ *in the case* $N = 1$. *If* $\Lambda \neq 0$, *then*

$$|\Lambda| > \exp\left(-c(N) \, d^{N+2}(W + \log(E \, dV_N^+)) \log(E \, dV_{N-1}^+)(\log E)^{-N-1} \prod_{\nu=1}^{N} V_\nu \right),$$

where $c(N) \leq 2^{8N+51} N^{2N}$.

In conjunction with Rieger's Theorem 9.7 this leads to

Theorem 9.9. *With the notation and the assumptions of Theorem 9.7 there exists an integer* h_0 *such that (9.31) holds and*

$$b \leq h_0 \leq b + 2 + \exp\left(2^{8N+51} N^{2N}(1 + 2\log p_N)(1 + \log p_{N-1}) \prod_{\nu=2}^{N} \log p_\nu \right)$$

$$\times ((3\omega U(N+2) \log p_N)^4 + 2)^{N+2};$$

in the special case that the primes $p_1 < p_2 < \ldots < p_N$ *are the first* N *successive prime numbers, we obtain the stronger inequality*

$$b \leq h_0 \leq b + (\omega U)^{(4+\varepsilon)N} \exp\left(N^{(2+\varepsilon)N} \right)$$

for any $\epsilon > 0$ *and sufficiently large* N.

Proof. For $t \in \mathbb{R}$ define

$$f(t) = 1 + \exp(2\pi i t) + \sum_{\nu=1}^{N} \exp\left(2\pi i \left(t \frac{u_\nu}{v} \log p_\nu - \beta_\nu \right) \right).$$

With $\gamma_{-1} := 0, \beta_{-1} := 0, \gamma_0 := 1, \beta_0 := 0$ and $\gamma_\nu := \frac{u_\nu}{v} \log p_\nu, 1 \leq \nu \leq N$, we have

$$f(t) = \sum_{\nu=-1}^{N} \exp(2\pi i(t\gamma_\nu - \beta_\nu)).$$

By the multinomial theorem,

$$f(t)^k = \sum_{\substack{j_\nu \geq 0 \\ j_{-1} + \cdots + j_N = k}} \frac{k!}{j_{-1}! \cdots j_N!} \exp\left(2\pi i \sum_{\nu=-1}^{N} j_\nu(t\gamma_\nu - \beta_\nu) \right).$$

Hence, for $0 < B \in \mathbb{R}$ and $k \in \mathbb{N}$,

$$
\begin{aligned}
J &:= \int_b^{b+B} |f(t)|^{2k}\, dt \\
&= \sum_{\substack{j_\nu \geq 0 \\ j_{-1}+\cdots+j_N = k}} \frac{k!}{j_{-1}!\cdots j_N!} \sum_{\substack{j_{\nu'} \geq 0 \\ j'_{-1}+\cdots+j'_N = k}} \frac{k!}{j'_{-1}!\cdots j'_N!} \\
&\quad \int_b^{b+B} \exp\left(2\pi i \left(\sum_{\nu=-1}^{N} (j_\nu - j'_\nu)\gamma_\nu t - \sum_{\nu=-1}^{N} (j_\nu - j'_\nu)\beta_\nu \right) \right) dt.
\end{aligned}
$$

Since the logarithms of prime numbers are linearly independent, the sum

$$
\sum_{\nu=-1}^{N} (j_\nu - j'_\nu)\gamma_\nu
$$

vanishes if and only if $j_\nu = j'_\nu$ for $-1 \leq \nu \leq N$. Thus, integration gives

$$
\int_b^{b+B} \exp\left(2\pi i \left(\sum_{\nu=-1}^{N} (j_\nu - j'_\nu)\gamma_\nu t - \sum_{\nu=-1}^{N} (j_\nu - j'_\nu)\beta_\nu \right) \right) dt = B
$$

if $j_\nu = j'_\nu$ for $-1 \leq \nu \leq N$, and

$$
\left| \int_b^{b+B} \exp\left(2\pi i \left(\sum_{\nu=-1}^{N} (j_\nu - j'_\nu)\gamma_\nu t - \sum_{\nu=-1}^{N} (j_\nu - j'_\nu)\beta_\nu \right) \right) dt \right|
$$
$$
\leq \frac{1}{\pi} \left| \sum_{\nu=-1}^{N} (j_\nu - j'_\nu)\gamma_\nu \right|^{-1}
$$

if $j_\nu \neq j'_\nu$ for some $\nu \in \{-1, 0, \ldots, N\}$. In the latter case, by Baker's estimate [13] for linear forms, there exists an effectively computable constant A such that

$$
\left| \sum_{\nu=-1}^{N} (j_\nu - j'_\nu)\gamma_\nu \right|^{-1} < A.
$$

Setting $\beta_0 = j_0 - j'_0, \beta_\nu = \frac{u_\nu}{v}(j_\nu - j'_\nu)$, and $a_\nu = p_\nu$ for $1 \leq \nu \leq N$, we find, with the notation of Theorem 9.8,

$$
\Lambda = \sum_{\nu=-1}^{N} (j_\nu - j'_\nu)\gamma_\nu.
$$

We may take $E = 1, W = \log p_N, V_1 = 1$ and $V_\nu = \log p_\nu$ for $2 \leq \nu \leq N$. If $N \geq 2$, Theorem 9.8 gives

$$|\Lambda| > \exp\left(-2^{8N+51}N^{2N}(1+2\log p_N)(1+\log p_{N-1})\prod_{\nu=2}^{N}\log p_\nu\right).$$

Thus we may take

$$A = \exp\left(2^{8N+51}N^{2N}(1+2\log p_N)(1+\log p_{N-1})\prod_{\nu=2}^{N}\log p_\nu\right). \qquad (9.32)$$

Hence we obtain

$$J \geq B \sum_{\substack{j_\nu \geq 0 \\ j_{-1}+\cdots+j_N=k}} \left(\frac{k!}{j_{-1}!\cdots j_N!}\right)^2$$
$$-\frac{A}{\pi} \sum_{\substack{j_\nu \geq 0 \\ j_{-1}+\cdots+j_N=k}} \frac{k!}{j_{-1}!\cdots j_N!} \sum_{\substack{j_{\nu'} \geq 0 \\ j'_{-1}+\cdots+j'_N=k}} \frac{k!}{j'_{-1}!\cdots j'_N!}. \qquad (9.33)$$

Since

$$\sum_{\substack{j_\nu \geq 0 \\ j_{-1}+\cdots+j_N=k}} 1 \leq (k+1)^{N+2},$$

an application of the Cauchy Schwarz-inequality to the first multiple sum and of the multinomial theorem to the second multiple sum on the right-hand side of (9.33) yields

$$J \geq \left(\frac{B}{(k+1)^{N+2}} - \frac{A}{\pi}\right)\left(\sum_{\substack{j_\nu \geq 0 \\ j_{-1}+\cdots+j_N=k}} \frac{k!}{j_{-1}!\cdots j_N!}\right)^2$$
$$\geq \left(\frac{B}{(k+1)^{N+2}} - \frac{A}{\pi}\right)(N+2)^{2k}.$$

Setting $B = A(k+1)^{N+2}$ and with $\tau \in [b, b+B]$ given by

$$|f(\tau)| = \max_{t\in[b,b+B]}|f(t)|,$$

we obtain

$$\frac{B(N+2)^{2k}}{2(k+1)^{N+2}} \leq J \leq B|f(\tau)|^{2k}.$$

This gives

$$|f(\tau)| > N+2-2\mu, \qquad \text{where} \quad \mu := \frac{(N+2)^2\log k}{3k}; \qquad (9.34)$$

note that $\mu < 1$ for $k \geq 11$. By definition

$$f(t) = 1 + \exp(2\pi i(t\gamma_\nu - \beta_\nu)) + \sum_{\substack{m=0 \\ m \neq \nu}}^{N} \exp(2\pi i(t\gamma_m - \beta_m)).$$

Therefore, using the triangle inequality,

$$|f(t)| \leq N + |1 + \exp(2\pi i(\tau\gamma_\nu - \beta_\nu))| \quad \text{for} \quad 0 \leq \nu \leq N$$

and arbitrary $t \in \mathbb{R}$. Thus in view of (9.34)

$$|1 + \exp(2\pi i(\tau\gamma_\nu - \beta_\nu))| > 2 - 2\mu \quad \text{for} \quad 0 \leq \nu \leq N.$$

If h_ν denotes the nearest integer to $\tau\gamma_\nu - \beta_\nu$, then

$$|\tau\gamma_\nu - \beta_\nu - h_\nu| < \sqrt{\frac{\mu}{2}} \quad \text{for} \quad 0 \leq \nu \leq N.$$

For $\nu = 0$ this implies $|\tau - h_0| < \sqrt{\mu}$. Thus, replacing τ by h_0 yields

$$|h_0\gamma_\nu - \beta_\nu - h_\nu| < \sqrt{\mu}\left(1 + \max_{1 \leq \nu \leq N}|\gamma_\nu|\right) \quad \text{for} \quad 1 \leq \nu \leq N.$$

Putting $k = [(3wU(N+2)\log p_N)^4] + 1$, we get

$$b - 1 \leq h_0 \leq b + 1 + B = b + 1 + A([(3\omega U(N+2)\log p_N)^4] + 2)^{N+2}.$$

Substituting (9.32) and replacing $b - 1$ by b, the first assertion of Theorem 9.7 follows with the estimate of Theorem 9.9; the second estimate can be proved by standard estimates involving the prime number Theorem 1.1.   □

Now we are in the position to present our effective result on the almost periodicity for the polynomial Euler products under consideration. For this purpose we shall apply Theorem 9.9 with $x = p_N$. Recall that m was the degree of each Euler factor in the polynomial Euler product for $\mathcal{L}$. For the sake of simplicity we use the rough estimate $N = \pi(x) \leq x$ which gives

$$M(x, \epsilon) = \epsilon^{-4x} \exp\left(x^{2x}\right).$$

Hence, by Theorem 9.6,

Corollary 9.10. *Let $\mathcal{L}(s)$ be given by (9.18) and $\delta > 0$. For sufficiently small $\epsilon > 0$,*

$$\ell(\mathcal{L}, \delta, \varepsilon) = (C\delta^{m+1}\varepsilon)^{-C\delta^{-(m+1)/\delta}\varepsilon^{-1/\delta}} \exp\left(\left(C\delta^{(m+1)/\delta}\varepsilon^{1/\delta}\right)^{-C\delta^{-(m+1)/\delta}\varepsilon^{-1/\delta}}\right),$$

where C is an absolute positive constant, depending only on m.

9.7 c-Values Revisited

We return now to a topic already touched in Chap. 7. Almost periodicity has a strong impact on the value distribution. In fact, if a value is taken by a polynomial Euler product in the half-plane of absolute convergence, then it is taken infinitely often, with a certain regularity which reflects almost periodicity.

Theorem 9.11. *Assume that $\mathcal{L}$ is given by (9.18). Let $N(T; a, b, c)$ count the number of c-values of $\mathcal{L}(s)$ (i.e., the roots of the equation $\mathcal{L}(s) = c$) in the rectangle $a < \sigma < b, 0 < t < T$ (counting multiplicities) and assume $a \geq 1$. If $N(T; a, b, c) \geq 1$, then, for sufficiently small positive ε and δ,*

$$\liminf_{T \to \infty} \frac{1}{T} N(T; a, b, c) \geq \frac{1}{\ell(\mathcal{L}, \delta, \varepsilon)} > 0.$$

Since the Euler products under observation have no zeros in their half-plane of convergence, we have $N(T; 1, \infty, 0) = 0$; however, we expect $N(T; 1, \infty, c) \geq 1$ for any $c \neq 0$ (as in the case of the zeta-function).

Again we shall use Rouché's theorem in order to localize c-values.

Proof. Suppose that $N(T; a, b, c) \geq 1$. Then there exists a complex number λ_c with $a < \operatorname{Re} \lambda_c < b$ and $0 < \operatorname{Im} \lambda_c < T$ such that $\mathcal{L}(\lambda_c) = c$. Locally, we have

$$\mathcal{L}(s) - c = \alpha(s - \lambda_c)^k + O\left(|s - \lambda_c|^{k+1}\right), \tag{9.35}$$

where α is a non-zero complex number and $k \in \mathbb{N}$. Now define

$$\mathcal{K}_\delta = \{s \in \mathbb{C} : |s - \lambda_c| \leq \delta\}.$$

For sufficiently small $\delta > 0$ the disk $\mathcal{K}_\delta$ is contained in the rectangle $a < \sigma < b, 0 < t < T$. Now assume that there is some real number τ for which

$$\max_{|s-\lambda_c|=\delta} |\mathcal{L}(s) - c - \{\mathcal{L}(s + i\tau) - c\}| < \varepsilon < \min_{|s-\lambda_c|=\delta} |\mathcal{L}(s) - c|; \tag{9.36}$$

the second inequality holds for sufficiently small ε and δ. By the maximum principle the left-hand side equals

$$\max_{s \in \mathcal{K}_\delta} |\mathcal{L}(s) - \mathcal{L}(s + i\tau)|.$$

By almost periodicity, the latter quantity can be made arbitrarily small for a set of τ with positive lower density. Now, for any τ satisfying (9.36), Rouché's Theorem 8.1 implies the existence of a zero ϱ_c of $\mathcal{L}(s) - c$ in

$$\mathcal{K}_\delta + i\tau := \{s \in \mathbb{C} : |s - i\tau - \lambda_c| \leq \delta\}.$$

In view of (9.35) the numbers λ_c and ϱ_c are intimately related. In fact,

$$\varepsilon > \max_{s \in \mathcal{K}_\delta} |\mathcal{L}(s) - \mathcal{L}(s + i\tau)|$$
$$\geq |\mathcal{L}(\varrho_c - i\tau) - \mathcal{L}(\varrho_c)| = |\mathcal{L}(\varrho_c - i\tau) - c|$$
$$= |\alpha| \cdot |\varrho_c - i\tau - \lambda_c|^k + \mathrm{O}(\delta^{k+1}).$$

Hence,

$$|\varrho_c - i\tau - \lambda_c| < \left(\frac{\varepsilon}{|\alpha|} \right)^{1/k} + \mathrm{O}\left(\delta^{1+(1/k)} \right).$$

In particular,

$$\mathrm{Re}\,\lambda_c - 2 \left(\frac{\varepsilon}{|\alpha|} \right)^{1/k} < \mathrm{Re}\,\varrho_c < \mathrm{Re}\,\lambda_c + 2 \left(\frac{\varepsilon}{|\alpha|} \right)^{1/k},$$

and

$$|\mathrm{Im}\,\varrho_c - (\tau + \mathrm{Im}\,\lambda_c)| < 2 \left(\frac{\varepsilon}{|\alpha|} \right)^{1/k}$$

for small $\delta = \mathrm{O}(\varepsilon^{1/(k+1)})$. Hence, for sufficiently small ε and δ, the root ϱ_c of the equation $\mathcal{L}(s) = c$ is contained in the open rectangle $a < \sigma < b, 0 < t < T$. By the almost periodicity of $\mathcal{L}(s)$, we can find a real number τ satisfying (9.36) in any interval of length $\ell(\mathcal{L}, \delta, \varepsilon)$. This implies the estimate of the theorem.

$\square$

Taking into account Bohr's results on the value-distribution of the zeta-function in the half-plane of absolute convergence (see Theorem 1.3), the condition of Theorem 9.11 is fulfilled for any non-zero complex number c, i.e., $N(T; 1, \infty, c) \geq 1$. Hence, for fixed $c \neq 0$ and sufficiently small $\varepsilon, \delta > 0$, we find for the number of c-values of $\zeta(s)$ in the half-plane $\sigma > 1$ the estimate

$$\liminf_{T \to \infty} \frac{1}{T} \sharp \{\varrho_c : \mathrm{Re}\,\varrho_c > 1, 0 < \mathrm{Im}\,\varrho_c < T, \mathcal{L}(\varrho_c) = c\} \geq \frac{1}{\ell(\zeta, \delta, \varepsilon)},$$

where

$$\ell(\zeta, \delta, \varepsilon) = (C\delta^2\varepsilon)^{-C\delta^{-2/\delta}\varepsilon^{-1/\delta}} \exp\left(\left(C\delta^{2/\delta}\varepsilon^{1/\delta} \right)^{-C\delta^{-2/\delta}\varepsilon^{-1/\delta}} \right) \quad (9.37)$$

by Corollary 9.10.

We apply Theorem 9.6 and (9.37) to prove the existence of *large* values of the zeta-function in some distance of the pole at $s = 1$. Recall that

$$\zeta(s) = \frac{1}{s-1} + \mathrm{O}(1)$$

for $s \to 1$. Let c be a large positive real number; then there exists a real number λ_c with

$$1 < \lambda_c \sim 1 + \frac{1}{c} \quad \text{for which} \quad c = \zeta(\lambda_c).$$

Put $\ell := \ell(\zeta, \delta, \delta)$. By Theorem 9.6, for sufficiently small δ, we can find a real number $\tau \in (\ell, 2\ell)$ such that

$$|\zeta(\sigma + i\tau)| \sim |\zeta(\lambda_c)| = c =: \frac{1}{\delta}$$

for some $\sigma > 1$. In view of (9.37) we obtain after a short computation

$$\frac{1}{\delta} \asymp \frac{\log\log\log \tau}{\log\log\log\log \tau}.$$

This leads to the estimate

$$\inf_{\sigma > 1, t > \ell} \frac{|\zeta(\sigma + it)|}{\log\log\log t} > 0.$$

It should be noted that this estimate for large values of the zeta-function in the half-plane of absolute convergence is quite far from the best possible estimate due to Bohr and Landau [34]; they proved the existence of infinitely many $s = \sigma + it$ with $\sigma \to 1+$ and $t \to +\infty$ for which

$$|\zeta(s)| \gg \log\log t.$$

By a standard argument based on the Phragmén–Lindelöf principle it follows that

$$\zeta(1 + it) = \Omega(\log\log t) \quad \text{and} \quad \frac{1}{\zeta(1 + it)} = \Omega(\log\log t). \qquad (9.38)$$

Under assumption of Riemann's hypothesis these estimates give the true order of growth (see [353, Sect. 14.8]).

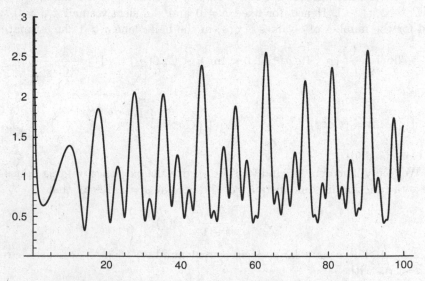

Fig. 9.3. $|\zeta(1 + it)|$ for $t \in (0, 100]$. The limit superior of the peaks grows with t to infinity, the limit inferior of the valleys tends to zero

Theorem 9.9 is due to Steuding [344] and was used to prove Ω-results for Dirichlet L-functions in the half-plane of absolute convergence. It was shown that for any real θ there are infinitely many values of $s = \sigma + it$ with $\sigma \to 1+$ and $t \to +\infty$ such that

$$\mathrm{Re}\left\{\exp(i\theta)\log L(s,\chi)\right\} \geq \log \frac{\log\log\log t}{\log\log\log\log t} + \mathrm{O}(1)$$

and

$$\mathrm{Re}\left\{\exp(i\theta)\log L(1+it,\chi)\right\} = \Omega(\log\log\log\log t).$$

It seems that also these estimates for $L(s,\chi)$ are one iterated logarithm off the true order. However, in contrast to Ω-results, these large values are taken frequently.

Consequences of Universality

> Betrachten wir beispielsweise die Klasse derjenigen Funktionen, die
> sich durch gewöhnliche oder partielle Differentialgleichungen charak-
> terisieren lassen. In dieser Klasse von Funktionen kommen, wie wir
> sofort bemerken, gerade solche Funktionen nicht vor, die aus der
> Zahlentheorie stammen und deren Erforschung für uns von höchster
> Wichtigkeit ist. Beispielsweise genügt die schon früher erwähnte Funk-
> tion $\zeta(s)$ keiner algebraischen Differentialgleichung, ..."
>
> David Hilbert

The phenomenon of universality has many interesting consequences. There
are the classic results due to Bohr, mentioned in the introduction, as well as
extensions of Ostrowski's theorem on the non-existence of algebraic differential
equations for Dirichlet series (which actually was part of the 18th Hilbert
problem). In this chapter, we prove generalizations for universal L-functions,
we discuss a disproof of a conjecture on certain mean-square estimates due to
Ramachandra by Andersson [1], and, finally, we study the value-distribution
of linear combinations of shifts of universal Dirichlet series.

10.1 Dense Sets in the Complex Plane

Let $\mathcal{L} \in \tilde{\mathcal{S}}$ and let c be a complex number. A standard argument yields that
the number of c-values up to level T is $\gg T$ provided $c \neq 0$. To see this, let
$N_{\mathcal{L}}^c(\sigma_1, \sigma_2, T)$ count the number of c-values of $\mathcal{L}(s)$ in the region $\sigma_1 < \sigma < \sigma_2$
and $|t| \leq T$, i.e., $N_{\mathcal{L}}^c(\sigma_1, \sigma_2, T) = N_{\mathcal{L}}^c(\sigma_1, T) - N_{\mathcal{L}}^c(\sigma_2, T)$ in the notation of
Sect. 7.2. We apply Theorem 5.14 to the function $g(s) = c \neq 0$ with a small
disk $\mathcal{K}$ inside the strip $\sigma_1 < \sigma < \sigma_2$, where $\max\{\frac{1}{2}, 1 - \frac{1}{d_{\mathcal{L}}}\} < \sigma_1 < \sigma_2 < 1$.
For any $\epsilon > 0$, we find

$$\liminf_{T \to \infty} \frac{1}{T} \operatorname{meas} \left\{ \tau \in [0, T] : \max_{s \in \mathcal{K}} |\mathcal{L}(s + i\tau) - c| < \epsilon \right\} > 0.$$

By Rouché's Theorem 8.1, it follows that $N_{\mathcal{L}}^c(\sigma_1, \sigma_2, T)$ is bounded below by a positive constant times T. In combination with the estimate (7.11) of Theorem 7.6 we obtain

Theorem 10.1. *Let $\mathcal{L} \in \tilde{\mathcal{S}}$ and let c be a complex number $\neq 0, 1$. Then, for any σ_1, σ_2 satisfying $\max\{\frac{1}{2}, 1 - \frac{1}{d_{\mathcal{L}}}\} < \sigma_1 < \sigma_2 < 1$,*

$$N_{\mathcal{L}}^c(\sigma_1, \sigma_2, T) \asymp T.$$

This result should be compared with Bohr's limit Theorem 1.5.

In Sect. 8.1, we have seen that $\zeta(s)$ cannot approximate functions having zeros. However, the same argument as above can be used to count zeros of the real part of $\zeta(s)$. Garaev [83] used Voronin's universality theorem in order to detect zeros of $\mathrm{Re}\,\zeta(s)$. He showed that, for fixed $\sigma \in (\frac{1}{2}, 1)$, the function $\mathrm{Re}\,\zeta(\sigma + it)$ has more than $\gg T$ zeros $t \in (0, T)$; this follows directly from an application of the universality theorem to the function $g(s) = \exp(Cs)$, where C is a sufficiently large constant. Moreover, the function $\mathrm{Re}\,\zeta(\sigma + it)$ has at least cT sign changes in the interval $(0, T)$, where c is some positive constant. Besides, Garaev proved also the upper bound $T \log T$ (the same statements also hold for $\mathrm{Im}\,\zeta(\sigma + it)$).

Next we consider the multidimensional analogue and extend Voronin's Theorem 1.6 to

Theorem 10.2. *Let $\mathcal{L} \in \tilde{\mathcal{S}}$ and suppose that $\sigma \in (\max\{\frac{1}{2}, 1 - \frac{1}{d_{\mathcal{L}}}\}, 1)$ is fixed. Then the set*

$$\{(\mathcal{L}(s), \mathcal{L}'(s), \ldots, \mathcal{L}^{(n-1)}(s)) : t \in \mathbb{R}\}$$

lies everywhere dense in $\mathbb{C}^n$.

Proof. First, for any vector $(a_0, a_1, \ldots, a_m) \in \mathbb{C}^{m+1}$ with $a_0 \neq 0$, we prove by induction on m, that there exists another vector $(b_0, b_1, \ldots, b_m) \in \mathbb{C}^{m+1}$ for which

$$\exp\left(\sum_{k=0}^{m} b_k s^k\right) \equiv \sum_{k=0}^{m} \frac{a_k}{k!} s^k \mod s^{m+1}.$$

For $m = 0$ one only has to choose $b_0 = \log a_0$. By the induction assumption we may assume that with some α

$$\exp\left(\sum_{k=0}^{m} b_k s^k\right) \equiv \sum_{k=0}^{m} \frac{a_k}{k!} s^k + \alpha s^{m+1} \mod s^{m+2}.$$

For any β,

$$\exp(\beta s^{m+1}) \equiv 1 + \beta s^{m+1} \mod s^{2m+2}.$$

Thus,

$$\exp\left(\sum_{k=0}^{m} b_k s^k + \beta s^{m+1}\right) \equiv (1 + \beta s^{m+1})\left(\sum_{k=0}^{m} \frac{a_k}{k!} s^k + \alpha s^{m+1}\right) \mod s^{m+2}.$$

Let $b_{m+1} = \beta$ be the solution of the equation

$$\beta a_0 + \alpha = \frac{a_{m+1}}{(m+1)!},$$

which exists since $a_0 \neq 0$. This shows

$$\exp\left(\sum_{k=0}^{m+1} b_k s^k\right) \equiv \sum_{k=0}^{m+1} \frac{a_k}{k!} s^k \quad \text{mod } s^{m+2},$$

proving the claim.

Now let

$$g(s) = \exp\left(\sum_{k=0}^{n-1} b_k s^k\right) \equiv \sum_{k=0}^{n-1} \frac{a_k}{k!} s^k \quad \text{mod } s^n.$$

Obviously, $g^{(k)}(0) = a_k$ for $0 \leq k \leq n-1$. Put

$$\sigma_0 = \frac{1}{2}\left(\max\left\{\frac{1}{2}, 1 - \frac{1}{d_{\mathcal{L}}}\right\} + 1\right).$$

By the universality Theorem 5.14, for any $\epsilon > 0$, there exists a τ such that

$$\max_{|s|\leq r} |\mathcal{L}(s + \sigma_0 + i\tau) - g(s)| < 2\pi \frac{\epsilon r^k}{k!}$$

for sufficiently small r. Applying Cauchy's formula,

$$f^{(k)}(0) = \frac{k!}{2\pi i} \oint_{|s|=r} \frac{f(s)}{s^{k+1}}\, ds$$

to the function

$$f(s) = \mathcal{L}(s + \sigma_0 + i\tau) - g(s)$$

with sufficiently small r, we obtain

$$|\mathcal{L}^{(k)}(\sigma_0 + i\tau) - a_k| = |\mathcal{L}^{(k)}(\sigma_0 + i\tau) - g^{(k)}(0)| < \epsilon$$

for $0 \leq k \leq n-1$. This proves the theorem. $\square$

10.2 Functional Independence

Hölder [135] proved in 1887 that the Gamma-function is hypertranscendental, i.e., $\Gamma(s)$ does not satisfy any algebraic differential equation: there exists no polynomial P in n variables, not identically zero, such that

$$P\left(z, \Gamma(z), \Gamma'(z), \ldots, \Gamma^{(n-1)}\right) = 0$$

for all $z \in \mathbb{C}$. In Hilbert's famous list of 23 challenging problems for the twentieth century [134], originated from his lecture given at the International Congress for Mathematicians in Paris 1900, he asked in problem 18 for a description of classes of function definable by differential equations. In this context Hilbert stated that the Riemann zeta-function is hypertranscendental; the first published proof for $\zeta(s)$ was written down by Stadigh in his dissertation (cf. [286]). Hilbert also asked for a proof of the hypertranscendence for the more general series

$$\zeta(s, x) := \sum_{n=1}^{\infty} \frac{x^n}{n^s}$$

and proposed to use the functional equation

$$x \frac{\partial}{\partial x} \zeta(s, x) = \zeta(s - 1, x).$$

This problem was solved by Ostrowski [286] as a particular case of a more general theorem; his argument relies on a comparison of the differential independence with the linear independence of its frequencies. By a similar reasoning Reich [307] proved that the Dedekind zeta-function does not satisfy any difference-differential equation. Popken [295] introduced a measure for the differential-transcendence of $\zeta(s)$ and other hypertranscendental Dirichlet series (similar to transcendence measures in the theory of transcendental numbers).

Following Voronin [362] we shall show now that universality implies the more general concept of functional independence. We say that the functions $f_1(s), \ldots, f_m(s)$ are functionally independent if for any continuous functions $F_0, F_1, \ldots, F_N : \mathbb{C}^m \to \mathbb{C}$, not all identically vanishing, the function

$$\sum_{k=0}^{N} s^k F_k(f_1(s), \ldots, f_m(s))$$

is non-zero for some values of s.

Theorem 10.3. *Let $\mathcal{L} \in \tilde{\mathcal{S}}$, $\mathbf{z} = (z_0, z_1, \ldots, z_{n-1}) \in \mathbb{C}^n$ and suppose that $F_0(\mathbf{z}), F_1(\mathbf{z}), \ldots, F_N(\mathbf{z})$ are given continuous functions, not all identically zero. Then*

$$\sum_{k=0}^{N} s^k F_k \left(\mathcal{L}(s), \mathcal{L}(s)', \ldots, \mathcal{L}(s)^{(n-1)} \right) \neq 0$$

for some $s \in \mathbb{C}$.

In particular, $\mathcal{L} \in \tilde{\mathcal{S}}$ does not satisfy any algebraic differential equation.

Proof. First, we show that if $F(\mathbf{z})$ is a continuous function and if

$$F \left(\mathcal{L}(s), \mathcal{L}(s)', \ldots, \mathcal{L}(s)^{(n-1)} \right) = 0$$

identically in $s \in \mathbb{C}$, then F vanishes identically.

Suppose the contrary, i.e., $F(\mathbf{z}) \not\equiv 0$. Then there exists $\mathbf{a} \in \mathbb{C}^n$ for which $F(\mathbf{a}) \neq 0$. Since F is continuous we can find a neighbourhood U of $\mathbf{a}$ and a positive ϵ such that

$$|F(\mathbf{z})| > \epsilon \quad \text{for} \quad \mathbf{z} \in U.$$

Choosing an arbitrary σ with $\max\{\frac{1}{2}, 1 - \frac{1}{d_{\mathcal{L}}}\} < \sigma < 1$, an application of Theorem 10.2 yields the existence of some t for which

$$\left(\mathcal{L}(\sigma + it), \mathcal{L}(\sigma + it)', \ldots, \mathcal{L}(\sigma + it)^{(n-1)}\right) \in U,$$

which contradicts our assumption. This proves our claim which is the assertion of the theorem with $N = 0$.

Without loss of generality we may assume that $F_0(\mathbf{z})$ is not identically zero. As above there exist an open bounded set U and a positive ϵ such that

$$|F_0(\mathbf{z})| > \epsilon \quad \text{for} \quad \mathbf{z} \in U.$$

Denote by M the maximum of all indices m for which

$$\sup_{\mathbf{z} \in U} |F_m(\mathbf{z})| \neq 0.$$

If $M = 0$, then the assertion of the theorem follows from the result proved above. Otherwise, we may take a subset $V \subset U$ such that:

$$\inf_{\mathbf{z} \in V} |F_M(\mathbf{z})| > \epsilon$$

for some positive ϵ. By Theorem 10.2, there exists a sequence t_j, tending with j to infinity, such that

$$\left(\mathcal{L}(\sigma + it_j), \mathcal{L}(\sigma + it_j)', \ldots, \mathcal{L}(\sigma + it_j)^{(n-1)}\right) \in V.$$

This implies

$$\lim_{j \to \infty} \left| \sum_{k=0}^{M} (\sigma + it_j)^k F_k(\mathcal{L}(\sigma + it_j), \mathcal{L}(\sigma + it_j)', \ldots, \mathcal{L}(\sigma + it_j)^{(n-1)}) \right| = \infty,$$

which proves the theorem. $\square$

The zeta-function does not satisfy any algebraic differential equation; however, another approximation property for zeta is related to a non-algebraic differential equation. Gauthier and Tarkhanov [95] proved that any meromorphic function on a compact subset $\mathcal{K}$ of $\mathbb{C}$ having at most simple poles can be approximated by linear combinations of certain translates of the Riemann zeta-function (not necessarily from the critical strip); if derivatives of $\zeta(s)$ are allowed, then arbitrary meromorphic functions can be approximated. The proof of this result does not rely on the universality property of the zeta-function but the fact that $\zeta(s)$ satisfies a certain (non-algebraic) inhomogeneous linear differential equation of order one for the Riemann–Cauchy operator.

10.3 Joint Functional Independence

Now we extend the previous results by restricting on some kind of minimal definition what *universality* might be. Assume that $F_1(s), \ldots, F_m(s)$ are meromorphic functions for $\sigma > \sigma_m$ with $\sigma_m < 1$ satisfying the following property: given arbitrary linearly independent continuous functions $f_1(s), \ldots, f_m(s)$ on a compact subset $\mathcal{K}$ of $\sigma_m < \sigma < 1$ with connected complement, which are analytic in the interior, for any $\epsilon > 0$ and all but finitely many points $s_0 \in \mathcal{K}$, there exists a neighbourhood $\tilde{\mathcal{K}}$ of s_0 with $\tilde{\mathcal{K}} \subset \mathcal{K}$ and a real number τ such that

$$\max_{1 \leq j \leq m} \max_{s \in \tilde{\mathcal{K}}} |F_j(s + i\tau) - f_j(s)| < \epsilon. \qquad (10.1)$$

We then say that the family $F_1, \ldots, F_m$ is *weakly jointly universal*. This notion was introduced by Sander and Steuding [315] in order to prove that families of Hurwitz zeta-functions with rational parameters satisfying certain side conditions are weakly jointly universal and functional independent.

Theorem 10.4. *Suppose that $F_1(s), \ldots, F_m(s)$ are weakly jointly universal. Then for almost all s_0 with $\operatorname{Re} s_0 \in (\sigma_m, 1)$ (i.e., all but from a discrete set) the image of the curve*

$$\Xi(t) := \left(F_1(s_0 + it), \ldots, F_m(s_0 + it), \ldots, F_1^{(\ell-1)}(s_0 + it), \ldots, F_m^{(\ell-1)}(s_0 + it) \right)$$

lies dense in $\mathbb{C}^{m\ell}$.

We only give a sketch of the proof since the argument is in principal the same as in the proof of Theorem 10.2. For $j = 1, \ldots, m$ and $0 \leq k < \ell$, to any given set of complex numbers a_{kj} with $a_{0j} \neq 0$, there exists a polynomial $g_j(s)$ such that

$$a_{kj} = (\exp(g_j(s)))^{(k)} \Big|_{s=s_0}. \qquad (10.2)$$

Let $\mathcal{K}$ be the closed disk with center s_0 and radius $r > 0$ such that $\mathcal{K}$ lies inside the strip $\sigma_m < \sigma < 1$. Now we make use of our assumption that the functions $F_j(s)$ are weakly jointly universal. We may assume that the functions $f_j(s) := \exp(g_j(s))$ are linearly independent (if not, then we may replace the $g_j(s)$ by $g_j(s) + \gamma_j(s)$ with suitable functions $\gamma_j(s)$ such that (10.2) still holds). Further, we may assume that the $F_j(s)$ are jointly universal with respect to the functions $f_j(s)$ for a small neighbourhood $\tilde{\mathcal{K}}$ of s_0, a disk with center s_0 of radius ϱ, say. In particular, for any $\epsilon > 0$, there exists a real number τ such that (10.1) holds. Now define $\eta_j(s) = F_j(s + i\tau) - g_j(s)$. Using Cauchy's integration formula,

$$\eta_j^{(k)}(s_0) = \frac{k!}{2\pi i} \oint_{|s-s_0|=\varrho} \frac{\eta_j(s)}{(s - s_0)^{k+1}} \, ds,$$

we obtain the estimate

$$\left| \eta_j^{(k)}(s_0) \right| \leq \frac{k!}{\varrho^k} \max_{s \in \tilde{\mathcal{K}}} |\eta_j(s)|.$$

This leads to

$$\max_{1 \leq j \leq m} \left| F_j^{(k)}(s_0 + i\tau) - a_{kj} \right| < \epsilon \frac{k!}{\varrho^k}$$

for $k = 0, 1, \ldots, \ell - 1$. Taking ϵ sufficiently small the assertion follows.

Now we may use the previous theorem to prove the functional independence for jointly universal families and their derivatives:

Theorem 10.5. *Suppose that $F_1(s), \ldots, F_m(s)$ are weakly jointly universal. Let $f_0, \ldots, f_n$ be continuous functions on $\mathbb{C}^{m\ell}$. If*

$$\sum_{k=0}^{n} s^k f_k \left(F_1(s), \ldots, F_m(s), \ldots, F_1^{(\ell-1)}(s), \ldots, F_1^{(\ell-1)}(s) \right) = 0$$

identically in $\sigma_m < \sigma < 1$, then $f_k(s) \equiv 0$ for $k = 0, 1, \ldots, n$.

Again we give only a sketch of the proof. First, we show that if f is a continuous function on $\mathbb{C}^{m\ell}$ and

$$f \left(F_1(s), \ldots, F_m(s), \ldots, F_1^{(\ell-1)}(s), \ldots, F_m^{(\ell-1)}(s) \right) = 0$$

in $\sigma_m < \sigma < 1$, then f vanishes identically.

Suppose the contrary, i.e., $f(z_0) \neq 0$ for some $z_0 \in \mathbb{C}^{m\ell}$. Since f is continuous, there exists a neighbourhood $\mathcal{U}$ of z_0 and some positive ϵ such that

$$|f(z)| > \epsilon \quad \text{for} \quad z \in \mathcal{U}. \tag{10.3}$$

By Theorem 10.4, for some fixed s_0 with $\operatorname{Re} s_0 \in (\sigma_m, 1)$, there exists a sequence (t_j) tending to $+\infty$ for which

$$\left(F_1(s_0 + it_j), \ldots, F_m(s_0 + it_j), \ldots, F_1^{(\ell-1)}(s_0 + it_j), \ldots, F_m^{(\ell-1)}(s_0 + it_j) \right) \in \mathcal{U}.$$

This gives the desired contradiction. So we have proved the assertion of the theorem in the case of $n = 0$.

To prove the full assertion we may now assume that f_0 is not identically zero. By the same reasoning as above, there exists an open bounded set $\mathcal{U}$ and a positive ϵ such that (10.3) holds with f_0 in place of f. Now denote by κ the maximum of all indices $1 \leq k \leq m$ such that the supremum of all values $|f_k(z)|$ for $z \in \mathcal{U}$ is positive. For sufficiently small $\epsilon > 0$, we may find a subset $\mathcal{V} \subset \mathcal{U}$ for which

$$\inf_{z \in \mathcal{V}} |f_\kappa(z)| > \epsilon.$$

By Theorem 10.4, for fixed s_0, there exists a sequence (t_j) tending to $+\infty$ such that

$$\left(F_1(s_0 + \mathrm{i}t_j), \ldots, F_m(s_0 + \mathrm{i}t_j), \ldots, F_1^{(\ell-1)}(s_0 + \mathrm{i}t_j), \ldots, F_m^{(\ell-1)}(s_0 + \mathrm{i}t_j)\right) \in \mathcal{V}.$$

This implies the divergence of

$$\sum_{k=0}^{\kappa} s^k f_k \left(F_1(s), \ldots, F_m(s), \ldots, F_1^{(\ell-1)}(s), \ldots, F_m^{(\ell-1)}(s)\right)\Big|_{s=s_0+\mathrm{i}t_j},$$

as $j \to \infty$, giving the contradiction. Theorem 10.5 is proved.

We conclude with an interesting result of Kaczorowski and Perelli. If P is an arbitrary entire function of minimum type of order one on $\mathbb{C}^n \times D$, where D is a region in $\mathbb{C}$ which contains the set

$$\left\{s \in \mathbb{C} : \sigma \geq \frac{1}{2}, |t| \geq T\right\} \cup \{s \in \mathbb{C} : \sigma > 1, |t| \leq T\}$$

for some positive T, and if the functions $\mathcal{L}_1, \ldots, \mathcal{L}_n \in \mathcal{S}$ are linearly independent over $\mathbb{Q}$, then Kaczorowski and Perelli [159] showed that the function

$$P(\log \mathcal{L}_1(s), \ldots, \log \mathcal{L}_n(s), s)$$

has infinitely many singularities in the half-plane $\sigma \geq \frac{1}{2}$.

10.4 Andersson's Disproof of a Mean-Square Conjecture

There are further applications of universality. First, we briefly discuss a disproof of a conjecture concerning the mean-square of Dirichlet polynomials. The theorem of Montgomery–Vaughan states that, for arbitrary complex numbers $a_1, \ldots, a_N$,

$$\int_0^T \left|\sum_{n=1}^N a_n n^{\mathrm{it}}\right|^2 \mathrm{d}t \ll \sum_{n=1}^N |a_n|^2 (T + n).$$

In the other direction, Ramachandra [299] conjectured that for any $N \geq 0$ and any complex numbers $a_1, \ldots, a_N$, there exists a positive constant C such that

$$\int_0^T \left|\sum_{n=1}^N a_n n^{\mathrm{it}}\right|^2 \mathrm{d}t \geq C \sum_{n \leq CT} |a_n|^2.$$

Andersson [1] gave a disproof of this and two related conjectures on moments of Dirichlet polynomials due to Ramachandra. Actually, Andersson proved the following statement: for any $H, \epsilon > 0$ and $0 < \delta < \frac{1}{2}$ there exists an $N \geq 2$

and complex numbers $a_2, \ldots, a_N$, satisfying the estimate $|a_n| \leq n^{\delta - 1/2}$, such that

$$\max_{0 \leq t \leq H} \left| 1 + \sum_{n=2}^{N} a_n n^{it} \right| < \epsilon. \tag{10.4}$$

The proof relies on an approximation of $\zeta(s)$ by a Dirichlet polynomial and Voronin's universality theorem. It follows from (1.3) that, for any $\delta \in (0, \frac{1}{2})$ and for sufficiently large T,

$$\left| \zeta(1 - \delta + it) - \sum_{n < 2T} n^{\delta - 1 - it} \right| < \frac{\epsilon}{3} \tag{10.5}$$

when $T \leq t \leq T + H$. Voronin's universality Theorem 1.9, applied to $g(s) = \frac{\epsilon}{3}$ and $\mathcal{K} = [1 - \delta, 1 - \delta + iH]$, yields

$$\max_{t \in \mathcal{K}} \left| \zeta(1 - \delta + it + iT) - \frac{\epsilon}{3} \right| < \frac{\epsilon}{3}$$

for some $T > 0$. Thus, if we replace t by $t - T$, then

$$|\zeta(1 - \delta + it)| < \frac{2\epsilon}{3}$$

for $T \leq t \leq T + H$. By (10.5), this implies the inequality

$$\left| \sum_{n < 2T} n^{\delta - 1 - it} \right| < \epsilon,$$

valid for $T \leq t \leq T + H$, which leads to Andersson's claim (10.4).

It follows from (10.4) that, for each $T \geq 0$ and any $\epsilon > 0$, there exist complex numbers $a_1, \ldots, a_N$, for which

$$\int_0^T \left| 1 + \sum_{n=2}^{N} a_n n^{it} \right|^2 dt \leq \epsilon.$$

This disproves Ramachandra's conjecture. Andersson also gave a simpler argument based on the Szasz–Müntz theorem.

10.5 Voronin's Theorems and Physics

We should mention applications of Voronin's results on the value distribution of the zeta-function to physics. Gutzwiller [112] discovered an interesting link between quantum theory, hyperbolic geometry and the Riemann zeta-function. On a closed surface of constant negative curvature the Selberg trace formula expresses the relation between classical and quantum states. Gutzwiller constructed a solution of Schrödinger's equation on the punctured torus which may be regarded as the quotient of the upper half-plane by a free subgroup of index six in $\mathsf{SL}_2(\mathbb{Z})$. This solution represents the scattering and reflection of a particle which enters the torus through the hole. The scattering phase shift is essentially the argument of $\zeta(s)$ on the line $\sigma = 1$. The universality theorem reflects the *chaotic* behaviour of the zeta-function inside the critical strip, and in particular the zero-distribution is expected to be closely related to the quantum chaos in this model.

Bitar, Khuri and Ren [24] found an interesting application of Voronin's Theorems 1.6 and 1.7 to Feynman's path integral in quantum physics. They obtained a formula for the partition function as a discrete sum over paths with each path labeled by an integer and given by a zeta-function evaluated at a fixed set of points in the critical strip. These points are the image of the space–time lattice resulting from a linear mapping.

10.6 Shifts of Universal Dirichlet Series

Recently, Kaczorowski, Laurinčikas and Steuding [157] studied shifts of universal Dirichlet series with respect to universality and their value-distribution. Assume that $\mathcal{F}(s)$ is analytic in some open strip, $\mathcal{D} = \{s \in \mathbb{C} : \sigma_1 < \sigma < \sigma_2\}$ say, and that $\mathcal{K}_1, \ldots, \mathcal{K}_n$ are disjoint compact subsets of $\mathcal{D}$ with connected complements. Let $g(s)$ be any (non-vanishing) continuous function, defined on $\bigcup_{j=1}^n \mathcal{K}_j$, which is analytic in the interior; here $g(s)$ is allowed to have zeros if and only if $\mathcal{F}(s)$ is strongly universal. If now for any $\epsilon > 0$, there exists a real number τ such that

$$\max_{s \in \cup_{j=1}^n \mathcal{K}_j} |\mathcal{F}(s + i\tau) - g(s)| < \epsilon,$$

then also

$$\max_{1 \leq j \leq n} \max_{s \in \mathcal{K}_j} |\mathcal{F}(s + i\tau) - g_j(s)| < \epsilon,$$

where the $g_j(s)$ are defined as restriction of $g(s)$ on $\mathcal{K}_j$; of course, equivalently, one can consider all $g_j(s)$ being defined on some compact subset $\mathcal{K}$ of $\mathcal{D}$ with connected complement and study shifts of $\mathcal{F}(s)$.

Let $\lambda_1, \ldots, \lambda_n$ be complex numbers and put

$$\lambda := \min_{1 \leq j \leq n} \operatorname{Re} \lambda_j \quad \text{and} \quad \Lambda := \max_{1 \leq j \leq n} \operatorname{Re} \lambda_j. \tag{10.6}$$

Further, let $\mathcal{K}$ be a compact set, and define $\mathcal{K}_j := \{s + \lambda_j : s \in \mathcal{K}\}$. Then the shifts

$$\mathcal{F}_{\lambda_j}(s) := \mathcal{F}(s + \lambda_j)$$

of $\mathcal{F}(s)$ are said to be jointly universal with respect to $\lambda_1, \ldots, \lambda_n$ if, for every compact $\mathcal{K}$ with connected complement and for which the sets $\mathcal{K}_j$ are disjoint subsets of $\mathcal{D}$, every family of (non-vanishing) continuous functions $g_j(s)$ defined on $\mathcal{K}$ which are analytic in the interior, and for any $\epsilon > 0$, we have

$$\liminf_{T \to \infty} \frac{1}{T} \text{meas} \left\{ \tau \in [0, T] : \max_{1 \le j \le n} \max_{s \in \mathcal{K}} : \mathcal{F}_{\lambda_j}(s + i\tau) - g_j(s)| < \epsilon \right\} > 0.$$

Clearly, the assumption on the $\mathcal{K}_j$ to be disjoint is necessary. Moreover, to have all shifts well defined on a non-empty subset of the complex plane we need the following necessary condition:

$$\Lambda - \lambda < \sigma_2 - \sigma_1 \qquad (10.7)$$

with λ and Λ defined in (10.6) and σ_1, σ_2 being the abscissae bounding $\mathcal{D}$. The condition $\mathcal{K}_k \cap \mathcal{K}_\ell = \emptyset$ for $k \ne \ell$ implies

$$\lambda_k \ne \lambda_\ell \quad \text{for} \quad k \ne \ell. \qquad (10.8)$$

According to the foregoing remarks one can formulate the following general

Shifts Universality Principle 1. *Every (strongly) universal function in $\mathcal{D}$ is jointly (strongly) universal with respect to every finite sequence of shifts $\lambda_1, \ldots, \lambda_n$ satisfying (10.7) and (10.8).*

For instance, $\zeta(s)$ and $\zeta(s + i\lambda)$ are jointly universal for any real $\lambda \ne 0$.

Now suppose that $\mathcal{F}(s)$ is universal in $\mathcal{D}$. By the same argument as in Sect. 10.1 it follows that then for every complex numbers $\lambda_1, \ldots, \lambda_n$ satisfying (10.7) and (10.8), for every $s_0 \in \mathcal{D}(\lambda_1, \ldots, \lambda_n) := \{ s \in \mathbb{C} : \sigma_1 - \lambda < \sigma < \sigma_2 - \Lambda \}$, the image of the curve

$$\left(\mathcal{F}_{\lambda_1}(s_0 + it), \ldots, \mathcal{F}_{\lambda_n}(s_0 + it), \ldots, \mathcal{F}_{\lambda_1}^{(\ell-1)}(s_0 + it), \ldots, \mathcal{F}_{\lambda_n}^{(\ell-1)}(s_0 + it) \right)$$

lies dense in $\mathbb{C}^{n\ell}$. It should be noticed that in the case of the zeta-function this implies and generalizes both statements of Voronin's Theorem 1.6. Moreover, as in Sect. 10.2, this statement can be used to deduce the functional independence for these shifts, i.e., if $f_0, \ldots, f_m$ are continuous functions on $\mathbb{C}^{n\ell}$, and if

$$\sum_{k=0}^{m} s^k f_k \left(\mathcal{F}_{\lambda_1}(s), \ldots, \mathcal{F}_{\lambda_n}(s), \ldots, \mathcal{F}_{\lambda_1}^{(\ell-1)}(s), \ldots, \mathcal{F}_{\lambda_n}^{(\ell-1)}(s) \right) = 0$$

identically for all $s \in \mathcal{D}(\lambda_1, \ldots, \lambda_n)$, then $f_k \equiv 0$ for $k = 0, 1, \ldots, m$.

Further, Kaczorowski, Laurinčikas and Steuding [157] considered linear combinations of jointly universal shifts $\mathcal{F}_{\lambda_j}(s)$. Let $n \ge 2$ and $a_1, \ldots, a_n$ be arbitrary, non-zero complex numbers and let

$$\mathcal{Z}(s) := \sum_{j=1}^{n} a_j \mathcal{F}_{\lambda_j}(s).$$

This linear combination is strongly universal:

Theorem 10.6. *Suppose that $n \geq 2$ and $\lambda_1, \ldots, \lambda_n$ are complex numbers satisfying (10.7), (10.8), and let $\mathcal{K}$ be a non-empty compact subset with the connected complement such that the sets $\mathcal{K}_j = \{s + \lambda_j : s \in \mathcal{K}\}$ are disjoint subsets of $\mathcal{D}$. Moreover, let $\mathcal{F}(s)$ be universal in $\mathcal{D}$ and let $g(s)$ be a continuous function on $\mathcal{K}$ which is analytic in the interior of $\mathcal{K}$. Then, for any $\epsilon > 0$,*

$$\liminf_{T \to \infty} \frac{1}{T} \operatorname{meas} \left\{ \tau \in [0, T] : \max_{s \in \mathcal{K}} |\mathcal{Z}(s + i\tau) - g(s)| < \epsilon \right\} > 0.$$

We sketch the proof. Define the functions

$$g_1(s) = \frac{1}{a_1}(g(s) + \ell + 1) \quad \text{and} \quad g_2(s) = -\frac{1}{a_2}(\ell + 1),$$

where $\ell := \max_{s \in \mathcal{K}} |g(s)|$. Then $g_1(s)$ does not vanish for $s \in \mathcal{K}$. Further, for $j = 3, \ldots, n$, define $g_j(s) \equiv \epsilon$ with some arbitrary positive ϵ. Then, for any τ,

$$\begin{aligned}
\mathcal{Z}(s + i\tau) &- g(s) \\
&= a_1 \left\{ \mathcal{F}_{\lambda_1}(s + i\tau) - g_1(s) \right\} + a_2 \left\{ \mathcal{F}_{\lambda_2}(s + i\tau) - g_2(s) \right\} \\
&\quad + \sum_{j=3}^{n} a_j \left\{ \mathcal{F}_{\lambda_j}(s + i\tau) - g_j(s) \right\} + \sum_{j=3}^{n} a_j \epsilon.
\end{aligned}$$

Now an application of the Shift Universality Principle yields the existence of a set of positive real numbers τ having positive lower density for each of which

$$\begin{aligned}
\max_{s \in \mathcal{K}} &|\mathcal{Z}(s + i\tau) - g(s)| \\
&\leq \sum_{j=1}^{n} |a_j| \left\{ \max_{1 \leq j \leq n} \max_{s \in \mathcal{K}} |\mathcal{F}_{\lambda_j}(s + i\tau) - g_j(s)| + \epsilon \right\} < 2\epsilon \sum_{j=1}^{n} |a_j|.
\end{aligned}$$

The right-hand side can be made as small as we please.

This theorem has interesting consequences on the zero-distribution. Under the conditions of the previous theorem, the function $\mathcal{Z}(s)$ can uniformly approximate functions $g(s)$ having zeros, so $\mathcal{Z}(s)$ is strongly universal. However, for a single shift $\mathcal{F}_{\lambda_j}(s)$ this is not true in general (e.g., the Riemann zeta-function is not strongly universal as shown in Sect. 8.1).

In the sequel we assume that $\mathcal{F}(s)$ is not vanishing identically, has a representation as an ordinary Dirichlet series, more precisely,

$$\mathcal{F}(s) = \sum_{n=1}^{\infty} \frac{f(n)}{n^s}$$

for $\sigma > \sigma_2$, and is regular and of finite order in the strip $\sigma_1 < \sigma < \sigma_2$ with bounded second moment:

$$\limsup_{T \to \infty} \frac{1}{T} \int_0^T |\mathcal{F}(\sigma + it)|^2 \, dt < \infty; \tag{10.9}$$

notice that these assumptions hold for the Riemann zeta-function, Dirichlet L-functions, Lerch zeta-functions, and many other naturally appearing Dirichlet series with $\sigma_1 = \frac{1}{2}$ and $\sigma_2 = 1$.

To obtain some information on the zero-distribution of $\mathcal{Z}(s)$ we denote by $N_{\mathcal{Z}}(\sigma, T)$ the number of zeros $\varrho = \beta + i\gamma$ of $\mathcal{Z}(s)$ satisfying $\beta > \sigma, 0 < \gamma < T$. Then

Theorem 10.7. *Let $\lambda_1, \ldots, \lambda_n$ be in pairs different and satisfy (10.7). Then for fixed $\sigma \in (\sigma_1 - \lambda, \sigma_2 - \Lambda)$, there exist positive constants c_1, c_2, depending only on σ and $\mathcal{Z}(s)$, such that*

$$c_1 \le \liminf_{T \to \infty} \frac{1}{T} N_{\mathcal{Z}}(\sigma, T) \le \limsup_{T \to \infty} \frac{1}{T} N_{\mathcal{Z}}(\sigma, T) \le c_2;$$

the constants c_1 and c_2 are explicitly given by (10.12) and (10.15) if condition (10.13) holds.

The method of proof relies on Theorem 10.6 and techniques from Sects. 7.1 and 8.2 (respectively, Sect. 9.2).

Proof. We start with the lower bound. For ℓ and $r > 0$ we shall consider the function $g(s) := s - \ell$ defined on the disk of radius r and center ℓ, that is $\mathcal{K} := \{\ell + s : |s| \le r\}$. We suppose that the disks $\mathcal{K}_j := \{\lambda_j + s : s \in \mathcal{K}\}$ are disjoint and that $\sigma_1 - \lambda + r < \ell < \sigma_2 - \Lambda - r$. An application of Theorem 10.6 yields the existence of some positive real numbers τ such that

$$\max_{s \in \mathcal{K}} |\mathcal{Z}(s + i\tau) - g(s)| < \epsilon. \tag{10.10}$$

This implies

$$\max_{|s|=r} |\mathcal{Z}(\ell + s + i\tau) - g(\ell + s)| < \epsilon \le r = \min_{|s|=r} |g(\ell + s)| \tag{10.11}$$

for sufficiently small ϵ. The function $g(\ell + s) = s$ has a zero inside the disk $|s| \le r$ and so, by Rouché's Theorem 8.1, there exists a zero of $\mathcal{Z}(s)$ inside any of the sets $\{\ell + s + i\tau : |s| \le r\}$ provided inequality (10.10) holds. The number of these zeros will give a lower bound for the zero-counting function $N_{\mathcal{Z}}$ in the formulation of the theorem. As in the proof of Theorem 8.3 (respectively, Theorem 9.1), we get

$$N_{\mathcal{Z}}(\sigma, T) \ge \frac{1}{2r} \operatorname{meas} \mathcal{I}(T),$$

where $\sigma := \ell - r > \sigma_1 - \lambda$, and $\mathcal{I}(T)$ is the set of $\tau \in [0, T]$ such that (10.10) is valid. In view of (10.11) we can take $r = \epsilon$. By Theorem 10.6,

$$\underline{d}(\epsilon) := \liminf_{T \to \infty} \frac{1}{T} \operatorname{meas} \left\{ \tau \in [0, T] : \max_{s \in \mathcal{K}} |\mathcal{Z}(s + i\tau) - g(s)| < \epsilon \right\} > 0$$

for any positive ϵ for which the sets $\mathcal{K}_j = \{\lambda_j + s : |s| \leq \epsilon\}$ lie disjoint. This yields the lower estimate of the theorem with the constant

$$c_1 = \frac{1}{2\epsilon} \, \underline{d}(\epsilon)$$

$$- \frac{1}{2\epsilon} \liminf_{T \to \infty} \frac{1}{T} \operatorname{meas} \left\{ \tau \in [0, T] : \max_{|s| \leq \epsilon} |\mathcal{Z}(\sigma + \epsilon + s + \mathrm{i}\tau) - s| < \epsilon \right\} \quad (10.12)$$

For the upper bound we suppose that

$$f(1) = 1 \quad \text{and} \quad a := \sum_{j=1}^{n} a_j \neq 0. \quad (10.13)$$

The general case can be treated analogously. We define the function

$$Z(s) := \frac{1}{a} \mathcal{Z}(s).$$

Obviously, the zeros of $Z(s)$ correspond one-to-one to the zeros of $\mathcal{Z}(s)$, and $Z(s)$ tends to 1 as $\sigma \to \infty$. It follows that $Z(s)$ has no zeros in some half-plane $\sigma \geq \sigma_3$. Littlewood's Lemma 7.2 yields

$$\int_{\kappa_1}^{\kappa_2} N_{\mathcal{Z}}(\sigma, T) \, \mathrm{d}\sigma = \frac{1}{2\pi\mathrm{i}} \int_{\mathcal{R}} \log Z(s) \, \mathrm{d}s + \mathrm{O}(1), \quad (10.14)$$

where $\mathcal{R}$ is the rectangular contour with vertices $\kappa_1, \kappa_2, \kappa_1 + \mathrm{i}T, \kappa_2 + \mathrm{i}T$ with $\sigma_1 - \lambda < \kappa_1 < \sigma_2 - \Lambda < \sigma_3 < \kappa_2$, and where the error term arises from possible poles at $s = 1 - \lambda_j$. The right-hand side of (10.14) is equal to

$$\frac{1}{2\pi} \int_0^T \log |Z(\kappa_1 + \mathrm{i}t)| \, \mathrm{d}t + \mathrm{O}(\log T);$$

the error term estimate comes from a standard use of Jensen's formula (as in the proof of Theorem 7.1). By Jensen's inequality, the appearing integral is

$$\leq \frac{T}{4\pi} \log \left(\frac{1}{T} \int_0^T |Z(\kappa_1 + \mathrm{i}t)|^2 \, \mathrm{d}t \right).$$

Since $\mathcal{F}(s)$ is regular and of finite order with bounded second moment (10.9), it follows from Carlson's Theorem 2.4 that, for $\sigma > \sigma_1$,

$$\lim_{T \to \infty} \frac{1}{T} \int_0^T |\mathcal{F}(\sigma + \mathrm{i}t)|^2 \, \mathrm{d}t = \sum_{n=1}^{\infty} \frac{|f(n)|^2}{n^{2\sigma}} =: F(2\sigma).$$

This leads to

$$\lim_{T\to\infty}\frac{1}{T}\int_0^T |Z(\kappa_1+\mathrm{i}t)|^2\,\mathrm{d}t \le \frac{n}{|a|^2}\sum_{j=1}^n |a_j|^2 \lim_{T\to\infty}\frac{1}{T}\int_0^T |\mathcal{F}_{\lambda_j}(\kappa_1+\mathrm{i}t)|^2\,\mathrm{d}t$$

$$= \frac{n}{|a|^2}\sum_{j=1}^n |a_j|^2 F(2(\kappa_1+\operatorname{Re}\lambda_j)).$$

Thus, we may replace (10.14) by

$$\sum_{\substack{\operatorname{Re}\varrho>\kappa_1\\0<\operatorname{Im}\varrho\le T}} (\operatorname{Re}\varrho-\kappa_1) \le \frac{T}{4\pi}\left(\log\left(\frac{n}{|a|^2}\sum_{j=1}^n |a_j|^2 F(2(\kappa_1+\operatorname{Re}\lambda_j))\right)+\mathrm{O}(1)\right);$$

here the sum on the left-hand side is taken over all zeros ϱ of $\mathcal{Z}(s)$ satisfying the condition of summation. Since, for $\sigma_1-\lambda<\kappa_1<\sigma$,

$$N_{\mathcal{Z}}(\sigma,T) \le \frac{1}{\sigma-\kappa_1}\sum_{\substack{\operatorname{Re}\varrho>\kappa_1\\0<\operatorname{Im}\varrho\le T}} (\operatorname{Re}\varrho-\kappa_1),$$

we obtain

$$N_{\mathcal{Z}}(\sigma,T) \le \frac{T}{4\pi(\sigma-\kappa_1)}\left(\log\left(\frac{n}{|a|^2}\sum_{j=1}^n |a_j|^2 F(2(\kappa_1+\operatorname{Re}\lambda_j))\right)+\mathrm{O}(1)\right).$$

Thus, taking $\kappa_1=\frac{1}{2}(\sigma_1-\lambda+\sigma)$ the upper estimate of the theorem holds with the constant

$$c_2 = \frac{1}{2\pi(\sigma+\lambda-\sigma_1)}\log\left(\frac{n}{|a|^2}\sum_{j=1}^n |a_j|^2 F(\sigma_1+\sigma+2\operatorname{Re}\lambda_j-\lambda)\right). \quad (10.15)$$

The theorem is proved. $\qquad\qquad\square$

We conclude with a few words concerning the zero-distribution of Dirichlet series. It seems that universal Dirichlet series have in their region of universality either *none* or $\asymp T$ *many zeros* with real part greater than σ for any fixed $\sigma\in[\frac{1}{2},1)$ up to level T. Selberg [323] investigated linear combinations of linearly independent elements of the Selberg class and conjectured that $\gg T\sqrt{\log\log T}$ *many of its zeros up to level T lie to the right of the critical line, but that almost all zeros lie on the critical line*. The latter conjecture was proved by Bombieri and Hejhal [38] in the case of Epstein zeta-functions attached to a rational quadratic form under the assumption of a well-spacing of the zeros. Gritsenko [106] showed that linear combinations of primitive degree 2-elements of the Selberg class have at least $T(\log\log T)^\eta$ many zeros on the critical line up to level T, where η is some constant. It is expected that a positive proportion of these zeros is located on the critical line.

11

Dirichlet Series with Periodic Coefficients

In Anschluß daran beweise ich u.a. eine Funktionalgleichung, der jede Dirichletsche Reihe mit periodischen Koeffizienten genügt; diese ist ganz überraschend einfach und scheint mir, wenn man sie auf die in der Zahlentheorie eine so große Rolle spielende Charakterreihen (*L*-Reihen) anwendet, die Quelle zu sein, von der aus die Funktionalgleichungen recht verständlich werden.   W. Schnee

In this chapter, we consider Dirichlet series associated with periodic arithmetical functions f, sometimes also called periodic zeta-functions. This class of Dirichlet series includes Dirichlet L-functions, but in general these functions do not have an Euler product; anyway, we shall denote them by $L(s, f)$. Such Dirichlet series are rather simple objects which have the advantage that many computations can be done explicitly. We prove universality for Dirichlet series attached to non-multiplicative periodic functions subject to some side restrictions. This leads to an interesting zero-distribution which is rather different to the one of Dirichlet L-functions. The results of this chapter are due to Steuding [340, 342, 343].

11.1 Zero-Distribution

Let q be a positive integer and let $f : \mathbb{Z} \to \mathbb{C}$ be a q-periodic arithmetical function, i.e.,

$$f(n + q) = f(n) \quad \text{for all} \quad n \in \mathbb{Z}.$$

We define the associated Dirichlet series by

$$L(s, f) = \sum_{n=1}^{\infty} \frac{f(n)}{n^s}.$$

Since the coefficients are bounded, this series converges for $\sigma > 1$. In the sequel, we assume that f does not vanish identically zero. However, we do not require that q is the minimal period; consequently, certain quantities appearing in asymptotic formulae below remain invariant if q is replaced by an integer multiple.

By the periodicity of f we find

$$L(s, f) = \frac{1}{q^s} \sum_{a=1}^{q} f(a) \zeta\left(s, \frac{a}{q}\right). \tag{11.1}$$

The analytic continuation of the Hurwitz zeta-function (1.42) leads immediately to an analytic continuation for $L(s, f)$. Thus, $L(s, f)$ is analytic throughout the whole complex plane except for at most one simple pole at $s = 1$; actually, $L(s, f)$ is regular at $s = 1$ if and only if $\sum_{a=1}^{q} f(a) = 0$ (respectively, $f^+(q) = 0$ in the notation (11.2)). By the identity

$$\zeta(1 - s, \alpha) = \frac{\Gamma(s)}{(2\pi)^s} \left\{ \exp\left(\frac{\pi i s}{2}\right) \sum_{n=1}^{\infty} \frac{\exp(-2\pi i \alpha n)}{n^s} \right.$$

$$\left. + \exp\left(-\frac{\pi i s}{2}\right) \sum_{n=1}^{\infty} \frac{\exp(2\pi i \alpha n)}{n^s} \right\},$$

valid for $\sigma > 1$ (this is a special case of the functional equation (1.44)), we get from (11.1)

$$L(1 - s, f) = \left(\frac{q}{2\pi}\right)^s \frac{\Gamma(s)}{\sqrt{q}} \left\{ \exp\left(\frac{\pi i s}{2}\right) \sum_{n=1}^{\infty} \frac{f^-(n)}{n^s} \right.$$

$$\left. + \exp\left(-\frac{\pi i s}{2}\right) \sum_{n=1}^{\infty} \frac{f^+(n)}{n^s} \right\}$$

for $\sigma > 1$, where

$$f^\pm(n) := \frac{1}{\sqrt{q}} \sum_{a=1}^{q} f(a) \exp\left(\pm 2\pi i \frac{an}{q}\right). \tag{11.2}$$

The functions $f^\pm$ may be interpreted as the discrete Fourier transforms of the functions $f_\pm : \mathbb{Z} \to \mathbb{C}$, $n \mapsto f(\pm n)$. Obviously, $f_+ = f$, and f_- is q-periodic as well. By analytic continuation we obtain the functional equation

$$L(1 - s, f_\pm) = \left(\frac{q}{2\pi}\right)^s \frac{\Gamma(s)}{\sqrt{q}} \left\{ \exp\left(\frac{\pi i s}{2}\right) L(s, f^\mp) \right.$$

$$\left. + \exp\left(-\frac{\pi i s}{2}\right) L(s, f^\pm) \right\}, \tag{11.3}$$

valid for all s, which was first proved by Schnee [316].

Denote the zeros of $L(s, f)$ by $\varrho = \beta + i\gamma$. As in the case of Dirichlet L-functions we have to distinguish between trivial and non-trivial zeros of

$L(s, f)$. Before we give a rigorous definition of trivial and non-trivial zeros we establish a zero-free region on the right. Let

$$C_f = \max\{|f(a)| : 1 \le a \le q\}$$

and

$$m_f = \min\{a : 1 \le a \le q, f(a) \ne 0\}.$$

Then, for $\sigma > 1$, we obtain

$$L(s, f) = \sum_{n=m_f}^{\infty} \frac{f(n)}{n^s} = \lambda(s, f) + \sum_{n=m_f+1}^{\infty} \frac{f(n)}{n^s},$$

where

$$\lambda(s, f) := \frac{f(m_f)}{m_f^s}.$$

This leads to the estimate

$$|L(s, f) - \lambda(s, f)| \le C_f \sum_{n=m_f+1}^{\infty} \frac{1}{n^\sigma} \le C_f \int_{m_f}^{\infty} \frac{\mathrm{d}x}{x^\sigma} \le \frac{C_f m_f^{1-\sigma}}{\sigma - 1}.$$

Hence, as $\sigma \to \infty$,

$$L(s, f) = \lambda(s, f) + \mathrm{O}\left(\frac{1}{\sigma m_f^\sigma}\right).$$

This implies

$$L(s, f) \ne 0 \quad \text{for} \quad \sigma > 1 + A(f),$$

where

$$A(f) := \frac{C_f}{|\lambda(1, f)|}.$$

By the functional equation (11.3) and the non-vanishing of the Gamma-function, $L(1 - s, f)$ vanishes if and only if

$$\exp\left(\frac{\pi \mathrm{i} s}{2}\right) L(s, f^-) = -\exp\left(-\frac{\pi \mathrm{i} s}{2}\right) L(s, f^+).$$

Define $B(f) = \max\{A(f^\pm)\}$. It follows that for $\sigma < -B(f)$ the function $L(s, f)$ can only have zeros close to the negative real axis if $m_{f^+} = m_{f^-}$, and close to the straight line, given by

$$\sigma = 1 + \frac{\pi t}{\log \frac{m_{f^-}}{m_{f^+}}},$$

if $m_{f^+} \ne m_{f^-}$. We call zeros $\varrho = \beta + \mathrm{i}\gamma$ of $L(s, f)$ with $\beta < -B(f)$ trivial. One can locate those zeros with the techniques of Spira [335] and Garunkštis [88]. It is easily seen that the number of trivial zeros ϱ with $|\varrho| \le R$ is $\sim cR$, where

c is some positive, computable constant depending on f. We call other zeros of $L(s, f)$ non-trivial. The non-trivial zeros lie in the vertical strip

$$-B(f) \leq \sigma \leq 1 + A(f). \tag{11.4}$$

If f is a Dirichlet character $\chi \mod q$, this definition of trivial and non-trivial zeros does not correspond with the traditional one in the theory of Dirichlet L-functions, where, for example, $L(s, \chi)$ has a trivial zero at $s = -1$ if $\chi(-1) = -1$. But note that there are only finitely many trivial zeros ϱ of $L(s, f)$ in any vertical strip of bounded width (independent on this or any other reasonable definition of *trivial*).

We start, similarly as in Chap. 7.1, with an asymptotic formula for a sum taken over the zeros.

Theorem 11.1. *Let f be a q-periodic arithmetic function and let b be a constant satisfying $-b \geq 3 + \max\{A(f), B(f)\}$. Then,*

$$\sum_{\substack{\beta > b \\ |\gamma| \leq T}} (\beta - b) = \left(\frac{1}{2} - b\right) \frac{T}{\pi} \log \frac{qT}{2\pi e m_f \sqrt{m_{f-} m_{f+}}}$$

$$+ \frac{T}{2\pi} \log \left| \frac{\lambda\left(\frac{1}{2}, f^-\right) \lambda\left(\frac{1}{2}, f^+\right)}{\lambda\left(\frac{1}{2}, f\right)^2} \right| + O(\log T).$$

Proof. Define

$$Z(s, f) = \frac{L(s, f)}{\lambda(s, f)}.$$

Of course, the zeros of $Z(s, f)$ are exactly the zeros of $L(s, f)$. Let $a = 2 + \max\{A(f), B(f)\}$. Then, by the condition on b, all non-trivial zeros of $L(s, f)$ and $L(s, f^{\pm})$ have real parts in the interval (b, a).

First, suppose that $L(s, f)$ is regular at $s = 1$. Then Littlewood's Lemma 7.2, applied to $Z(s, f)$ and the rectangle $\mathcal{R}$ with vertices $a \pm iT, b \pm iT$, yields

$$2\pi \sum_{\substack{\beta > b \\ |\gamma| \leq T}} (\beta - b) = \int_{-T}^{T} \log |Z(b + it, f)| \, dt - \int_{-T}^{T} \log |Z(a + it, f)| \, dt +$$

$$- \int_{b}^{a} \arg Z(\sigma - iT, f) \, d\sigma + \int_{b}^{a} \arg Z(\sigma + iT, f) \, d\sigma$$

$$= \sum_{j=1}^{4} I_j, \tag{11.5}$$

say. To define $\log Z(s, f)$ we choose the principal branch of the logarithm on the real axis, as $\sigma \to \infty$; for other points s the value of the logarithm is obtained by analytic continuation in a standard manner. If $L(s, f)$ is not regular at $s = 1$, we have to replace $Z(s, f)$ by $(s - 1)Z(s, f)$ at the expense of an error $O(1)$. To see this one applies Littlewood's lemma to $(s - 1)Z(s, f)$ and the function $s - 1$ separately and subtracts the resulting formulas. Since

$$\mathrm{i} \int_{\partial \mathcal{R}} \log(s - 1) \, \mathrm{d}s = 2\pi(b + 1) \ll 1,$$

we obtain (11.5) with an additional error term $O(1)$.

To evaluate I_1 note that

$$\log |Z(b + \mathrm{i}t, f)| = -\log |\lambda(b, f)| + \log |L(b + \mathrm{i}t, f)|.$$

By the functional equation (11.3) we get

$$\log |L(b + \mathrm{i}t, f)| = \log \left| \left(\frac{q}{2\pi}\right)^{1-b-\mathrm{i}t} \frac{\Gamma(1 - b - \mathrm{i}t)}{\sqrt{q}} \right| \tag{11.6}$$

$$+ \log \left| \exp \left(\frac{\pi\mathrm{i}(1 - b) + \pi t}{2}\right) L(1 - b - \mathrm{i}t, f^-) \right.$$

$$+ \exp \left(-\frac{\pi\mathrm{i}(1 - b) + \pi t}{2}\right) L(1 - b - \mathrm{i}t, f^+) \Big|.$$

For $|t| > 1$, Stirling's formula (2.17) implies

$$\log \left| \left(\frac{q}{2\pi}\right)^{1-b-\mathrm{i}t} \frac{\Gamma(1 - b - \mathrm{i}t)}{\sqrt{q}} \right| = \left(\frac{1}{2} - b\right) \log \frac{q|t|}{2\pi} - \frac{\pi|t|}{2} + O(|t|^{-1}).$$

For $t > 1$, the second term on the right-hand side of (11.6) equals

$$\log \left| \exp \left(\frac{\pi\mathrm{i}(1 - b) + \pi t}{2}\right) L(1 - b - \mathrm{i}t, f^-) \right.$$

$$\times \left. \left(1 + \exp(-\pi\mathrm{i}(1 - b) - \pi t) \frac{L(1 - b - \mathrm{i}t, f^+)}{L(1 - b - \mathrm{i}t, f^-)}\right) \right|$$

$$= \frac{\pi t}{2} + \log |\lambda(1 - b, f^-)| + \log \left| \frac{L(1 - b - \mathrm{i}t, f^-)}{\lambda(1 - b - \mathrm{i}t, f^-)} \right| + O(\exp(-\pi|t|)),$$

and, for $t < -1$,

$$-\frac{\pi t}{2} + \log |\lambda(1 - b, f^+)| + \log \left| \frac{L(1 - b - \mathrm{i}t, f^+)}{\lambda(1 - b - \mathrm{i}t, f^+)} \right| + O(\exp(-\pi|t|)).$$

Collecting, we obtain

$$I_1 = 2 \left(\frac{1}{2} - b\right) T \log \frac{qT}{2\pi e} + T \left(\log |\lambda(1 - b, f^-)\lambda(1 - b, f^+)| \right.$$

$$- 2 \log |\lambda(b, f)|) + \int_{-T}^{-1} \log \left| \frac{L(1 - b - \mathrm{i}t, f^+)}{\lambda(1 - b - \mathrm{i}t, f^+)} \right| \, \mathrm{d}t$$

$$+ \int_{1}^{T} \log \left| \frac{L(1 - b - \mathrm{i}t, f^-)}{\lambda(1 - b - \mathrm{i}t, f^-)} \right| \, \mathrm{d}t + O(1).$$

The appearing integrals look similar to I_2. We estimate them all as follows. For $\sigma > 1$,

$$Z(s,f) = 1 + \sum_{n=m_f+1}^{\infty} \frac{f(n)}{f(m_f)} \left(\frac{m_f}{n}\right)^s = 1 + \mathrm{O}\left(\left(\frac{m_f}{m_f+1}\right)^\sigma\right).$$

This yields $I_2 \ll 1$ by the same argument as in the proof of the estimate (7.8). Hence, we get

$$I_1 + I_2 = 2\left(\frac{1}{2} - b\right) T \log \frac{T}{2\pi e m_f \sqrt{m_{f^-} m_{f^+}}}$$

$$+ T \log \left|\frac{\lambda\left(\frac{1}{2}, f^-\right) \lambda\left(\frac{1}{2}, f^+\right)}{\lambda\left(\frac{1}{2}, f\right)^2}\right| + \mathrm{O}(1).$$

The horizontal integrals I_3, I_4 can be estimated as in the proof of Theorem 7.1. Thus the theorem is proved.   □

Denote by $N(T, f)$ the number of non-trivial zeros ϱ of $L(s, f)$ with $|\gamma| \le T$ (according multiplicities). The formula of Theorem 11.1 implies

Corollary 11.2. *Let f be a q-periodic arithmetic function. Then*

$$N(T, f) = \frac{T}{\pi} \log \frac{qT}{2\pi e m_f \sqrt{m_{f^-} m_{f^+}}} + \mathrm{O}(\log T),$$

and

$$\sum_{\substack{\varrho \text{ non-trivial} \\ |\gamma| \le T}} \left(\beta - \frac{1}{2}\right) = \frac{T}{2\pi} \log \left|\frac{\lambda\left(\frac{1}{2}, f^-\right) \lambda\left(\frac{1}{2}, f^+\right)}{\lambda\left(\frac{1}{2}, f\right)^2}\right| + \mathrm{O}(\log T).$$

The latter sum is $\mathrm{O}(N(T, f))$. So most of the non-trivial zeros are either approximately symmetrically distributed or lie close to the critical line. Indeed, setting

$$N_+(\sigma, T, f) = \sharp\{\varrho : \beta > \sigma, |\gamma| \le T\}$$

and

$$N_-(\sigma, T, f) = \sharp\{\varrho : \beta < \sigma, |\gamma| \le T\},$$

one can prove by the techniques which we used for the proof of Theorem 7.6:

Theorem 11.3. *Let f be a q-periodic arithmetic function. Then, uniformly in $\delta > 0$,*

$$N_+\left(\frac{1}{2} + \delta, T, f\right) + N_-\left(\frac{1}{2} - \delta, T, f\right) \ll \frac{1}{\delta} T \log \log T$$

$$\ll \frac{\log \log T}{\delta \log T} N(T, f).$$

Moreover, for every fixed $\sigma > \frac{1}{2}$,

$$N_+(\sigma, T, f) \ll T.$$

Thus, the non-trivial zeros of any $L(s, f)$ are clustered around the critical line. It seems that the clustering is a common pattern for Dirichlet series with a Riemann-type functional equation.

Now define

$$\Sigma(f) = \lim_{T \to \infty} \frac{2\pi}{T} \sum_{\substack{\varrho \text{ non-trivial} \\ |\gamma| \le T}} \left(\beta - \frac{1}{2} \right).$$

It follows from Corollary 11.2 that $L(s, f)$ has infinitely many zeros off the critical line if

$$\Sigma(f) = \log \left| \frac{\lambda\left(\frac{1}{2}, f^-\right) \lambda\left(\frac{1}{2}, f^+\right)}{\lambda\left(\frac{1}{2}, f\right)^2} \right| \ne 0.$$

Hence, a non-zero value of $\Sigma(f)$ indicates an asymmetrical distribution of the non-trivial zeros of $L(s, f)$ (with respect to the critical line), and in this case the number of non-trivial zeros of $L(s, f)$ with $\beta \ne \frac{1}{2}$ and $|\gamma| \le T$ is

$$\ge \frac{|\Sigma(f)|}{1 + 2 \max\{A(f), B(f)\}} \frac{T}{\pi}.$$

For example, the Ramanujan sum

$$c_q(n) := \sum_{\substack{a \bmod q \\ (a,q)=1}} \exp\left(2\pi \mathrm{i} \frac{an}{q} \right)$$

is q-periodic (in the n-aspect). Since $c_q(1) = \mu(q)$ and $c_q(1)^{\pm} = \sqrt{q}$, where μ is the Möbius μ-function, we have $\Sigma(c_q) = \log q$ by Corollary 11.2 if q is squarefree. It turns out that there are more than

$$\frac{\log q}{1 + 2\varphi(q)} \frac{T}{\pi}$$

many zeros off the critical line up to level T. However, in this special example we have Ramanujan's identity [301]

$$L(s, c_q) = \zeta(s) \sum_{d \mid q} \mu(q/d) d^{1-s}, \tag{11.7}$$

which gives precise information on the asymmetrically distributed zeros off the critical line.

We conclude this section by a look on the zero distributions of $L(s, f_{\pm})$ and $L(s, f^{\pm})$ with respect to each other. Note that with $f_{\pm}$ also the Fourier transforms $f^{\pm}$ are q-periodic. By the inversion formula for Fourier transforms,

$$(f^+)^{\pm} = f_{\mp} \quad \text{and} \quad (f^-)^{\pm} = f_{\pm},$$

we deduce from (11.3) the functional equation

$$L(1 - s, f^{\pm}) = \left(\frac{q}{2\pi} \right)^s \frac{\Gamma(s)}{\sqrt{q}} \left\{ \exp\left(\frac{\pi \mathrm{i} s}{2} \right) L(s, f_{\mp}) + \exp\left(-\frac{\pi \mathrm{i} s}{2} \right) L(s, f_{\pm}) \right\}.$$

Now we deduce from Corollary 11.2.

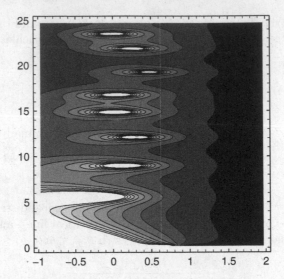

Fig. 11.1. The reciprocal of the absolute value of $L(s, \mathbf{1}_{1,5})$ for $\sigma \in [-1, 2], t \in [0, 25]$ as a contour plot. In each *white island* there is a zero of $\zeta(s, 1/5)$

Corollary 11.4. *For every q-periodic function f,*

$$\Sigma(f_+ f_- f^+ f^-) = \Sigma(f_+) + \Sigma(f_-) + \Sigma(f^+) + \Sigma(f^-) = 0.$$

For example, define for positive integers a, q with $1 \le a \le q$

$$\mathbf{1}_{a,q}(n) = \begin{cases} 1 & \text{if } n \equiv a \bmod q, \\ 0 & \text{otherwise.} \end{cases}$$

Obviously, $\mathbf{1}_{a,q}$ is q-periodic. One easily calculate

$$\Sigma((\mathbf{1}_{a,q})_+) + \Sigma((\mathbf{1}_{a,q})_-) + \Sigma((\mathbf{1}_{a,q})^+) + \Sigma((\mathbf{1}_{a,q})^-)$$
$$= \log \frac{a}{q} + \log \frac{q-a}{q} + \log \frac{q}{\sqrt{(q-a)a}} + \log \frac{q}{\sqrt{a(q-a)}}$$
$$= 0.$$

Note that

$$L(s, \mathbf{1}_{a,q}) = \frac{1}{q^s} \zeta\left(s, \frac{a}{q}\right). \tag{11.8}$$

Hence, the zeros of $L(s, \mathbf{1}_{a,q})$ are exactly the zeros of $\zeta(s, \frac{a}{q})$ (see Fig. 11.1 for an example). This special case should be compared with analogous results of Garunkštis and Steuding [91] for the Lerch zeta-function.

11.2 A Link to the Selberg Class

In this section, we are interested in the class $\mathcal{S}^\sharp$ consisting of all Dirichlet series

$$\mathcal{L}(s) := \sum_{n=1}^{\infty} \frac{a(n)}{n^s},$$

not identically zero, satisfying the axioms:

(ii') *Analytic continuation.* There exists a non-negative integer k such that $(s-1)^k \mathcal{L}(s)$ is an entire function of finite order.

(iii') *Functional equation.* $\mathcal{L}(s)$ satisfies a functional equation of type

$$\Lambda_{\mathcal{L}}(s) = \omega \overline{\Lambda_{\mathcal{L}}(1-\bar{s})},$$

where

$$\Lambda_{\mathcal{L}}(s) := \mathcal{L}(s) Q^s \prod_{j=1}^{f} \Gamma(\lambda_j s + \mu_j)$$

with positive real numbers Q, λ_j, and complex numbers μ_j, ω with $\operatorname{Re} \mu_j \geq 0$ and $|\omega| = 1$.

These axioms are the analytic hypotheses of the Selberg class $\mathcal{S}$. Therefore, $\mathcal{S} \subset \mathcal{S}^\sharp$. We refer to $\mathcal{S}^\sharp$ as the extended Selberg class. We define the degree of elements of $\mathcal{S}^\sharp$ in the same way as for $\mathcal{S}$ (see (6.2)) and denote by $\mathcal{S}^\sharp_{\mathrm{d}}$ the set of $\mathcal{L} \in \mathcal{S}^\sharp$ of degree d.

Kaczorowski and Perelli [160] proved

Theorem 11.5. *If $\mathcal{L} \in \mathcal{S}^\sharp_0$, then $\mathcal{L}(s)$ is a Dirichlet polynomial and*

$$\mathcal{L}(s) = \sum_{n|Q^2} \frac{a(n)}{n^s} \quad \text{with} \quad a(n) = \omega \frac{n}{Q} \overline{a(Q^2/n)}.$$

This follows more or less directly from the functional equation for $\mathcal{L}(s)$ which might be viewed as an identity between absolutely convergent Dirichlet series (see the proof of Theorem 6.1).

Our aim is to give a different classification for the degree zero-elements in the Selberg class, namely as entire quotients of Dirichlet series with periodic coefficients and the Riemann zeta-function $\zeta(s)$. One example we have already met in (11.7). The general result is

Theorem 11.6. *A function $\mathcal{L}$ lies in $\mathcal{S}^\sharp_0$ if and only if there exists a periodic arithmetical function $f : \mathbb{N} \to \mathbb{C}$ which satisfies*

$$f^+ = \omega \overline{f} \quad \text{with} \quad \omega \in \mathbb{C}, \ |\omega| = 1, \tag{11.9}$$

and the quotient $\mathcal{L}(s) = L(s,f)/\zeta(s)$ defines an entire function. In this case, $L(s,f) \in \mathcal{S}^\sharp_1$.

This result should be compared with a related theorem of Kaczorowski and Perelli [161] on a characterization of $\mathcal{S}^\sharp_1$-functions as $\mathcal{S}^\sharp_0$-linear combinations of Dirichlet L-functions and with the construction of Dirichlet series with periodic coefficients having zeros off the critical line by Balanzario [14].

Before we give the proof of the theorem we note

Lemma 11.7. *Let f be a q-periodic arithmetic function. If $L(s, f)/\zeta(s)$ is an entire function, then*

$$L(1 - s, f) = \frac{2}{\sqrt{q}} \left(\frac{q}{2\pi}\right)^s \Gamma(s) \cos\left(\frac{\pi s}{2}\right) L(s, f^+). \qquad (11.10)$$

Proof. To compensate the trivial zeros of $\zeta(s)$ in $s = -2m$ with $m \in \mathbb{N}$, the function $L(s, f)$ has to vanish there too. From the functional Equation (11.3) we deduce

$$L(2m + 1, f^+) = L(2m + 1, f^-)$$

and

$$\sum_{n=1}^\infty \frac{f^+(n) - f^-(n)}{n^{2m+1}} = 0 \quad \text{for all} \quad m \in \mathbb{N}.$$

Since any convergent Dirichlet series, which does not vanish identically, has a zero-free half-plane (see Sect. 2.1), it follows that $f^+ = f^-$, and, since $f = f_+$, we can replace (11.3) by (11.10). The lemma is proved. $\square$

Now we are in the position to give the

Proof of Theorem 11.6. First, assume that $\mathcal{L}(s) = L(s, f)/\zeta(s)$ is an entire function. By Lemma 11.7 we see that $L(s, f)$ satisfies the functional equation (11.10). If f satisfies additionally (11.9), then we can rewrite (11.10) as

$$L(1 - s, f) = \frac{2\omega}{\sqrt{q}} \left(\frac{q}{2\pi}\right)^s \Gamma(s) \cos\left(\frac{\pi s}{2}\right) L(s, \overline{f}), \qquad (11.11)$$

or, using well-known identities for the Gamma-function,

$$\left(\frac{q}{\pi}\right)^{(1-s)/2} \Gamma\left(\frac{1-s}{2}\right) L(1 - s, f) = \omega \left(\frac{q}{\pi}\right)^{s/2} \Gamma\left(\frac{s}{2}\right) L(s, \overline{f}).$$

We note that $L(s, f) \in \mathcal{S}_1^\sharp$. Using (11.11) also with f constant equal to 1 (the functional equation for $\zeta(s)$), we obtain

$$q^{s/2} \mathcal{L}(s) = \omega q^{(1-s)/2} \overline{\mathcal{L}(1 - \overline{s})}.$$

Therefore, $\mathcal{L}(s)$, being the entire quotient of two Dirichlet series in $\mathcal{S}_1^\sharp$, lies in $\mathcal{S}_0^\sharp$. This proves the sufficiency.

Now assume that $\mathcal{L} \in \mathcal{S}_0^\sharp$. By Theorem 11.5 there exist a positive integer Q^2 and complex numbers $a(n)$ such that the coefficients $a(n)$ in the Dirichlet series expansion of $\mathcal{L}(s)$ vanish if n does not divide Q^2. Now define $f(n) = \sum_{d|n} a(d)$, then

$$f(n + Q^2) = \sum_{\substack{d|(n+Q^2) \\ (d|Q^2)}} a(d) = \sum_{\substack{d|n \\ (d|Q^2)}} a(d) = f(n),$$

and thus f is a Q^2-periodic function. Further, $L(s, f) = \zeta(s)\mathcal{L}(s)$, and it remains to show (11.9). Using the functional equation for $\zeta(s)$, we see that $L(s, f)$ satisfies the functional equation (11.11), after replacing q by Q^2. Moreover, since $L(s, f)/\zeta(s)$ is entire, Lemma 11.7 implies that further (11.10) holds with Q^2 instead of q. Comparing both functional equations, we find $L(s, f^+) = \omega L(s, \overline{f})$. In view of the uniqueness of Dirichlet series expansions we obtain (11.9). Theorem 11.6 is proved.   □

It is interesting to investigate the difference between the Selberg class and the extended Selberg class, or which functions lie in $\mathcal{S}^\sharp \setminus \mathcal{S}$? The descriptions of these classes are complete for degree less than or equal to one. For degree two Kaczorowski et al. [158] gave examples with Dirichlet series associated with cusp forms of certain Hecke groups. Note that for a positive real number λ, the Hecke group $\mathsf{G}(\lambda)$ is defined as the subgroup of $\mathsf{PSL}_2(\mathbb{R})$ given by

$$\mathsf{G}(\lambda) := \left\langle \begin{pmatrix} 1 & \lambda \\ 0 & 1 \end{pmatrix}, \begin{pmatrix} 0 & 1 \\ -1 & 0 \end{pmatrix} \right\rangle.$$

Kaczorowski et al. showed that the associated Dirichlet series are elements of $\mathcal{S}^\sharp$ or a related class of Dirichlet series where the axiom on the functional equation is appropriately adjusted. Moreover, they showed that the Dirichlet series associated to newforms for $\mathsf{G}(\lambda)$ have an Euler product representation if and only if $\mathsf{G}(\lambda)$ can be arithmetically defined, i.e., if $\lambda \in \{1, \sqrt{2}, \sqrt{3}, 2\}$. Their result is based on Hecke's famous theorem on the correspondence of Dirichlet series with functional equation and modular forms, and a result of Wolfart on the arithmetic nature of the Fourier coefficients of the associated modular forms. For $\lambda \leq 2$, Molteni and Steuding [257] proved that all these Dirichlet series are almost primitive (i.e., primitive up to factors of degree zero) and primitive if $\lambda \notin \{\sqrt{2}, \sqrt{3}, 2\}$; if the latter condition is not fulfilled, there are examples of non-primitive functions.

11.3 Strong Universality

As a more or less immediate consequence of the joint universality of Dirichlet L-functions we obtain a strong universality result for $L(s, f)$:

Theorem 11.8. *Suppose that $q > 2$ and f is a q-periodic arithmetical function, not a multiple of a character $\mathrm{mod}\, q$, satisfying $f(n) = 0$ for n with $(n, q) > 1$. Let $\mathcal{K}$ be a compact subset of the strip $\frac{1}{2} < \sigma < 1$ with connected complement and let $g(s)$ be a continuous function on $\mathcal{K}$ which is analytic in the interior. Then, for any $\epsilon > 0$,*

$$\liminf_{T \to \infty} \frac{1}{T} \, \mathrm{meas} \left\{ \tau \in [0, T] : \max_{s \in \mathcal{K}} |L(s + i\tau, f) - g(s)| < \epsilon \right\} > 0.$$

For example, the Davenport–Heilbronn zeta-function (7.1) satisfies the conditions and is therefore strongly universal.

The assumptions on f from Theorem 11.8 imply that f is not multiplicative; this follows from a characterization of q-periodic multiplicative functions due to Leitmann and Wolke [213]. In particular, the associated Dirichlet series $L(s, f)$ has no Euler product representation.

Proof. For the $\varphi(q)$ pairwise non-equivalent characters χ_j mod q define the matrix

$$\Xi = (\chi_j(n))_{\substack{1 \le n \le q, (n,q)=1 \\ 1 \le j \le \varphi(q)}}.$$

By the properties of characters the matrix Ξ is invertible (since pairwise non-equivalent characters are linearly independent over $\mathbb{C}$, and $\det \Xi$ is a van der Monde-determinant). Hence there exist uniquely determined complex numbers $c_1, \ldots, c_{\varphi(q)}$ such that

$$f(n) = \sum_{j=1}^{\varphi(q)} c_j \chi_j(n) \quad \text{for} \quad 1 \le n \le q, (n,q) = 1,$$

where at least two distinct coefficients c_j are non-vanishing (since f is not a multiple of a character mod q and $\varphi(q) \ge 2$). In view of $f(n) = 0$ for integers n which are not coprime with q, we obtain the representation

$$L(s, f) = \sum_{j=1}^{\varphi(q)} c_j L(s, \chi_j). \tag{11.12}$$

Put $\ell_g = 1 + \max_{s \in \mathcal{K}} |g(s)|$. Then $g(s) + \ell_g \ne 0$ for any $s \in \mathcal{K}$. Without loss of generality, we may assume that $c_1, c_2 \ne 0$. Now define

$$g_{\chi_1}(s) = \frac{g(s) + \ell_g}{c_1}, \quad g_{\chi_2} = -\frac{\ell_g}{c_2},$$

and

$$g_{\chi_j}(s) = \eta \quad \text{for} \quad 3 \le j \le \varphi(q),$$

where η is a small positive parameter, chosen later. Using the joint universality Theorem 1.10 for Dirichlet L-functions, we obtain, for any $\delta > 0$,

$$\liminf_{T \to \infty} \frac{1}{T} \text{meas} \left\{ \tau \in [0, T] : \max_{1 \le j \le \varphi(q)} \max_{s \in \mathcal{K}} |L(s + i\tau, \chi_j) - g_{\chi_j}(s)| < \delta \right\} > 0.$$

Note that

$$g(s) = c_1 g_{\chi_1}(s) + c_2 g_{\chi_2}(s).$$

Consequently, (11.12) implies

$$\max_{s \in \mathcal{K}} |L(s + i\tau, f) - g(s)|$$

$$\leq \sum_{j=1}^{\varphi(q)} |c_j| \max_{s \in \mathcal{K}} |L(s + i\tau, \chi_j) - g_{\chi_j}(s)| + \eta \sum_{j=3}^{\varphi(q)} |c_j|$$

$$\leq \varphi(q)(\delta \max_{1 \leq j \leq \varphi(q)} |c_j| + \eta).$$

With sufficiently small δ and η the assertion of the theorem follows.   □

Theorem 11.8 is due to Bagchi [9]; however, his argument differs slightly from ours. Generalizations were given by Sander and Steuding [315], which include universality for arbitrary Dirichlet series with periodic coefficients and elements of degree 1 of the extended Selberg class in particular. The latter case was independently proved by Kaczorowski and Kulas [156] in a stronger form.

Next we shall give explicit upper bounds for the density of universality. With regard to Theorem 9.1 we get, using the same notation as in Sect. 9.2,

Theorem 11.9. *Suppose that $q > 2$ and f is a q-periodic arithmetical function, not a multiple of a character mod q, satisfying $f(1) = 1$ and $f(n) = 0$ for n with $(n, q) > 1$. Further, assume that $g \in \mathcal{A}_r$. Then, for any $\epsilon \in \left(0, \frac{1}{2r}\left(\frac{1}{4} + \operatorname{Re}|\xi|\right)\right)$ (where ξ is the zero of $g(s)$ in $|s| \leq r$) and any $\sigma_1 \in \left(\frac{1}{2}, \frac{3}{4} + \operatorname{Re}\xi - 2r\epsilon\right)$,*

$$\overline{\mathrm{d}}(\epsilon, g, L(s, f)) \leq \frac{2r^3 \epsilon}{\pi(r^2 - |\xi|^2)\left(\frac{3}{4} + \operatorname{Re}\xi - 2r\epsilon - \sigma_1\right)}$$

$$\times \log\left(\frac{1}{q^{2\sigma_1}} \sum_{a=1}^{q} |f(a)|^2 \zeta\left(2\sigma_1, \frac{a}{q}\right)\right).$$

The proof is quite similar to some parts of the proof for Theorem 10.7.

Proof. Denote by $N(\sigma, T, f)$ the number of zeros $\varrho = \beta + i\gamma$ of $L(s, f)$ with $\beta > \sigma$ and $0 < \gamma \leq T$. Then Littlewood's Lemma 7.2 yields

$$\int_{\sigma_1}^{\sigma_2} N(\sigma, T, f)\, d\sigma = \frac{1}{2\pi i} \int_{\mathcal{R}} \log L(s, f)\, ds + O(1), \qquad (11.13)$$

where $\mathcal{R}$ is a rectangular contour with vertices $\sigma_1, \sigma_2, \sigma_1 + iT, \sigma_2 + iT$, where $\frac{1}{2} < \sigma_1 < 1 < \sigma_2$. Here the error term arises from the possible pole of $L(s, f)$ at $s = 1$ and $\log L(s, f)$ is defined in a standard manner (as in the proof of Theorem 11.1). Since

$$L(s, f) = 1 + O(1)$$

as $\sigma \to \infty$, we choose σ_2 so large that $L(s, f)$ has no zeros in the half-plane $\sigma \geq \sigma_2$. A standard application of Jensen's formula (as in the proof of Theorem 7.1) shows that the right-hand side of (11.13) is equal to

$$\frac{1}{2\pi} \int_0^T \log |L(\sigma_1 + it, f)| \, dt + O(\log T)$$

$$\leq \frac{T}{4\pi} \log \left(\frac{1}{T} \int_0^T |L(\sigma_1 + it, f)|^2 \, dt \right) + O(\log T).$$

Therefore, using the mean-square formula of Kačėnas and Laurinčikas [152],

$$\frac{1}{T} \int_0^T |L(\sigma + it, f)|^2 \, dt \sim \frac{1}{q^{2\sigma}} \sum_{a=1}^q |f(a)|^2 \zeta \left(2\sigma, \frac{a}{q} \right),$$

valid for $\sigma > \frac{1}{2}$, and the identity (7.4), we may replace (11.13) by

$$\sum_{\substack{\beta > \sigma_1 \\ 0 < \gamma \leq T}} (\beta - \sigma_1) \leq \frac{T}{4\pi} \log \left(\frac{1}{q^{2\sigma_1}} \sum_{a=1}^q |f(a)|^2 \zeta \left(2\sigma_1, \frac{a}{q} \right) \right) \qquad (11.14)$$

$$+ O(\log T).$$

Now let $\sigma_3 > \sigma_1$. Since

$$N(\sigma_3, T, f) \leq \frac{1}{\sigma_3 - \sigma_1} \sum_{\substack{\beta > \sigma_1 \\ 0 < \gamma \leq T}} (\beta - \sigma_1),$$

we obtain the estimate

$$N \left(\frac{3}{4} + \operatorname{Re} \xi - 2r\epsilon, T, f \right)$$

$$\leq \frac{T}{4\pi \left(\frac{3}{4} + \operatorname{Re} \xi - 2r\epsilon - \sigma_1 \right)} \log \left(\frac{1}{q^{2\sigma_1}} \sum_{a=1}^q |f(a)|^2 \zeta \left(2\sigma_1, \frac{a}{q} \right) \right)$$

$$+ O(\log T).$$

In view of (9.6) this implies the estimate in the theorem.   □

We give an example. For an odd prime q, consider the function $L(s, \mathbf{1}_{1,q})$, given by (11.8). By Theorem 11.8, $L(s, \mathbf{1}_{1,q})$ has the universality property, and, using Theorem 11.9, we obtain after some lengthy calculations, for sufficiently small $\epsilon > 0$,

$$\overline{d}(\epsilon, g, L(s, \mathbf{1}_{1,q})) \leq \frac{2\zeta \left(\frac{9}{8} + \frac{|\xi|}{2} \right) r^3 \epsilon}{q^{9/8 + |\xi|/2} \pi (r^2 - |\xi|^2)}.$$

Since the right-hand side tends to zero as $q \to \infty$, the problem to approximate an analytic isomorphism $g(s)$ by a function $L(s, \mathbf{1}_{1,q})$ becomes, in some sense, more and more difficult with increasing q.

Following the reasoning of Chap. 10, universality for Dirichlet series with periodic coefficients leads to results on the denseness of the set of values taken on vertical lines, and functional independence. Here we only consider the particular case of the distribution of zeros. Strong universality leads to the existence of *many* zeros off the critical line. Via Rouché's Theorem 8.1 the universality property implies the existence of $\gg T$ zeros in each strip $\frac{1}{2} < \sigma_1 < \sigma < \sigma_2 < 1$ up to level T. This reasoning is very similar to the proof of Theorem 10.1. In particular, it follows that the real parts of the zeros lie dense in $[\frac{1}{2}, 1]$. In conjunction with the estimates of Theorem 11.3 we get the exact order of size for the zero-counting function $N(\sigma, T, f)$, introduced in the proof of Theorem 11.9.

Corollary 11.10. *Suppose that $L(s, f)$ satisfies the conditions of Theorem 11.8. Then, for $\frac{1}{2} < \sigma < 1$,*

$$N(\sigma, T, f) \asymp T.$$

There is another interesting result in this context. Kaczorowski and Kulas [156] proved that any element $\mathcal{L}$ of degree 1 of the extended Selberg class has $o(T)$ many zeros in the same range as in Theorem 11.9 if and only if $\mathcal{L}(S)$ has a factorization

$$\mathcal{L}(s) = P(s)L(s, \chi),$$

where $P(s)$ is a Dirichlet polynomial from $\mathcal{S}_0^\sharp$ and χ is a primitive Dirichlet character. Their reasoning is based on the structure theorem for $\mathcal{S}_1^\sharp$ from Kaczorowski and Perelli [161] and a variation of the joint universality theorem for Dirichlet L-functions 1.10. The result of Kaczorowski and Kulas gives a partial answer to the question whether any Dirichlet series satisfying Riemann's hypothesis or, at least, a density estimate has necessarily an Euler product.

11.4 Hurwitz Zeta-Functions

We consider Hurwitz zeta-functions with rational parameters (for their definition see (1.42)). In view of (11.8) and Theorem 11.8 we see that the function

$$\frac{1}{q^s} \zeta\left(s, \frac{a}{q}\right) = L(s, 1_{a,q}). \tag{11.15}$$

is strongly universal for $q > 2$. For $\alpha \in \{\frac{1}{2}, 1\}$, we have

$$\zeta\left(s, \frac{1}{2}\right) = 2^s L(s, \chi) \quad \text{and} \quad \zeta(s, 1) = \zeta(s), \tag{11.16}$$

where χ is the unique character mod 2. Thus, it follows from Voronin's results, respectively, the refinements due to Bagchi and Gonek, Theorem 1.9, 1.10, and their discrete variant Theorem 5.17) that the functions

$$\frac{1}{2^s}\zeta\left(s,\frac{1}{2}\right) \quad \text{and} \quad \zeta(s,1)$$

are universal but not strongly universal (by the results from Chap. 8). The appearance of the factor 2^{-s} in the case $\alpha = \frac{1}{2}$ is a bit unpleasant here but it can be removed as we shall explain now.

The first argument is by an application of a universality theorem for Dirichlet L-functions due to Gonek [104]. Incorporating some ideas from Good [105], Gonek proved

Theorem 11.11. *Let q be a positive integer and let $\mathcal{K}$ be a simply connected compact set in $\frac{1}{2} < \sigma < 1$. Suppose that for each prime $p|q$ we have $0 \leq \theta_p < 1$ and that for each character χ mod q, the function $g_\chi(s)$ is continuous on $\mathcal{K}$ and analytic in the interior. Then, for every $\epsilon > 0$, there is a real number τ such that*

$$\min_{z \in \mathbb{Z}} |\tau \log p - 2\pi(z + \theta_p)| < \epsilon \quad \text{for all} \quad p|q, \tag{11.17}$$

and

$$\max_{\chi \bmod q} \max_{s \in \mathcal{K}} |L(s + i\tau, \chi) - g_\chi(s)| < \epsilon.$$

Gonek's theorem does not show that the set of these τ has positive lower density as in Voronin's universality theorem for the zeta-function and also the set $\mathcal{K}$ on which these approximations are realized is more restricted than usually in universality theorems (however, it seems to be possible to extend Gonek's approach in order to obtain these stronger concepts of universality). But this result has the option that the diophantine condition (11.17) can be used to *remove* the factor 2^{-s} in (11.16) as well as the factor q^{-s} in (11.15). Bagchi [9] gave a different argument based on statistical independence. A third possibility is to use the joint discrete universality Theorem 5.17 for Dirichlet L-functions due to Bagchi [9].

Sander and Steuding [315] used the latter argument to obtain a joint universality theorem for Hurwitz zeta-functions with rational parameters. We sketch the simple argument which mimics the proof of Theorem 11.8. Define vectors

$$\mathsf{Z} := \left(\frac{1}{q^s}\zeta\left(s,\frac{a}{q}\right)\right)^t_{1 \leq a \leq q; \gcd(a,q)=1} \quad \text{and} \quad \mathsf{L} := \left(L(s,\chi_j)\right)^t_{1 \leq j \leq \varphi(q)}.$$

Then we deduce from

$$\frac{1}{q^s}\zeta\left(s,\frac{a}{q}\right) = \frac{1}{\varphi(q)}\sum_{\chi \bmod q} \overline{\chi}(a)L(s,\chi)$$

(this is the inverse of (11.1) for $f = \chi$) that

$$\mathsf{Z} = \mathsf{M}\mathsf{L} \quad \text{with} \quad \mathsf{M} := \left(\frac{\overline{\chi}_j(a)}{\varphi(q)}\right)_{\substack{1 \leq j \leq \varphi(q) \\ 1 \leq a \leq q; \gcd(a,q)=1}},$$

where the χ_j denote the $\varphi(q)$ pairwise non-equivalent characters modulo q. Clearly, M is a square matrix. Imagine we want to approximate a family of *nice* target functions $f_a : \mathcal{K} \to \mathbb{C}$ with $1 \leq a \leq q$ and a coprime with q. The idea is to find appropriate functions $g_j : \mathcal{K} \to \mathbb{C}, 1 \leq j \leq \varphi(q)$ such that $g_j(s)$ is approximated by $L(s, \chi_j)$ and

$$ \mathsf{f} = \mathsf{Mg}, \quad \text{where} \quad \mathsf{f} := \left(\frac{1}{q^s} f_a(s)\right)^{\mathsf{t}}_{1 \leq a \leq q; \gcd(a,q)=1}, \quad \mathsf{g} := \left(g_j(s)\right)^{\mathsf{t}}_{1 \leq j \leq \varphi(q)}. $$

The functions $g_j(s)$ are uniquely determined since M is invertible. We have $\mathsf{g} = \mathsf{M}^{-1}\mathsf{f}$. To approximate the functions $g_j(s)$ by Dirichlet L-functions, we need to assume that the $g_j(s)$ are non-vanishing continuous functions on $\mathcal{K}$ which are analytic in the interior. Then, given $\epsilon > 0$, for any real $\Delta \neq 0$, by Theorem 5.17 we can find a set of positive integers n with positive lower density such that

$$ \max_{1 \leq j \leq \varphi(q)} \max_{s \in \mathcal{K}} |L(s + i\Delta n, \chi_j) - g_j(s)| < \epsilon. $$

Hence, we may deduce from the identity $\mathsf{Z} - \mathsf{f} = \mathsf{M}(\mathsf{L} - \mathsf{g})$ that

$$ \max_{\substack{1 \leq a \leq q \\ \gcd(a,q)=1}} \max_{s \in \mathcal{K}} \left| \frac{1}{q^{i\Delta n}} \zeta\left(s + i\Delta n, \frac{a}{q}\right) - f_a(s) \right| < \epsilon. $$

Now, choosing Δ appropriately, we may get rid of the factor $q^{-i\Delta n}$ and arrive at the following *weak* joint universality theorem:

Theorem 11.12. *Let $\mathcal{K}$ be a compact subset of $\frac{1}{2} < \sigma < 1$ with connected complement, let q be a positive integer, and for each $1 \leq a \leq q$ with $\gcd(a, q) = 1$ let $f_a(s)$ be a continuous functions on $\mathcal{K}$ which is analytic in the interior. If all components of $\mathsf{M}^{-1}\mathsf{f}$ are non-vanishing on $\mathcal{K}$ (in the notation from above), then, for any $\epsilon > 0$,*

$$ \liminf_{N \to \infty} \frac{1}{N} \sharp \left\{ n \leq N : \max_{\substack{1 \leq a \leq q \\ \gcd(a,q)=1}} \max_{s \in \mathcal{K}} \left| \zeta\left(s + i\Delta n, \frac{a}{q}\right) - f_a(s) \right| < \epsilon \right\} > 0, $$

where $\Delta = \frac{2\pi k}{\log q}$ with any $k \in \mathbb{N}$ if $q \geq 2$, and $0 \neq \Delta \in \mathbb{R}$ otherwise.

Notice that q is allowed to be equal to 1 or 2; however, in these cases we do not obtain strong universality since then $\varphi(q) = 1$ and $a = 1$ and so the condition on the non-vanishing of the components of $\mathsf{M}^{-1}\mathsf{f}$ is nothing but the restriction for $f_1(s)$ to be free of zeros in $\mathcal{K}$. This corresponds to our remarks concerning (11.16).

How strong is Theorem 9.5? By (11.1) with $f = \chi$,

$$ L(s, \chi) = \frac{1}{q^s} \sum_{a=1}^{q} \chi(a) \zeta\left(s, \frac{a}{q}\right), $$

it follows that joint universality for Dirichlet L-functions is equivalent to joint universality for Hurwitz zeta-functions to rational parameters. If the non-vanishing condition for the components of $M^{-1}f$ would be dropped, it would follow that Dirichlet L-functions could approximate functions with zeros which is impossible.

Sander and Steuding [315] also investigated to which extent the condition on the parameters of the Hurwitz zeta-functions to be reduced fractions with a common denominator can be skipped. However, this is not always possible. Assume that we are given a set $\mathcal{A}$ of integers $1 \leq a \leq q$. By relations as for example

$$\zeta(s) = \frac{1}{q^s} \sum_{1 \leq a \leq q} \zeta\left(s, \frac{a}{q}\right)$$

it is easily seen that there exist dependence relations among the Hurwitz zeta-functions $\zeta(s, \frac{a}{q})$ with $a \in \mathcal{A}$ if $\sharp\mathcal{A} > \varphi(q)$. However, if $\sharp\mathcal{A} \leq \varphi(q)$, then there exists *restricted* joint universality whenever the target functions are linearly independent; here the conditions are rather technical. Nevertheless, using the notion of *weak joint universality* from Sect. 10.3, it follows that many families of Hurwitz zeta-functions with rational parameters are jointly functional independent.

There is an interesting related, recent result due to Nakamura [276] who proved that generalizations of Lerch zeta-functions are *not jointly* universal. This is a rather surprising result. For $\lambda, \alpha, \beta \in (0, 1]$ and $\operatorname{Re}\gamma \in [0, \frac{1}{2})$ the generalized Lerch zeta-function $\mathfrak{L}(\lambda, \alpha, \beta, \gamma, s)$ is for $\sigma > 1$ defined by

$$\mathfrak{L}(\lambda, \alpha, \beta, \gamma, s) = \sum_{n=0}^{\infty} \frac{\exp(2\pi i \lambda n)}{(n+\alpha)^{s-\gamma}(n+\beta)^{\gamma}}$$

and by analytic continuation for $\sigma > 0$ except for at most a simple pole at $s = 1$. We note that for $\alpha = \beta$ or $\gamma = 0$ this function is indeed a Lerch zeta-function:

$$\mathfrak{L}(\lambda, \alpha, \alpha, \gamma, s) = L(\lambda, \alpha, s) \quad \text{and} \quad \mathfrak{L}(\lambda, \alpha, \beta, 0, s) = L(\lambda, \alpha, s).$$

In fact, Nakamura [276] showed joint universality for any finite family $\mathfrak{L}(\lambda_j, \alpha_j, \beta_j, \gamma, s)$ provided the α_j are algebraically independent transcendental numbers; however, he also proved that any family $\mathfrak{L}(\lambda_j, \alpha, \beta_j, \gamma, s)$ with transcendental α is *not* jointly universal in the sense of Voronin if at least two of the λ_j are equal. The latter statement follows from the fact that the set

$$(\mathfrak{L}(\lambda, \alpha, \beta_j, \gamma, s + i\tau), \mathfrak{L}(\lambda, \alpha, \beta_k, \gamma, s + i\tau))$$

is not dense for any admissible pair β_j, β_k in the according function space as τ varies in $\mathbb{R}$. However, this does not exclude the possibility of joint universality for a large class of functions.

We conclude with another open problem in universality theory. In [243], Matsumoto asked whether multiple zeta-functions are universal; for example, the Barnes multiple zeta-function defined by

$$\zeta_\ell(s; \alpha, \omega_1, \ldots, \omega_\ell) = \sum_{m_\ell=0}^{\infty} \cdots \sum_{m_\ell}^{\infty} \frac{1}{(\alpha + m_1\omega_1 + \ldots + m_\ell\omega_\ell)^s},$$

or the double Hurwitz–Lerch zeta-function given by

$$\zeta_2(s_1, s_2; \alpha, \beta, \omega) = \sum_{m_1=0}^{\infty} \frac{1}{(\alpha + m_1)^{s_1}} \sum_{m_2=0}^{\infty} \frac{\exp(2\pi i m_2 \beta)}{(\alpha + m_1 + m_2\omega)^{s_2}}.$$

Joint Universality

What we know is not much. What we do not know is immense.
P.S. Laplace

In this chapter, we shall prove a conditional joint universality theorem for functions in $\tilde{\mathcal{S}}$. *Joint* universality means that we are concerned with simultaneous uniform approximation, a topic invented by Voronin [362, 364]. Of course, such a result cannot hold for an arbitrary family of L-functions: e.g., $\zeta(s)$ and $\zeta(s)^2$ cannot be jointly universal. The L-functions need to be sufficiently independent to possess this joint universality property. We formulate sufficient conditions for a family of L-functions in order to be jointly universal and give examples when these conditions are fulfilled; for instance, Dirichlet L-functions to pairwise non-equivalent characters (this is an old result of Voronin) or twists of L-functions in the Selberg class subject to some condition on uniform distribution.

12.1 A Joint Limit Theorem

Our first aim is to prove a joint limit theorem for L-functions from $\tilde{\mathcal{S}}$; this generalizes the limit Theorem 4.3 to a multi-dimensional limit theorem for a family of elements $\mathcal{L}_k \in \tilde{\mathcal{S}}$, $1 \le k \le \ell$, which we consider as given by the vector

$$\underline{\mathcal{L}}(s) := (\mathcal{L}_1(s), \ldots, \mathcal{L}_\ell(s)).$$

As far as possible we shall use the same notation as in the previous chapters. As for the proof of the one-dimensional limit theorem we will not make use of axiom (v). Recall that $\sigma_m(\mathcal{L})$ is the abscissa of the mean-square half-plane for $\mathcal{L} \in \tilde{\mathcal{S}}$. We denote by $\mathcal{D}$ the intersection of all mean-square half-planes $\sigma > \sigma_m(\mathcal{L}_k)$ for $\mathcal{L}_k$, $1 \le k \le \ell$, and the critical strip:

$$\mathcal{D} = \left\{ s \in \mathbb{C} : \max_{1 \le k \le \ell} \sigma_m(\mathcal{L}_k) < \sigma < 1 \right\}. \tag{12.1}$$

In principal, we could also deal with different strips for the different $\mathcal{L}_k(s)$, however, to simplify the presentation we may restrict on one for all. By Theorem 2.4 (respectively, (4.2)), the set $\mathcal{D}$ is not empty. We write $\mathcal{H}(\mathcal{D})^\ell$ for the ℓ-dimensional cartesian product of the spaces of analytic functions $\mathcal{H}(\mathcal{D})$. Further, we define a probability measure $\mathbf{P}_T^{\underline{\mathcal{L}}}$ by setting

$$\mathbf{P}_T^{\underline{\mathcal{L}}}(A) = \frac{1}{T} \operatorname{meas}\{\tau \in [0,T] : \underline{\mathcal{L}}(s+i\tau) \in A\}$$

for $A \in \mathcal{B}(\mathcal{H}(\mathcal{D})^\ell)$. For $s \in \mathcal{D}$ and $\omega \in \Omega$, we put

$$\underline{\mathcal{L}}(s,\omega) = (\mathcal{L}_1(s,\omega), \dots, \mathcal{L}_\ell(s,\omega)),$$

where $\mathcal{L}_k(s,\omega)$ is given by (4.6). By Lemma 4.1, any component $\mathcal{L}_k(s,\omega)$ of the vector $\underline{\mathcal{L}}(s,\omega)$ is an $\mathcal{H}(\mathcal{D})$-valued random element. Hence, $\underline{\mathcal{L}}(s,\omega)$ is an $\mathcal{H}(\mathcal{D})^\ell$-valued random element on the probability space $(\Omega, \mathcal{B}(\Omega), \mathrm{m})$ with Ω as in Sect. 4.1. Let $\mathbf{P}^{\underline{\mathcal{L}}}$ denote the distribution of $\underline{\mathcal{L}}(s,\omega)$. We shall prove

Theorem 12.1. *Let $\mathcal{L}_1, \dots, \mathcal{L}_\ell \in \tilde{\mathcal{S}}$. The probability measure $\mathbf{P}_T^{\underline{\mathcal{L}}}$ converges weakly to $\mathbf{P}^{\underline{\mathcal{L}}}$, as $T \to \infty$.*

The proof relies in the main part on the one-dimensional limit Theorem 4.3 which applies individually to any component $\mathcal{L}_k(s)$ of the vector $\underline{\mathcal{L}}(s)$.

We start with

Lemma 12.2. *The family of probability measures $\{\mathbf{P}_T^{\underline{\mathcal{L}}} : T > 0\}$ is relatively compact.*

Proof. To any $\mathcal{L}_k(s)$ we attach a probability measure $\mathbf{P}_T^{\mathcal{L}_k}$ by putting

$$\mathbf{P}_T^{\mathcal{L}_k}(A) = \frac{1}{T} \operatorname{meas}\{\tau \in [0,T] : \mathcal{L}_k(s+i\tau) \in A\}$$

for $A \in \mathcal{B}(\mathcal{H}(\mathcal{D}))$. By Theorem 4.3, restricted to $\mathcal{D}$, the probability measure $\mathbf{P}_T^{\mathcal{L}_k}$ converges weakly to the distribution of the corresponding random element $\mathcal{L}_k(s,\omega)$, i.e, as $T \to \infty$,

$$\mathbf{P}_T^{\mathcal{L}_k} \Rightarrow \mathbf{P}^{\mathcal{L}_k},$$

say. Hence, the family of probability measures $\{\mathbf{P}_T^{\mathcal{L}_k} : T > 0\}$ is relatively compact for any $1 \le k \le \ell$. Since $\mathcal{H}(\mathcal{D})$ is a complete separable space, we obtain by Prokhorov's Theorem 3.5 that this family is also tight, i.e., for every $T > 0$ and every $\epsilon > 0$, there exists a compact set $K_k \subset \mathcal{H}(\mathcal{D})$ such that

$$\mathbf{P}_T^{\mathcal{L}_k}(\mathcal{H}(\mathcal{D}) \setminus K_k) < \frac{\epsilon}{\ell}. \tag{12.2}$$

Now let θ_T be a random variable defined on some probability space $(\widetilde{\Omega}, \mathcal{F}, \mathbf{Q})$ with the probability measure

$$\mathbf{Q}(\theta_T \in A) = \frac{1}{T} \int_0^T \mathbf{1}(t; A)\, \mathrm{d}t,$$

for $A \in \mathcal{B}(\mathbb{R})$, where $\mathbf{1}(t; A)$ is the indicator function of the set A. Consider the $\mathcal{H}(\mathcal{D})^\ell$-valued random element

$$\underline{\mathcal{L}}^{\theta_T}(s) = (\mathcal{L}_1(s + i\theta_T), \dots, \mathcal{L}_\ell(s + i\theta_T)).$$

Then, by (12.2),

$$\mathbf{Q}(\mathcal{L}_k(s + i\theta_T) \in \mathcal{H}(\mathcal{D}) \setminus K_k) < \frac{\epsilon}{\ell}$$

for $1 \le k \le \ell$. Hence, for $K := K_1 \times \cdots \times K_\ell$,

$$\mathbf{P}_T^{\underline{\mathcal{L}}}(\mathcal{H}(\mathcal{D})^\ell \setminus K) = \mathbf{Q}(\underline{\mathcal{L}}^{\theta_T}(s) \in \mathcal{H}(\mathcal{D})^\ell \setminus K)$$

$$= \mathbf{Q}\left(\bigcup_{k=1}^\ell (\mathcal{L}_k(s + i\theta_T) \in \mathcal{H}(\mathcal{D}) \setminus K_k) \right)$$

$$\le \sum_{k=1}^\ell \mathbf{Q}(\mathcal{L}_k(s + i\theta_T) \in \mathcal{H}(\mathcal{D}) \setminus K_k) < \varepsilon$$

for all $T > 0$. Consequently, the family $\{\mathbf{P}_T^{\underline{\mathcal{L}}}\}$ is tight and so, by Prokhorov's Theorem 3.4, it is relatively compact. The lemma is proved. $\square$

Let $s_1, \dots, s_r$ be arbitrary points in $\mathcal{D}$. We put

$$\sigma_0 = \min_{1 \le m \le r} \operatorname{Re} s_m \quad \text{and} \quad \sigma_1 = \frac{1}{2} - \sigma_0.$$

Clearly, $\sigma_0 > \frac{1}{2}$ and $\sigma_1 < 0$. Now let $\mathcal{D}_1 = \{s \in \mathbb{C} : \sigma > \sigma_1\}$. For given complex numbers u_{km} for $1 \le k \le \ell$ and $1 \le m \le r$, define a function $h : \mathcal{H}(\mathcal{D})^\ell \to \mathcal{H}(\mathcal{D}_1)$ for $s \in \mathcal{D}_1$ by

$$h(f_1(s), \dots, f_\ell(s)) = \sum_{k=1}^\ell \sum_{m=1}^r u_{km} f_k(s_m + s). \tag{12.3}$$

Lemma 12.3. *Let*

$$W(s) = h(\underline{\mathcal{L}}(s)).$$

Then $W(s + i\theta_T)$ converges to $h(\underline{\mathcal{L}}(s, \omega))$ in distribution, where θ_T is the random variable, defined in the proof of Lemma 12.2.

Proof. We denote the nth Dirichlet coefficient of $\mathcal{L}_k \in \tilde{\mathcal{S}}$ by $a_{\mathcal{L}_k}(n)$. We have

$$W(s) = \sum_{k=1}^\ell \sum_{m=1}^r u_{km} \mathcal{L}_k(s_m + s) = \sum_{k=1}^\ell \sum_{m=1}^r u_{km} \sum_{n=1}^\infty \frac{a_{\mathcal{L}_k}(n)}{n^{s+s_m}}$$

$$= \sum_{n=1}^\infty \frac{a_n}{n^s} \quad \text{with} \quad a_n = \sum_{k=1}^\ell \sum_{m=1}^r \frac{u_{km} a_{\mathcal{L}_k}(n)}{n^{s_m}}.$$

Since $a_n \ll 1$ by (i), the Dirichlet series $\sum_n \frac{a_n}{n^s}$ converges for $\sigma > 1$. Moreover,

$$\sum_{n \le x} |a_n|^2 \ll x.$$

In view of this estimate, we may apply Theorem 4.12 (with any $\sigma_0 > \frac{1}{2}$) and obtain the weak convergence of the probability measure given by

$$\frac{1}{T} \operatorname{meas} \{\tau \in [0,T] : W(s + i\tau) \in A\}$$

for $A \in \mathcal{B}(\mathcal{H}(\mathcal{D}_1))$ to the distribution of the associated random element

$$\sum_{n=1}^{\infty} \frac{a_n \omega(n)}{n^s}$$

for $\omega \in \Omega$ and $s \in \mathcal{D}_1$, as $T \to \infty$. Since

$$\sum_{n=1}^{\infty} \frac{a_n \omega(n)}{n^s} = \sum_{k=1}^{\ell} \sum_{m=1}^{r} u_{km} \sum_{n=1}^{\infty} \frac{a_{\mathcal{L}_j}(n)\omega(n)}{n^{s+s_m}}$$

$$= \sum_{k=1}^{\ell} \sum_{m=1}^{r} u_{km} \mathcal{L}_k(s, \omega) = h(\underline{\mathcal{L}}(s, \omega)),$$

the assertion of the lemma follows.   □

Proof of Theorem 12.1. By Lemma 12.2, there exists a sequence $T_1 \to \infty$ such that $\mathbf{P}_{T_1}^{\underline{\mathcal{L}}}$ converges weakly to some probability measure $\mathbf{P}$. Suppose that $\mathbf{P}$ is the distribution of an $\mathcal{H}(\mathcal{D})^{\ell}$-valued random element

$$\underline{\widetilde{\mathcal{L}}}(s) := (\widetilde{\mathcal{L}}_1(s), \ldots, \widetilde{\mathcal{L}}_{\ell}(s)).$$

Then, obviously,

$$\underline{\mathcal{L}}(s + i\theta_{T_1}) \xrightarrow[T_1 \to \infty]{\mathcal{D}} \underline{\widetilde{\mathcal{L}}}(s). \tag{12.4}$$

By the continuity of the function h, defined by (12.3), we deduce

$$h(\underline{\mathcal{L}}^{\theta_{T_1}}(s)) \xrightarrow[T_1 \to \infty]{\mathcal{D}} h(\underline{\widetilde{\mathcal{L}}}(s)). \tag{12.5}$$

By Lemma 12.3,

$$h(\underline{\mathcal{L}}^{\theta_{T_1}}(s)) \xrightarrow[T_1 \to \infty]{\mathcal{D}} h(\underline{\mathcal{L}}(s, \omega)).$$

This and (12.5) yield

$$h(\underline{\widetilde{\mathcal{L}}}(s)) \overset{\mathcal{D}}{=} h(\underline{\mathcal{L}}(s, \omega)). \tag{12.6}$$

For $f \in \mathcal{H}(\mathcal{D}_1)$, define $h_1 : \mathcal{H}(\mathcal{D}_1) \to \mathbb{C}$ by $h_1(f) = f(0)$. Then it follows from (12.6) that:

$$h_1(h(\widetilde{\mathcal{L}}(s))) \stackrel{\mathcal{D}}{=} h_1(h(\mathcal{L}(s,\omega))),$$

and in particular

$$h(\widetilde{\mathcal{L}}(0)) \stackrel{\mathcal{D}}{=} h(\mathcal{L}(0,\omega)).$$

Taking into account the definition of h this yields

$$\sum_{k=1}^{\ell} \sum_{m=1}^{r} u_{km} \mathcal{L}_k(s_m, \omega) \stackrel{\mathcal{D}}{=} \sum_{k=1}^{\ell} \sum_{m=1}^{r} u_{km} \widetilde{\mathcal{L}}_k(s_m). \qquad (12.7)$$

The hyperplanes in the space $\mathbb{R}^{2\ell r}$ form a determining class and therefore, the same is true for the hyperplanes in $\mathbb{C}^{\ell r}$. Taking into account (12.7), we obtain that, for $1 \le k \le \ell$ and $1 \le m \le r$, the random elements $\mathcal{L}_k(s_m, \omega)$ and $\widetilde{\mathcal{L}}_k(s_m)$ have the same distribution.

Now let K be a compact subset of $\mathcal{D}$, and let $f_1, \ldots, f_\ell \in \mathcal{H}(\mathcal{D})$. For an arbitrary $\epsilon > 0$ we set

$$U = \left\{ (g_1, \ldots, g_\ell) \in \mathcal{H}(\mathcal{D})^\ell : \max_{1 \le k \le \ell} \max_{s \in K} |g_k(s) - f_k(s)| \le \epsilon \right\}.$$

Now choose a sequence $\{s_m\}$, which is dense in K and define, for $r \in \mathbb{N}$,

$$U_r = \left\{ (g_1, \ldots, g_\ell) \in \mathcal{H}(\mathcal{D})^\ell : \max_{1 \le k \le \ell} \max_{1 \le m \le r} |g_k(s_m) - f_k(s_m)| \le \epsilon \right\}.$$

From the properties of the random elements $\mathcal{L}_k(s_m, \omega)$ and $\widetilde{\mathcal{L}}_k(s_m)$ it follows that:

$$\mathrm{m}\{\omega \in \Omega : \mathcal{L}(s, \omega) \in U_r\} = \mathbf{P}^{\underline{\mathcal{L}}}(\widetilde{\mathcal{L}}(s) \in U_r). \qquad (12.8)$$

Since the sequence $\{s_m\}$ was chosen to be dense in K, we have $U_1 \supset U_2 \supset \ldots$, and $U_r \to U$, as $r \to \infty$. Thus, letting $r \to \infty$ in (12.8), we find

$$\mathrm{m}\{\omega \in \Omega : \mathcal{L}(s, \omega) \in U\} = \mathbf{P}^{\underline{\mathcal{L}}}(\widetilde{\mathcal{L}}(s) \in U).$$

Since the class of all sets U form a determining class, we obtain

$$\mathcal{L}(s, \omega) \stackrel{\mathcal{D}}{=} \widetilde{\mathcal{L}}(s).$$

This and (12.4) gives

$$\mathcal{L}(s + i\theta T_1) \xrightarrow[T_1 \to \infty]{\mathcal{D}} \mathcal{L}(s, \omega).$$

This means that the probability measure $\mathbf{P}_{T_1}^{\mathcal{L}}$ converges weakly to the distribution of the random element $\mathcal{L}(s, \omega)$ as $T_1 \to \infty$. By Lemma 12.2, the family $\{\mathbf{P}_T^{\mathcal{L}}\}$ is relatively compact. Since the random element $\mathcal{L}(s, \omega)$ does not depend on the choice of the sequence T_1, the assertion of the theorem follows. $\qquad \square$

Let $\tilde{\sigma}_m$ denote the abscissa limiting the intersection of all mean-square half-planes for the $\mathcal{L}_k(s)$. Then, for $M > 0$, define

$$\mathcal{D}_M = \{s \in \mathbb{C} : \tilde{\sigma}_m < \sigma < 1, |t| < M\}. \qquad (12.9)$$

Since $\mathcal{D}_M \subset \mathcal{D}$ (see (12.1)), we obtain, by the induced topology, that $\underline{\mathcal{L}}(s, \omega)$ is an $\mathcal{H}(\mathcal{D}_M)^\ell$-valued random element on the probability space $(\Omega, \mathcal{B}(\Omega), m)$. Now denote by $\mathbf{Q}^{\underline{\mathcal{L}}}$ the distribution of $\underline{\mathcal{L}}(s, \omega)$ on $(\mathcal{H}(\mathcal{D}_M)^\ell, \mathcal{B}(\mathcal{H}(\mathcal{D}_M)^\ell))$. Further, define the probability measure $\mathbf{Q}_T^{\underline{\mathcal{L}}}$ by

$$\mathbf{Q}_T^{\underline{\mathcal{L}}}(A) = \frac{1}{T} \operatorname{meas} \{\tau \in [0, T] : \underline{\mathcal{L}}(s + i\tau) \in A\} \qquad (12.10)$$

for $A \in \mathcal{B}(\mathcal{H}(\mathcal{D}_M)^\ell))$. Then, by Theorem 12.1, we deduce

Corollary 12.4. *Let $\mathcal{L}_1, \dots, \mathcal{L}_\ell \in \tilde{S}$. The probability measure $\mathbf{Q}_T^{\underline{\mathcal{L}}}$ converges weakly to $\mathbf{Q}^{\underline{\mathcal{L}}}$, as $T \to \infty$.*

This is the multi-dimensional analogue of Corollary 5.11.

12.2 A Transfer Theorem

Now, we shall prove a conditional joint universality theorem for elements $\mathcal{L}_k \in \tilde{S}$. Here, we shall make use of axiom (v). We cannot prove that any two L-functions $\mathcal{L}_1(s), \mathcal{L}_2(s)$ can approximate uniformly a nice function $g(s)$ simultaneously, since $\mathcal{L}_1(s)$ and $\mathcal{L}_2(s)$ need not be independent (see the remark after Theorem 1.10 in Sect. 1.4). Therefore, we have to assume a certain independence.

Given $\mathcal{L}_1, \dots, \mathcal{L}_\ell \in \tilde{S}$, we write their Euler product in the sequel

$$\mathcal{L}_k(s) = \prod_p \prod_{j=1}^m \left(1 - \frac{\alpha_{jk}(p)}{p^s}\right)^{-1}$$

for $1 \leq k \leq \ell$. In principle, the functions $\mathcal{L}_1(s), \dots, \mathcal{L}_\ell(s)$ may have polynomial Euler products of different degree; however, without loss of generality we may assume that they all have degree equal to m by adding factors $(1 - \frac{\alpha_{jk}(p)}{p^s})^{-1}$ with local roots $\alpha_{jk} = 0$ if necessary. Recall that $\mathcal{D}$ denotes the intersection of all mean-square half-planes of $\mathcal{L}_k(s)$ for $1 \leq k \leq \ell$ with the critical strip (see (12.1)). In view of Theorem 2.4 (respectively, (4.2)), the strip of universality $\mathcal{D}$ contains at least the strip

$$\max_{1 \leq k \leq \ell} \max \left\{\frac{1}{2}, 1 - \frac{1 - \sigma_{\mathcal{L}_k}}{1 + 2\mu_{\mathcal{L}_k}}\right\} < \sigma < 1,$$

where $\sigma_{\mathcal{L}_k}$ and $\mu_{\mathcal{L}_k}$ are defined by axioms (ii) and (iii).

Further, we define

$$g_{pk}(s, b(p)) = -\sum_{j=1}^{m} \log\left(1 - \frac{\alpha_{jk}(p)b(p)}{p^s}\right) \qquad (12.11)$$

for $1 \le k \le \ell$, where $b(p) \in \gamma := \{s \in \mathbb{C} : |s| = 1\}$. Finally, we write

$$\underline{g}_p(s, b(p)) = (g_{p1}(s, b(p)), \ldots, g_{p\ell}(s, b(p))). \qquad (12.12)$$

Then our transfer theorem takes the form:

Theorem 12.5. *Let* $\mathcal{L}_1(s), \ldots, \mathcal{L}_\ell(s)$ *be elements of* $\tilde{S}$, *let* $\mathcal{K}_1, \ldots, \mathcal{K}_\ell$ *be compact subsets of* $\mathcal{D}$ *with connected complements (where* $\mathcal{D}$ *is defined by (12.1)), and, for each* $1 \le k \le \ell$, *let* $g_k(s)$ *be a continuous non-vanishing function on* $\mathcal{K}_k$ *which is analytic in the interior of* $\mathcal{K}_k$. *Suppose that the set of all convergent series*

$$\sum_p \underline{g}_p(s, b(p)) \quad with \quad b(p) \in \gamma \qquad (12.13)$$

is dense in $\mathcal{H}(\mathcal{D})^\ell$ *(where* $\underline{g}_p$ *is defined by (12.11) and (12.12)). Then, for any* $\epsilon > 0$,

$$\liminf_{T \to \infty} \frac{1}{T} \text{meas} \left\{ \tau \in [0, T] : \max_{1 \le k \le \ell} \max_{s \in \mathcal{K}_k} |\mathcal{L}_k(s + i\tau) - g_k(s)| < \epsilon \right\} > 0.$$

Condition (12.13) is the multi-dimensional analogue of the assertion of Theorem 5.10. This condition turns out to be essential; however, it would be nice to replace it by a more lucid statement as, for example, statistical independence.

Proof. Since all $\mathcal{K}_k$ are compact subsets of $\mathcal{D}$, there exists a number M such that $\mathcal{K}_k \subset \mathcal{D}_M$ for $1 \le k \le \ell$, where the rectangle $\mathcal{D}_M$ is defined by (12.9).

Our first aim is to determine the support of the measure $\mathbf{Q}_T^{\underline{\mathcal{L}}}$, defined by (12.10). This will be done in a similar way as in the one-dimensional case, Lemma 5.12. Recall that the minimal closed set $S_{\mathbf{Q}_T^{\underline{\mathcal{L}}}} \subset \mathcal{H}(\mathcal{D}_M)^\ell$ with

$$\mathbf{Q}_T^{\underline{\mathcal{L}}}(S_{\mathbf{Q}_T^{\underline{\mathcal{L}}}}) = 1$$

is called the support of $\mathbf{Q}_T^{\underline{\mathcal{L}}}$; it consists of all $\underline{f} := (f_1, \ldots, f_\ell) \in \mathcal{H}(\mathcal{D}_M)^\ell$ such that for every neighbourhood $\mathcal{G}$ of $\underline{f}$ the inequality $\mathbf{Q}_T^{\underline{\mathcal{L}}}(\mathcal{G}) > 0$ holds.

Lemma 12.6. *The support of the measure* $\mathbf{Q}_T^{\underline{\mathcal{L}}}$ *is the set*

$$S_M^\ell := \{\underline{\varphi} \in \mathcal{H}(\mathcal{D}_M)^\ell : \underline{\varphi}(s) \ne 0 \text{ for } s \in \mathcal{D}_M, \text{ or } \underline{\varphi}(s) \equiv 0\}.$$

For the proof we make use of

Lemma 12.7. *Let $\{\underline{X}_n\}$ be a sequence of independent $\mathcal{H}(\mathcal{G})^\ell$-valued random elements, where $\mathcal{G}$ is a region in $\mathbb{C}$, and suppose that the series $\sum_{n=1}^{\infty} \underline{X}_n$ converges almost everywhere. Then the support of the sum of this series is the closure of the set of all $\underline{f} \in \mathcal{H}(\mathcal{G})^\ell$ which may be written as a convergent series*

$$\underline{f} = \sum_{n=1}^{\infty} \underline{f}_n \quad \text{with} \quad \underline{f}_n \in S_{\underline{X}_n},$$

where $S_{\underline{X}_n}$ is the support of the (distribution of the) random element $\underline{X}_n$.

This is a multi-dimensional version of Theorem 3.16; for its proof, we refer to Laurinčikas [186] (Theorem 1.7.10).

Proof of Lemma 12.6. Recall that $\omega(p)$ is the projection of $\omega \in \Omega$ on the coordinate space $\gamma_p = \gamma$. The sequence $\{\omega(p)\}$ is a sequence of independent random variables on the probability space $(\Omega, \mathcal{B}(\Omega), m)$. Thus

$$\underline{g}_p(s, \omega(p)) := \{(g_{p1}(s, \omega(p)), \dots, g_{p\ell}(s, \omega(p)))\}$$

is a sequence of independent $\mathcal{H}(\mathcal{D}_M)^\ell$-valued random elements. The support of each $\omega(p)$ is the unit circle γ, and therefore the support of the random elements $\underline{g}_p(s, \omega(p))$ is the set

$$\{\underline{f} \in \mathcal{H}(\mathcal{D}_M)^\ell : \underline{f}(s) = (g_{p1}(s, b), \dots, g_{p\ell}(s, b))\},$$

where

$$g_{pk}(s, b) = -\sum_{j=1}^{m} \log\left(1 - \frac{\alpha_{jk}(p)b}{p^s}\right) \quad \text{with} \quad b \in \gamma.$$

Consequently, by Lemma 3.16, the support of the $\mathcal{H}(\mathcal{D}_M)^\ell$-valued random element

$$\log \underline{\mathcal{L}}(s, \omega) = \sum_p \underline{g}_p(s, \omega(p))$$

is the closure of the set of all convergent series $\sum_p \underline{f}_p(s)$. By condition (12.13), the set of these series is dense in $\mathcal{H}(\mathcal{D}_M)^\ell$. The mapping $h : \mathcal{H}(\mathcal{D}_M)^\ell \to \mathcal{H}(\mathcal{D}_M)^\ell$, defined by

$$\underline{f} = (f_1, \dots, f_\ell) \mapsto \exp(\underline{f}) = (\exp(f_1), \dots, \exp(f_\ell)),$$

is a continuous function sending $\log \underline{\mathcal{L}}(s, \omega)$ to $\underline{\mathcal{L}}(s, \omega)$ and $\mathcal{H}(\mathcal{D}_M)^\ell$ to $S_M^\ell \setminus \{0\}$. Therefore, the support $S_{\underline{\mathcal{L}}}$ of $\underline{\mathcal{L}}(s, \omega)$ contains $S_M^\ell \setminus \{0\}$. On the other hand, the support of $\underline{\mathcal{L}}(s, \omega)$ is closed. By Hurwitz's Theorem 5.13, it follows that $\overline{S_M^\ell \setminus \{0\}} = S_M^\ell$. Thus, $S_M^\ell \subset S_{\underline{\mathcal{L}}}$. By axiom (iv) on the polynomial Euler product for elements in $\tilde{\mathcal{S}}$, the functions

$$\exp(g_{pk}(s,\omega(p))) = \prod_{j=1}^{m} \left(1 - \frac{\alpha_{jk}(p)\omega(p)}{p^s}\right)$$

are non-zero for $s \in \mathcal{D}_M$ and $\omega \in \Omega$. Hence, the functions $\mathcal{L}_k(s,\omega)$, $1 \leq k \leq \ell$, are almost surely convergent products of non-vanishing factors. If we apply the Hurwitz theorem again, we conclude that $\underline{\mathcal{L}}(s,\omega) \in S_M^{\ell}$ almost surely. Therefore, $S_{\underline{\mathcal{L}}} \subset S_M^{\ell}$. The lemma is proved. $\square$

Now we have finished all preliminaries and can start with the main part of the proof of Theorem 12.5.

First, we suppose that the functions $g_1(s), \ldots, g_{\ell}(s)$ all have a non-vanishing analytic continuation to $\mathcal{D}_M$. By Lemma 12.6, the g_k's are contained in the support $S_{\underline{\mathcal{L}}}$ of the random element $\underline{\mathcal{L}}(s,\omega)$. Denote by Φ the set of functions $\underline{\varphi} = (\varphi_1, \ldots, \varphi_{\ell}) \in \mathcal{H}(\mathcal{D}_M)^{\ell}$ such that

$$\max_{1 \leq k \leq \ell} \max_{s \in \mathcal{K}_k} |\varphi_k(s) - g_k(s)| < \epsilon.$$

Since by Corollary 12.4 the measure $\mathbf{Q}_T^{\underline{\mathcal{L}}}$ converges weakly to $\mathbf{Q}^{\underline{\mathcal{L}}}$, as $T \to \infty$, and since the set Φ is open, it follows from Theorem 3.1 and properties of the support that:

$$\liminf_{T \to \infty} \frac{1}{T} \, \mathrm{meas} \left\{ \tau \in [0,T] : \max_{1 \leq k \leq \ell} \max_{s \in \mathcal{K}_k} |\mathcal{L}_k(s + i\tau) - g_k(s)| < \epsilon \right\}$$

$$= \liminf_{T \to \infty} \mathbf{P}_T^{\underline{\mathcal{L}}}(\Phi) \geq \mathbf{Q}^{\underline{\mathcal{L}}}(\Phi) > 0. \tag{12.14}$$

This proves the theorem in the case of functions $g_k(s)$ which have a non-vanishing analytic continuation to $\mathcal{D}_M$.

Now assume that the functions $g_k(s)$ are as in the statement of the theorem. By Mergelyan's approximation Theorem 5.15, for $1 \leq k \leq \ell$, there exists a sequence of polynomials $G_{kn}(s)$ which converges uniformly on $\mathcal{K}_k$ to $g_k(s)$ as $n \to \infty$. Since the $g_k(s)$ are non-vanishing on $\mathcal{K}_k$, it follows that, for sufficiently large m, the functions $G_{km}(s)$ do not vanish on $\mathcal{K}_k$ and

$$\max_{1 \leq k \leq \ell} \max_{s \in \mathcal{K}_k} |g_k(s) - G_{km}(s)| < \frac{\epsilon}{4}. \tag{12.15}$$

Since any of the polynomials $G_{km}(s)$ has only finitely many zeros, there exists a region $\mathcal{G}$ whose complement is connected such that $\mathcal{K}_k \subset \mathcal{G}$ for $1 \leq k \leq \ell$ and $G_{km}(s) \neq 0$ on $\mathcal{G}$. Hence, there exist continuous branches $\log G_{km}(s)$ on $\mathcal{G}$, and the logarithms $\log G_{km}(s)$ are analytic in the interior of $\mathcal{G}$. Thus, by Mergelyan's Theorem 5.15, for $1 \leq k \leq \ell$, there exists a sequence of polynomials $F_{kn}(s)$ which converges uniformly on $\mathcal{K}_k$ to $\log G_{km}(s)$ as $n \to \infty$. Thus, for all $1 \leq k \leq \ell$ and all sufficiently large n,

$$\max_{s \in \mathcal{K}_k} |G_{km}(s) - \exp(F_{kn}(s))| < \frac{\epsilon}{4}.$$

This and (12.15) imply

$$\max_{1 \le k \le \ell} \max_{s \in \mathcal{K}_k} |g_k(s) - \exp(F_{kn}(s))| < \frac{\epsilon}{2}. \tag{12.16}$$

From (12.14) we deduce

$$\liminf_{T \to \infty} \frac{1}{T} \operatorname{meas} \left\{ \tau \in [0, T] : \max_{1 \le k \le \ell} \max_{s \in \mathcal{K}_k} |\mathcal{L}_k(s + i\tau) - \exp(F_{kn}(s))| < \frac{\epsilon}{2} \right\} > 0.$$

In combination with (12.16) this proves Theorem 12.5.   □

12.3 Twisted L-Functions

Next, we want to give an example for a family of L-functions which satisfies Condition (12.13) of Theorem 12.5 and thus is jointly universal.

For $\mathcal{L} \in \tilde{\mathcal{S}}$ and a Dirichlet character $\chi \bmod q$ define

$$\mathcal{L}_\chi(s) = \sum_{n=1}^{\infty} \frac{a(n)}{n^s} \chi(n). \tag{12.17}$$

It is obvious that such a twist satisfies the Ramanujan hypothesis (i) (since $|\chi(n)| \le 1$). By the complete multiplicativity of Dirichlet characters, we obtain the identity

$$\mathcal{L}_\chi(s) = \sum_{n=1}^{\infty} \frac{a(n)}{n^s} \chi(n) = \prod_p \prod_{j=1}^{m} \left(1 - \frac{\alpha_j(p)}{p^s} \chi(p) \right)^{-1};$$

hence, also the axiom on the polynomial Euler product (iv) holds for $\mathcal{L}_\chi(s)$. Next we observe that

$$\sum_{p \le x} |a(p)\chi(p)|^2 = \sum_{\substack{p \le x \\ p \nmid q}} |a(p)|^2 = \sum_{p \le x} |a(p)|^2 + O(1), \tag{12.18}$$

since there are only finitely many primes dividing q. Hence, $\mathcal{L}_\chi(s)$ satisfies the axiom on the mean-square with the same constant $\kappa > 0$ as for $\mathcal{L}(s)$. However, we cannot show that the analytic axioms (ii) and (iii) hold in this general frame – analytic continuation is a difficult problem in analytic number theory. Besides the assumption of axioms (ii) and (iii), we shall need one more condition on the distribution of the values of the Dirichlet coefficients $a(p)$ for prime p in arithmetic progressions.

Theorem 12.8. *Let $\mathcal{L} \in \tilde{\mathcal{S}}$. Let $q_1, \ldots, q_\ell$ be positive integers, let χ_1 mod $q_1, \ldots, \chi_\ell$ mod q_ℓ be pairwise non-equivalent characters. Further, for $1 \leq k \leq \ell$, let g_k be a continuous non-vanishing function on $\mathcal{K}_k$ which is analytic in the interior, where $\mathcal{K}_k$ is a compact subset of the strip $\mathcal{D}$ with connected complement (where $\mathcal{D}$ is defined by (12.1)). For $1 \leq k \leq \ell$, assume that $\mathcal{L}_{\chi_k}(s)$ satisfies the axioms (ii) and (iii), and that*

$$\lim_{x \to \infty} \frac{1}{\pi(x)} \sum_{\substack{p \leq x \\ p \equiv h \bmod q}} |a(p)|^2 = \frac{\kappa}{\varphi(q)} \tag{12.19}$$

holds for any prime residue class h mod q, where q is the least common multiple of $q_1, \ldots, q_k$, and κ is the quantity from axiom (v). Then, for any $\epsilon > 0$,

$$\liminf_{T \to \infty} \frac{1}{T} \operatorname{meas} \left\{ \tau \in [0,T] : \max_{1 \leq k \leq \ell} \max_{s \in \mathcal{K}_k} |\mathcal{L}_{\chi_k}(s + i\tau) - g_k(s)| < \varepsilon \right\} > 0.$$

Notice that the additional condition (12.19) implies axiom (v).

Proof. Because of Theorem 12.5 it suffices to prove that the set of all convergent series

$$\sum_p \underline{g}_p(s, b(p)) \quad \text{with} \quad b(p) \in \gamma$$

is dense in $\mathcal{H}(\mathcal{D})^\ell$, where, according to (12.11),

$$g_{pk}(s, b(p)) = -\sum_{j=1}^m \log\left(1 - \frac{\alpha_j \chi_k(p)b(p)}{p^s}\right), \tag{12.20}$$

and $\underline{g}_p(s,b(p)) = (g_{p1}(s,b(p)), \ldots, g_{p\ell}(s,b(p)))$. We follow the proof of Theorem 5.10. Almost all arguments can be easily extended to our multi-dimensional case; however, in one instance the linear independence of our characters plays a crucial role. The main tool for the proof will be a multi-dimensional analogue of Theorem 5.7:

Theorem 12.9. *Let D be a simply connected domain in the complex plain. Suppose that the sequence of functions $\{\underline{f}_n = (f_{n1}, \ldots, f_{n\ell})\}$ be a sequence in $\mathcal{H}(D)^\ell$ satisfies:*

- *If $\mu_1, \ldots, \mu_\ell$ are complex Borel measures on $(\mathbb{C}, \mathcal{B}(\mathbb{C}))$ with compact supports contained in D such that*

$$\sum_{n=1}^\infty \left| \sum_{k=1}^\ell \int_{\mathbb{C}} f_{nk} \, d\mu_k \right| < \infty,$$

then

$$\int_{\mathbb{C}} s^r \, d\mu_k(s) = 0 \quad \text{for} \quad r \in \mathbb{N} \cup \{0\}, \quad k = 1, \ldots, \ell.$$

- *The series $\sum_{n=1}^{\infty} \underline{f}_n$ converges on $\mathcal{H}(D)^{\ell}$.*
- *For every compact $K \subset D$*

$$\sum_{n=1}^{\infty} \max_{s \in K} |\underline{f}_n(s)|^2 < \infty.$$

Then the set of all convergent series $\sum_{n=1}^{\infty} b(n)\underline{f}_n$ with $b(n) \in \gamma$ is dense in $\mathcal{H}(D)^{\ell}$.

A proof can be found in Bagchi [10]; it follows along the lines of Theorem 5.7.

We continue with the proof of Theorem 12.8. For $1 \leq k \leq \ell$, define

$$\tilde{g}_{pk}(s) = g_{pk}(s, 1) = -\sum_{j=1}^{m} \log\left(1 - \frac{\alpha_j(p)\chi_k(p)}{p^s}\right),$$

and $\underline{\tilde{g}}_p(s) = (\tilde{g}_{p1}(s), \ldots, \tilde{g}_{p\ell}(s))$. Moreover, for $1 \leq k \leq \ell$ and a parameter $N \in \mathbb{N}$ which will be chosen later, define

$$\hat{g}_{pk}(s) = \begin{cases} \tilde{g}_{pk}(s) & \text{if } p > N, \\ 0 & \text{if } p \leq N, \end{cases}$$

and let $\underline{\hat{g}}_p(s) = (\hat{g}_{p1}(s), \ldots, \hat{g}_{p\ell}(s))$. In the proof of Theorem 5.10, we have shown that there exists a sequence $\{\hat{b}(p) : \hat{b}(p) \in \gamma\}$ such that the series

$$\sum_{p} \hat{b}(p)\hat{g}_{pk}(s)$$

converges in $\mathcal{H}(D)$ (see (5.12)). Similarly as in the one-dimensional case, one can show the existence of a sequence $\{\hat{b}(p) : \hat{b}(p) \in \gamma\}$ such that the series

$$\sum_{p} \hat{b}(p)\underline{\hat{g}}_p(s) \tag{12.21}$$

converges in $\mathcal{H}(D)$. Next, we prove that the set of all convergent series

$$\sum_{p} \tilde{b}(p)\underline{\hat{g}}_p(s) \quad \text{with} \quad \tilde{b}(p) \in \gamma \tag{12.22}$$

is dense in $\mathcal{H}(D)^{\ell}$. It suffices to show that the set of all convergent series

$$\sum_{p} \tilde{b}(p)\underline{f}_p(s) \quad \text{with} \quad \tilde{b}(p) \in \gamma \tag{12.23}$$

is dense, where $f_{pk}(s) := \hat{b}(p)\hat{g}_{pk}(s)$ and $\underline{f}_p(s) := (f_{p1}(s), \ldots, f_{p\ell}(s))$. For this purpose we apply Theorem 12.9. In view of the convergence of (12.21), it follows that the series $\sum_{p} \underline{f}_p(s)$ converges in $\mathcal{H}(D)^{\ell}$; this implies the second assumption of Theorem 12.9. The third assumption follows immediately from the fact that for any component

$$\sum_p |f_{pk}(s)|^2 \ll \sum_p \frac{1}{p^{2\sigma}},$$

as in the one-dimensional case.

To verify the first assumption, let $\mu_1, \ldots, \mu_\ell$ be complex Borel measures on $(\mathbb{C}, \mathcal{B}(\mathbb{C}))$ with compact support contained in $\mathcal{D}$ such that

$$\sum_p \left| \sum_{k=1}^{\ell} \int_{\mathbb{C}} f_{pk}(s) \, d\mu_k(s) \right| < \infty. \tag{12.24}$$

On account of the Euler product representation, we find

$$\tilde{g}_{pk}(s) = \sum_{j=1}^{m} \sum_{\nu=1}^{\infty} \frac{\alpha_j(p)\chi_k(p)}{\nu p^{\nu s}} = \frac{a(p)\chi_k(p)}{p^s} + r_{pk}(s)$$

for $1 \le k \le \ell$, where $r_{pk}(s) \ll p^{-2\sigma}$. Then,

$$\sum_p \left| f_{pk}(s) - \frac{a(p)\chi_k(p)\hat{b}(p)}{p^s} \right| = \sum_{\substack{p \le N \\ p \nmid q_k}} \frac{|a(p)|}{p^\sigma} + \sum_{p > N} |r_p(s)| \ll 1$$

uniformly on compact subsets of $\mathcal{D}$. Hence, (12.24) implies

$$\sum_p \left| a(p) \sum_{k=1}^{\ell} \int_{\mathbb{C}} \chi_k(p) p^{-s} \, d\mu_k(s) \right| < \infty. \tag{12.25}$$

Recall that q denotes the least common multiple of the moduli $q_1, \ldots, q_\ell$ and, for $1 \le k \le \ell$, denote by $\tilde{\chi}_k$ the character mod q induced by χ_k mod q_k. Now let h be an integer satisfying $1 \le h \le q$ and coprime with q. If $p \equiv h \bmod q$, then

$$\chi_k(p) = \tilde{\chi}_k(p) = \tilde{\chi}_k(h).$$

Hence, we may rewrite (12.25) as

$$\sum_{p \equiv h \bmod q} \left| a(p) \sum_{k=1}^{\ell} \tilde{\chi}_k(h) \int_{\mathbb{C}} p^{-s} \, d\mu_k(s) \right| < \infty$$

or, equivalently,

$$\sum_{p \equiv h \bmod q} |a(p)| \varrho_h(\log p)| < \infty, \tag{12.26}$$

where

$$\varrho_h(z) := \int_{\mathbb{C}} \exp(-sz)\, d\nu_h(s) \quad \text{and} \quad \nu_h(s) := \sum_{k=1}^{\ell} \tilde{\chi}_k(h)\mu_k(s) \qquad (12.27)$$

for $1 \leq h \leq q$ coprime with q. Recall our argument in the one-dimensional case. For any prime p we have defined angles $\phi_p \in [0, \frac{\pi}{2}]$ by

$$|a(p)| = \left| \sum_{j=1}^{m} \alpha_j(p) \right| = m \cos\phi_p.$$

Hence, we can replace (12.26) by

$$\sum_{p \equiv h \bmod q} \cos\phi_p |\varrho_h(\log p)| < \infty, \qquad (12.28)$$

Our next aim is to show that $\varrho_h(z)$ vanishes identically. We apply Theorem 5.9. Now, we choose a sufficiently large positive constant M such that the support of all μ_k is contained in the region $\{s \in \mathbb{C} : \sigma_m < \sigma < 1, |t| < M\}$. Then, by the definition of $\varrho_h(s)$,

$$|\varrho_h(\pm iy)| \leq \exp(My) \sum_{k=1}^{\ell} \int_{\mathbb{C}} |d\mu_k(s)|$$

for $y > 0$. Therefore,

$$\limsup_{y \to \infty} \frac{\log|\varrho_h(\pm iy)|}{y} \leq M,$$

and the first condition of Theorem 5.9 is valid with $\lambda = M$. Fix a number ϕ with $0 < \phi < \min\left\{1, \frac{\sqrt{\kappa}}{m}\right\}$, where m is the degree of the polynomial defining the local Euler factors, and κ is the quantity appearing in the axiom on the mean-square for $\mathcal{L} \in \tilde{\mathcal{S}}$. Next, denote by $\mathbb{P}(h)$ the set of all prime numbers $p \equiv h \bmod q$ and define

$$\mathbb{P}^\phi(h) = \{p : p \equiv h \bmod q \text{ prime and } \cos\phi_p > \phi\}.$$

Then (12.28) yields

$$\sum_{p \in \mathbb{P}^\phi(h)} |\varrho_h(\log p)| < \infty \qquad (12.29)$$

for any h with $1 \leq h \leq q$, coprime with q. Further, fix a number η with $0 < \eta < \frac{\pi}{M}$, and define

$$A = \left\{ n \in \mathbb{N} : \exists r \in \left(\left(n - \frac{1}{4}\right)\eta, \left(n + \frac{1}{4}\right)\eta\right] \text{ with } |\varrho_h(r)| \leq \exp(-r) \right\}.$$

Now let

$$\alpha := \alpha(n) := \exp\left(\left(n - \frac{1}{4}\right)\eta\right) \quad \text{and} \quad \beta := \beta(n) := \exp\left(\left(n + \frac{1}{4}\right)\eta\right).$$

Then

$$\sum_{p \in \mathbb{P}^\phi(h)} |\varrho_h(\log p)| \geq \sum_{n \notin A} \sum_{\substack{p \in \mathbb{P}^\phi(h) \\ \alpha < p \leq \beta}} |\varrho_h(\log p)| \geq \sum_{n \notin A} \sum_{\substack{p \in \mathbb{P}^\phi(h) \\ \alpha < p \leq \beta}} \frac{1}{p}$$

and, in view of (12.29),

$$\sum_{n \notin A} \sum_{\substack{p \in \mathbb{P}^\phi(h) \\ \alpha < p \leq \beta}} \frac{1}{p} < \infty. \tag{12.30}$$

Let

$$\pi_\phi(x; h \bmod q) = \sharp\{p \leq x : p \in \mathbb{P}^\phi(h)\}.$$

Then we obtain, for $\alpha < u \leq \beta$,

$$\sum_{\substack{p \in \mathbb{P}(h) \\ \alpha < p \leq u}} (\cos\phi_p)^2 \leq \sum_{\substack{p \in \mathbb{P}^\phi(h) \\ \alpha < p \leq u}} 1 + \phi^2 \sum_{\substack{p \in \mathbb{P}(h) \setminus \mathbb{P}^\phi(h) \\ \alpha < p \leq u}} 1$$

$$= (1 - \phi^2)(\pi_\phi(u; h \bmod q) - \pi_\phi(\alpha; h \bmod q))$$
$$+ \phi^2(\pi(u; h \bmod q) - \pi(\alpha; h \bmod q). \tag{12.31}$$

By condition (12.19),

$$\sum_{\substack{p \leq x \\ p \equiv h \bmod q}} (\cos\phi_p)^2 = \frac{1}{m^2} \sum_{\substack{p \leq x \\ p \equiv h \bmod q}} |a(p)|^2 \sim \frac{\kappa}{m^2} \pi(x; h \bmod q).$$

Let δ be a small positive constant. Substituting the latter formula in (12.31) gives

$$\pi_\phi(u; h \bmod q) - \pi_\phi(\alpha; h \bmod q)$$
$$\geq \left(\frac{\frac{\kappa}{m^2} - \phi^2}{1 - \phi^2} + O(1)\right)(\pi(u; h \bmod q) - \pi(\alpha; h \bmod q))$$

for $u \geq \alpha(1 + \delta)$, as $n \to \infty$. Thus, we get by partial summation

$$\sum_{\substack{p \in \mathbb{P}^\phi(h) \\ \alpha < p \leq \beta}} \frac{1}{p} = \int_\alpha^\beta \frac{d\pi_\phi(u; h \bmod q)}{u} \geq \left(\frac{\frac{\kappa}{m^2} - \phi^2}{1 - \phi^2} + O(1)\right) \int_\alpha^\beta \frac{d\pi(u; h \bmod q)}{u}$$

$$\geq \left(\frac{\frac{\kappa}{m^2} - \phi^2}{1 - \phi^2} + O(1)\right) \sum_{\substack{p \in \mathbb{P}(h) \\ \alpha(1+\delta) < p \leq \beta}} \frac{1}{p}, \tag{12.32}$$

as $n \to \infty$. By the prime number theorem for arithmetic progressions (that is (1.29) with a suitable remainder term), we find for the prime residue class $h \bmod q$

$$\sum_{\substack{p \le x \\ p \equiv h \bmod q}} \frac{1}{p} = \frac{1}{\varphi(q)} \log\log x + C_h + \mathrm{O}((\log x)^{-2})$$

with some constant C_h, depending only on $h \bmod q$. Thus

$$\sum_{\substack{p \in \mathbb{P}(h) \\ \alpha(1+\delta) < p \le \beta}} \frac{1}{p} = \frac{1}{\varphi(q)} \left(\frac{1}{2} - \frac{\log(1+\delta)}{\eta} \right) \frac{1}{n} + \mathrm{O}\left(\frac{1}{n^2} \right).$$

This gives in (12.32)

$$\sum_{\substack{p \in \mathbb{P}^\phi(h) \\ \alpha < p \le \beta}} \frac{1}{p} \ge \frac{1}{\varphi(q)} \frac{\frac{\kappa}{m^2} - \phi^2}{1 - \phi^2} \left(\frac{1}{2} - \frac{\log(1+\delta)}{\eta} \right) \frac{1}{n} + \mathrm{O}\left(\frac{1}{n} \right),$$

as $n \to \infty$. Hence, it follows from (12.30) that:

$$\sum_{n \notin A} \frac{1}{n} < \infty.$$

Let $A = \{a_j : j \in \mathbb{N}\}$ with $a_1 < a_2 < \cdots$. This implies

$$\lim_{j \to \infty} \frac{a_j}{j} = 1. \tag{12.33}$$

By the definition of the set A, there exists a sequence $\{\xi_j\}$ such that

$$\left(a_j - \frac{1}{4} \right) \eta < \xi_j \le \left(a_j + \frac{1}{4} \right) \eta \quad \text{and} \quad |\varrho_h(\xi_j)| \le \exp(-\xi_j).$$

Hence, from (12.33) it follows that:

$$\lim_{j \to \infty} \frac{\xi_j}{j} = \eta \quad \text{and} \quad \limsup_{j \to \infty} \frac{\log |\varrho_h(\xi_j)|}{\xi_j} \le -1.$$

Applying Theorem 5.9, we obtain

$$\limsup_{r \to \infty} \frac{\log |\varrho_h(r)|}{r} \le -1. \tag{12.34}$$

However, by Lemma 5.8, if $\varrho_h(z)$ does not vanish identically, then

$$\limsup_{r \to \infty} \frac{\log |\varrho_h(r)|}{r} > 0,$$

contradicting (12.34). Therefore, $\varrho_h(z)$ vanishes identically. By (12.27), differentiation yields

$$\int_{\mathbb{C}} s^r \, d\nu_h(s) \equiv 0,$$

respectively

$$\sum_{k=1}^{\ell} \tilde{\chi}_k(h) \int_{\mathbb{C}} s^r \mu_k(s) \equiv 0,$$

for $r = 0, 1, 2, \ldots$. The latter identity holds for any h coprime with q. Multiplying with $\overline{\tilde{\chi}_i}(h)$ for some arbitrary i with $1 \le i \le \ell$ and summing up over all prime residue classes $h \bmod q$, we get

$$0 \equiv \sum_{k=1}^{\ell} \int_{\mathbb{C}} s^r \, d\mu_k(s) \sum_{\substack{1 \le h \le q \\ (h,q)=1}} \overline{\tilde{\chi}_i}(h)\tilde{\chi}_k(h)$$

By the orthogonality relation for characters (1.28), this gives

$$0 \equiv \int_{\mathbb{C}} s^r \mu_i(s),$$

and so $\mu_i(s) \equiv 0$. Since i was arbitrary, all $\mu_1(s), \ldots, \mu_\ell(s)$ vanish identically. Thus, the first assumption of Theorem 12.9 is also satisfied and we deduce the denseness of all convergent series (12.22). It remains to show that this does not change when we add terms of the form $\sum_{p \le N} \tilde{\underline{g}}_p(s)$. However, this can be done in a similar way as in the one-dimensional case. The theorem is proved. □

12.4 First Applications

As a first application of Theorem 12.8 we extend Voronin's joint universality Theorem 1.10 for Dirichlet L-functions to pairwise non-equivalent characters. Dirichlet L-functions to primitive characters χ lie in the Selberg class (see Sect. 6.1) and so they are universal by Theorem 6.12; however, this is not true for imprimitive characters since the axiom on the functional equation is violated. What about Dirichlet L-functions to imprimitive characters with respect to universality? Assume that χ is an imprimitive character $\bmod q$, induced by a primitive character $\chi^* \bmod q^*$. Then, by (1.27),

$$L(s, \chi) = L(s, \chi^*) f_\chi(s), \quad \text{where} \quad f_\chi(s) := \prod_{p|q} \left(1 - \frac{\chi^*(p)}{p^s}\right).$$

Obviously, $f_\chi(s)$ is analytic in the whole complex plane, all its zeros are located on the line $\sigma = 0$, and it satisfies $|f_\chi(\sigma + it)| \le 1$ for $\sigma \ge 0$. Thus, $L(s, \chi)$ is an entire function. This yields axiom (ii) and axiom (iii). In case of a primitive character, the mean-square half-plane of $L(s, \chi^*)$ is $\sigma > \frac{1}{2}$ by Corollary 6.11. For Dirichlet L-functions to imprimitive characters, we can do better than applying Theorem 2.4. Taking into account the properties of the function $f_\chi(s)$ outlined above, we find

$$\int_1^T |L(\sigma + \mathrm{i}t, \chi)|^2 \, \mathrm{d}t \leq \int_1^T |L(\sigma + \mathrm{i}t, \chi^*)|^2 \, \mathrm{d}t$$

for $\sigma \geq 0$. Thus, by Corollary 6.11 and Carlson's Theorem 2.1, the asymptotic mean-square of $L(s, \chi)$ exists in the same region as the mean-square of $L(s, \chi^*)$. It remains to verify condition (12.19) which we may rewrite as

$$\lim_{x \to \infty} \frac{1}{\pi(x)} \sum_{\substack{p \leq x \\ p \equiv h \bmod q}} 1 = \frac{\kappa}{\varphi(q)},$$

where h is coprime with q. This is the prime number theorem in arithmetic progressions (1.29). The same reasoning holds for principal characters. Thus, we have proved Theorem 1.10: Dirichlet L-functions to pairwise non-equivalent characters are jointly universal.

Another example for an application of Theorem 12.8 is provided by L-functions associated with newforms (see Sect. 1.5). The joint universality for twists of these L-functions, Theorem 1.12, was obtained by Laurinčikas and Matsumoto [202]. Here, condition (12.19) is satisfied because of the asymptotic formula (1.38).

12.5 A Conjecture

We conclude this chapter with a conjecture. So far, all known explicit families of jointly universal Dirichlet series were constructed in a very special way, namely as twists of a single universal Dirichlet series twisted by pairwise non-equivalent characters. Examples are Dirichlet L-functions (see Theorem 1.10 for the continuous and Theorem 5.17 for discrete joint universality) and L-functions to newforms twisted by Dirichlet characters (Theorem 1.12); Theorem 12.8 may be considered as a conditional attempt to generalize this type of joint universality. The knowledge of these jointly universal L-functions can be used to prove joint universality for other Dirichlet series as, for example, Dirichlet series with periodic coefficients (see Theorem 11.8) or Hurwitz zeta-functions with rational parameters (Theorem 11.12), Dedekind zeta-functions (see Sect. 13.1) or further L-functions to number fields (Sect. 13.8). In all these examples, the result is obtained by suitable representations of the functions in question as linear combinations or products of jointly universal L-functions. Here, the necessary independence for these joint universality theorems is deduced from the linear independence of Dirichlet characters (or ray class group characters in Sect. 13.8) over $\mathbb{C}$. We should mention that there exist also analogous results for additive characters: Matsumoto and Laurinčikas [200, 203] obtained joint universality for certain classes of Lerch zeta-functions, which may be regarded as twists of strongly universal Hurwitz zeta-functions by additive characters $\exp(2\pi i \frac{a}{q})$ with rational $\frac{a}{q}$ (in addition with some algebraic side restrictions).

To replace the orthogonality relation of (multiplicative or additive) characters by a more general statement, Laurinčikas [191], Laurinčikas and Matsumoto [199], respectively, introduced an interesting matrix condition. To explain this matrix condition in a few words, given a collection of Matsumoto zeta-functions, it is assumed that there are finitely many distinct subclasses $\mathbb{P}_j$ of the set of prime numbers each of which of positive density such that the Dirichlet coefficients $a_k(p)$ for prime p multiplied with some weight are constant B_{kj} on the $\mathbb{P}_j$. If the rank of the matrix built from the values (B_{kj}) is as large as possible, the coefficients $a_k(p)$ are linearly independent which allows to deduce the joint universality of the Matsumoto zeta-functions. Since the subsets $\mathbb{P}_j$ are not necessarily related to the multiplicative structure of prime residue classes, this approach allows generalizations of joint universality theorems for Dirichlet series twisted by multiplicative characters.

We may ask whether this is all? The methods sketched above fail to prove, for example, joint universality for $\zeta(s)$ and an L-function to some newform. More generally, we ask for a necessary and sufficient condition that a given finite family of L-functions is jointly universal? In the context of the Selberg class, we expect that Selberg's Conjecture B (or a suitable quantitative extension) could be used to answer this question. By Selberg's Conjecture B, primitive functions are expected to form an orthonormal system. Recall that Bombieri and Hejhal [39] proved, assuming a stronger version of Selberg's conjecture B, the statistical independence of any collection of independent L-functions in any family of independent elements of $\mathcal{S}$. With regard to this statistical independence, predicted by Selberg's Conjecture B, it seems reasonable to conjecture that *any finite collection of distinct primitive functions in the Selberg class is jointly universal*. Moreover, we expect that *any two functions* $\mathcal{L}_1(s), \mathcal{L}_2(s) \in \mathcal{S}$ *are jointly universal, if and only if*

$$\sum_{p \leq x} \frac{a_1(p)\overline{a_2(p)}}{p} = O(1), \qquad (12.35)$$

where the Dirichlet series coefficients of $\mathcal{L}_j(s)$ are denoted by $a_j(n)$ and the summation is over the prime numbers. We shall briefly illustrate how this fits to some special cases of pairs of jointly universal functions.

Assume that we are given functions $\mathcal{L}_j(s)$ which are twists of an element of $\mathcal{S}$ by distinct primitive Dirichlet characters $\chi_j \bmod q$, i.e., $a_j(n) = a(n)\chi_j(n)$ for $j = 1, 2$. Then

$$\sum_{p \le x} \frac{a_1(p)\overline{a_2(p)}}{p} = \sum_{p \le x} \frac{|a(p)|^2}{p} \chi_1(p)\overline{\chi_2(p)}$$

$$= \sum_{h \bmod q} \chi_1(h)\overline{\chi_2}(h) \sum_{\substack{p \le x \\ p \equiv h \bmod q}} \frac{|a(p)|^2}{p}.$$

Now suppose a certain equidistribution: for any prime residue class $h \bmod q$, the inner sum on the right is asymptotically equal to $c \log \log x + O(1)$, where c is some positive constant, independent on $h \bmod q$. By the orthogonality relation for characters, it follows that:

$$\sum_{p \le x} \frac{a_1(p)\overline{a_2(p)}}{p} = c \sum_{h \bmod q} \chi_1(h)\overline{\chi_2}(h) \log \log x + O(1) = O(1),$$

which is (12.35). On the other hand, with regard to Theorem 12.8 we expect $\mathcal{L}_1(s)$ and $\mathcal{L}(s)$ to be universal (which is known in particular cases).

Finally, let us consider an example where we do not have joint universality, $\zeta(s)$ and $\zeta(s)^2$, say. Here, since the Dirichlet coefficients of $\zeta(s)^2$ are given by the divisor function $d(n)$, the left-hand side of (12.35) is obviously unbounded.

L-Functions of Number Fields

The zeta-function of a field is like the atom of physics.
H.M. Stark

In this chapter, we shall obtain universality for many classical *L*-functions, including Dedekind zeta-functions as well as Hecke and Artin *L*-functions. Further, we shall briefly discuss the arithmetic axioms in the definition of $\tilde{S}$ with respect to the Langlands program. We give only a sketch of the analytic theory of all these *L*-functions and refer to Bump et al. [47] for further details. For details from algebraic number theory we refer to Heilbronn's survey [129], the monographs of Murty and Murty [270], of Neukirch [279], and, last but not least, Stark's article [337].

13.1 Dedekind Zeta-Functions

Let $\mathbb{K}$ be an algebraic number field (i.e., a finite extension of $\mathbb{Q}$). The associated Dedekind zeta-function is for $\sigma > 1$ defined by

$$\zeta_{\mathbb{K}}(s) = \sum_{\mathfrak{a}} \frac{1}{N(\mathfrak{a})^s} = \prod_{\mathfrak{p}} \left(1 - \frac{1}{N(\mathfrak{p})^s} \right)^{-1};$$

here the sum is taken over all non-zero integral ideals, the product is taken over all prime ideals of the ring of integers of $\mathbb{K}$ and $N(\mathfrak{a})$ denotes the norm of the ideal $\mathfrak{a}$. This generalization of the Riemann zeta-function was introduced by Dedekind [66] in order to obtain information about the multiplicative structure of number fields. The identity between series and product is an analytic version of the unique factorization of integral ideals into prime ideals. In an algebraic number field $\mathbb{K}$ over $\mathbb{Q}$ of degree d any rational prime number p has a unique factorization into a product of prime ideals

$$(p) = \prod_{j=1}^{r} \mathfrak{p}_j^{e_j} \quad \text{with} \quad N(\mathfrak{p}_j) = p^{f_j} \quad \text{and} \quad \sum_{j=1}^{r} e_j f_j = d; \qquad (13.1)$$

of course, the non-negative integers e_j, f_j, and r depend on p (which is not indicated here for simplicity). Hence we can write

$$\zeta_{\mathbb{K}}(s) = \prod_{\mathfrak{p}} \left(1 - \frac{1}{\mathrm{N}(\mathfrak{p})^s}\right)^{-1} = \prod_{p} \prod_{\substack{j=1 \\ \mathfrak{p}_j \mid (p)}}^{r} \left(1 - \frac{1}{p^{s f_j}}\right)^{-1}. \qquad (13.2)$$

Assume that $\mathbb{K} = \mathbb{Q}(\alpha)$ with an algebraic integer α and $F(X)$ is the minimal polynomial of α. Then for all but finitely many primes p, the factorization of $F(X)$ modulo p determines the arithmetic of the local Euler factor: if

$$F(X) \equiv \prod_{j=1}^{r} F_j(X) \bmod p \qquad (13.3)$$

is the factorization into irreducible factors, then $f_j = \deg F_j$ in (13.2). Of course, we have $r \leq [\mathbb{K} : \mathbb{Q}] = d$. Thus, the Dedekind zeta-function has a representation as a polynomial Euler product. Since the norm of an integral ideal is a positive rational integer, the series can be rewritten as an ordinary Dirichlet series

$$\sum_{\mathfrak{a}} \frac{1}{\mathrm{N}(\mathfrak{a})^s} = \sum_{n=1}^{\infty} \frac{f_{\mathbb{K}}(n)}{n^s},$$

where $f_{\mathbb{K}}(n)$ counts the number of integral ideals $\mathfrak{a}$ with norm $\mathrm{N}(\mathfrak{a}) = n$. We see that the Riemann zeta-function is the Dedekind zeta-function of $\mathbb{Q}$ and, as a matter of fact, Dedekind zeta-functions share many properties with Riemann's zeta-function. The Dirichlet series defining the Dedekind zeta-function $\zeta_{\mathbb{K}}(s)$ converges for $\sigma > 1$, independent of the field $\mathbb{K}$. To see this note that, for $\sigma > 1$,

$$|\zeta_{\mathbb{K}}(s)| \leq \prod_{\mathfrak{p}} \left|1 - \frac{1}{\mathrm{N}(\mathfrak{p})^s}\right|^{-1} \leq \prod_{p} \left(1 - \frac{1}{p^\sigma}\right)^{-d} = \zeta(\sigma)^d,$$

since there are at most $d = [\mathbb{K} : \mathbb{Q}]$ many primes $\mathfrak{p}$ lying above each rational prime p and $\mathrm{N}(\mathfrak{p})$ is smallest if (p) splits completely.

Landau [176] proved the analogue of the prime number theorem for number fields, that is the prime ideal theorem

$$\pi_{\mathbb{K}}(x) := \#\{\mathfrak{p} \subset \mathcal{O}_{\mathbb{K}} \text{ prime} : \mathrm{N}(\mathfrak{p}) \leq x\} \sim \frac{x}{\log x}. \qquad (13.4)$$

Hecke [124] obtained the first deeper results concerning the analytic behaviour of Dedekind zeta-functions. He showed that $\zeta_{\mathbb{K}}(s)$ has an analytic continuation to $\mathbb{C}$ except for a simple pole at $s = 1$ and satisfies a Riemann-type functional equation. Especially the residue of the simple pole contains important information about the underlying number field. Hecke proved that

$$\lim_{s \to 1+} (s - 1)\zeta_{\mathbb{K}}(s) = \frac{2^{r_1}(2\pi)^{r_2} h R}{\omega \sqrt{|d_{\mathbb{K}}|}}, \qquad (13.5)$$

where r_1 is the number of real conjugate fields, $2r_2$ is the number of complex conjugate fields, h is the class number, R is the regulator, ω is the number of roots of unity in the group of units, and $d_{\mathbb{K}}$ is the discriminant of $\mathbb{K}$. The class number is the number of equivalence classes of fractional ideals of $\mathbb{K}$ modulo principal ideals, and so it measures the deviation of $\mathcal{O}_{\mathbb{K}}$ from having unique prime factorization. Gauss conjectured that the class numbers $h = h(d)$ of imaginary quadratic number fields $\mathbb{K} = \mathbb{Q}(\sqrt{D})$ with discriminant $d < 0$ tend with $-d$ to infinity; note that $d = D$ if $D \equiv 1 \bmod 4$, and $d = 4D$ if $D \equiv 2, 3 \bmod 4$. This was first proved by Heilbronn [128] and in refined form by Siegel [330]. The problem of finding an effective algorithm to determine all imaginary quadratic fields with a given class number h is known as the Gauss class number h problem. This problem is of interest with respect to the non-existence of exceptional real zeros of Dirichlet L-functions off the critical line. The general Gauss class number problem was solved by Goldfeld, Gross & Zagier [102, 107]. A complete determination of the imaginary quadratic fields with class number one was first given by Heegner [127] (but his solution was not completely accepted due to a number of gaps), Baker [12], and Stark [336] (independently):

$$h = 1 \quad \Longleftrightarrow \quad d \in \{-3, -4, -7, -8, -11, -19, -43, -67, -163\}.$$

Note that class number one is equivalent to unique prime factorization in the corresponding ring of algebraic integers.

Voronin [362] proved functional independence for any finite collection of

- Dedekind zeta-functions to distinct quadratic number fields and
- Dedekind zeta-functions to distinct cyclotomic number fields $\mathbb{Q}(\zeta_r)$, where ζ_r is a primitive rth root of unity and r is assumed to be odd and square-free.[1]

Gonek [104] showed that a single Dedekind zeta-function to an abelian number field is universal in the strip $1/2 < \sigma < 1$ and Reich [305, 306] succeeded in proving discrete universality for Dedekind zeta-functions to arbitrary number fields (Theorem 5.16); however, here the strip of universality is restricted to $1 - [\mathbb{K} : \mathbb{Q}]^{-1} < \sigma < 1$. We shall explain the reason for this difference.

If we are dealing with a Dedekind zeta-function $\zeta_{\mathbb{K}}(s)$ to a non-abelian field, we may apply our universality Theorem 6.12 (this is easily deduced from the properties listed above). Since $\zeta_{\mathbb{K}}(s)$ is an element of degree $d = [\mathbb{K} : \mathbb{Q}]$ in the

---

[1] Voronin's paper is published in the same year, 1975, as his universality theorem for $\zeta(s)$ and his reasoning is based on the very same ideas as in his approach to universality, however, while writing [362], the step to approximating functions rather than numbers was not done (actually, the rearrangement theorem for series in Hilbert spaces was then not available; cf. [198]). Later Voronin included the joint universality for Dedekind zeta-functions for such number fields in his thesis [364] (and also the monograph [166] of Karatsuba & Voronin contains these results).

Selberg class, this yields universality in the same strip as obtained by Reich. Now assume that we are given an abelian number field $\mathbb{K}$ (i.e., a finite Galois extension with abelian Galois group). Then

$$\zeta_{\mathbb{K}}(s) = \prod_{k=1}^{\ell} L(s, \chi_k^*), \tag{13.6}$$

where χ_k^* denotes the primitive character which induces χ_k, resp. the principal character $\chi \bmod 1$ which induces the trivial character of the Galois group, and ℓ is the number of primitive characters. The simplest example provide quadratic number fields $\mathbb{K} = \mathbb{Q}(\sqrt{D})$ with

$$\zeta_{\mathbb{K}}(s) = \zeta(s) L(s, \chi_d),$$

where χ_d is the Jacobi symbol and d is the discriminant. (For these facts we refer to Washington [368].)

Now it is clear how to apply the joint universality for Dirichlet L-functions in order to obtain universality for $\zeta_{\mathbb{K}}(s)$. The product (13.6) consists of Dirichlet L-functions to pairwise non-equivalent characters and we may apply the joint universality Theorem 1.10. For all but one of the factors $L(s, \chi_k^*)$ we define the target function to be $g_k(s) \equiv 1$, whereas the last one may be chosen freely, $g(s)$ say. Then

$$g(s) = \prod_{k=1}^{\ell} g_k(s).$$

If $g(s)$ is assumed to be non-vanishing and continuous on some compact subset $\mathcal{K}$ of $\frac{1}{2} < \sigma < 1$ with connected complement, and analytic in the interior, then Theorem 1.10 implies

$$\liminf_{T \to \infty} \frac{1}{T} \operatorname{meas} \left\{ \tau \in [0, T] : \max_{1 \le k \le \ell} \max_{s \in \mathcal{K}} |L(s + i\tau, \chi_k) - g_k(s)| < \varepsilon \right\} > 0.$$

Hence, we find for the product (13.6)

$$\liminf_{T \to \infty} \frac{1}{T} \operatorname{meas} \left\{ \tau \in [0, T] : \max_{s \in \mathcal{K}} |\zeta_{\mathbb{K}}(s + i\tau) - g(s)| < \varepsilon \right\} > 0.$$

However, we can do slightly more. In the sequel we shall sketch how to generalize this idea in order to obtain joint universality.

Dependence relations between Dedekind zeta-functions are of special interest with respect to so-called class number relations (see Brauer [43]). Besides functional independence we shall consider the concept of multiplicative independence. The functions $f_1(s), \ldots, f_m(s)$ are said to be multiplicatively independent if the identity

$$\prod_{j=1}^{m} f_j(s)^{a_j} = 1 \qquad \text{with} \quad a_j \in \mathbb{Z}$$

holds only trivially, i.e., $a_1 = \cdots = a_m = 0$.

Recently, Marszałek [238] showed that the multiplicative independence of Dedekind zeta-functions to abelian fields is equivalent to their functional independence; he also gave a sufficient condition for their independence as well as all possible multiplicative dependence relations. His argument relies essentially on the representation of Dedekind zeta-functions as products of Dirichlet L-functions.

Assume that we are given abelian number fields $\mathbb{K}_1, \ldots, \mathbb{K}_m$. Then, by the theorem of Kronecker–Weber, their compositum $\mathbb{K}$ is a subfield of some cyclotomic field $\mathbb{Q}(\zeta_r)$, where ζ_r is an rth root of unity. We denote by $\hat{\mathsf{G}} = \{\chi_1, \ldots, \chi_\ell\}$ the group of Dirichlet characters associated with $\mathbb{K}$ and by $\hat{\mathsf{G}}_1, \ldots, \hat{\mathsf{G}}_\ell$ the subgroups of $\hat{\mathsf{G}}$ corresponding to $\mathbb{K}_1, \ldots, \mathbb{K}_m$, respectively. We put

$$\delta_{jk} = \begin{cases} 1 & \text{if } \chi_k \in \hat{\mathsf{G}}_j, \\ 0 & \text{otherwise.} \end{cases} \tag{13.7}$$

Then, for $1 \leq j \leq m$, by class field theory we have the multiplicative representation

$$\zeta_{\mathbb{K}_j}(s) = \prod_{k=1}^{\ell} L(s, \chi_k^*)^{\delta_{jk}}, \tag{13.8}$$

where χ_k^* denotes the primitive character which induces χ_k. (For these facts we once more refer to Washington [368].)

Sander and Steuding [315] showed that both concepts of independence, functional independence and multiplicative independence, are equivalent to the discrete joint universality for the Dedekind zeta-functions in question.

Theorem 13.1. *Suppose that $\mathbb{K}_1, \ldots, \mathbb{K}_m$ are abelian number fields. Then the following statements are equivalent:*

(i) *The associated Dedekind zeta-functions are jointly universal: Let $\mathcal{K}$ be a compact subset of $\frac{1}{2} < \sigma < 1$ with connected complement, let q be a positive integer, and for each $1 \leq j \leq m$ let $f_j(s)$ be a continuous non-vanishing function on $\mathcal{K}$ which is analytic in the interior. Then, for any real $\Delta \neq 0$ and any $\epsilon > 0$,*

$$\liminf_{N \to \infty} \frac{1}{N} \sharp \left\{ n \leq N : \max_{1 \leq j \leq m} \max_{s \in \mathcal{K}} |\zeta_{\mathbb{K}_j}(s + \mathrm{i}\Delta n) - f_j(s)| < \epsilon \right\} > 0.$$

(ii) *The matrix $\mathsf{M} := (\delta_{jk})_{\substack{1 \leq j \leq m \\ 1 \leq k \leq \ell}}$ with δ_{jk} given by (13.7) has rank $m \leq \ell$.*

(iii) *The associated Dedekind zeta-functions are multiplicatively independent.*

(iv) *The associated Dedekind zeta-functions are functionally independent.*

The proof is quite simple. The equivalence of (ii)–(iv) is from Marszałek [238]. We sketch the argument. Suppose that the matrix M has rank $m \leq \ell$ and that

$$\prod_{j=1}^{m} \zeta_{\mathbb{K}_j}(s)^{a_j} = 1$$

for some integers a_j. In view of (13.8) this implies

$$1 = \prod_{j=1}^{m} \prod_{k=1}^{\ell} L(s, \chi_k^*)^{a_j \delta_{jk}} = \prod_{k=1}^{\ell} L(s, \chi_k^*)^{\sum_{j=1}^{m} a_j \delta_{jk}},$$

resp.

$$\sum_{j=1}^{m} a_j \delta_{jk} = 0 \qquad \text{for} \quad k = 1, \ldots, \ell$$

by the functional independence of Dirichlet *L*-functions. However, since the rank of M is equal to m, the latter system of linear equations has a unique solution: $a_1 = \cdots = a_m = 0$. Conversely, if we would have $m > \ell$ or the rank would be less than m, the system would have non-trivial solutions. Hence, (ii) is equivalent to the multiplicative independence (iii). Clearly, functional independence implies multiplicative independence. Next we show that the converse is also true.

Assume that the Dedekind zeta-functions are multiplicatively independent and that, for continuous functions $F_0, F_1, \ldots, F_N : \mathbb{C}^m \to \mathbb{C}$, not all identically vanishing, and any s,

$$\sum_{k=0}^{N} s^k F_k(\zeta_{\mathbb{K}_1}(s), \ldots, \zeta_{\mathbb{K}_m}(s)) = 0. \tag{13.9}$$

For each k define $G_k = F_k(\underline{P}(s))$, where $\underline{P} := (P_1, \ldots, P_m) : \mathbb{C}^\ell \to \mathbb{C}^m$ is given by

$$P_j(s_1, \ldots, s_\ell) = \prod_{k=1}^{\ell} s_k^{\delta_{jk}}.$$

Now it follows from (13.9) that

$$\sum_{k=0}^{N} s^k G_k(L(s, \chi_1^*), \ldots, L(s, \chi_\ell^*)) = 0.$$

The functional independence of Dirichlet *L*-functions implies that the G_k are vanishing identically and it is easy to deduce the same for the F_k. This shows that multiplicative and functional independence are equivalent. In particular, we have shown (ii) $\iff$ (iv).

In Sect. 10.3, we have shown that joint universality implies functional independence: (i) $\Rightarrow$ (iv). It remains to prove that (ii) implies (i). Given the condition on the rank of the matrix M, the system of linear equations

$$\mathsf{F} = \mathsf{M}\mathsf{G}, \qquad \text{where} \quad \mathsf{F} := \left(F_j(s)\right)_{1 \leq j \leq m}^{\mathsf{t}}, \quad \mathsf{G} := \left(G_k(s)\right)_{1 \leq k \leq \ell}^{\mathsf{t}}$$

with $F_j(s) := \log f_j(s)$ is solvable in G; note that the logarithms $F_j(s) = \log f_j(s)$ for $s \in \mathcal{K}$ exist by the condition of the theorem. Now we switch from the identities

$$F_j(s) = \sum_{k=1}^{\ell} \delta_{jk} G_k(s)$$

by exponentiation to

$$f_j(s) = \prod_{k=1}^{\ell} g_k(s)^{\delta_{jk}},$$

where $g_k(s) := \exp(G_k(s))$; here the functions $g_k(s)$ are continuous non-vanishing functions on $\mathcal{K}$ which are analytic in the interior. By Theorem 5.17 we may approximate any $g_k(s)$ by certain shifts $L(s + i\Delta n, \chi_k^*)$, i.e.,

$$\max_{1 \le k \le \ell} \max_{s \in \mathcal{K}} |L(s + i\Delta n, \chi_k^*) - g_k(s)| < \epsilon. \qquad (13.10)$$

In view of (13.8) we have

$$\zeta_{\mathbb{K}_j}(s + i\Delta n) = \prod_{k=1}^{\ell} L(s + i\Delta n, \chi_k^*)^{\delta_{jk}}$$

$$= \prod_{k=1}^{\ell} (g_k(s) + O(\epsilon))^{\delta_{jk}} = f_j(s) + O(\epsilon)$$

for any $s \in \mathcal{K}$ and any j. Since (13.10) holds for a set of positive integers n with positive lower density, the assertion follows.

Of course, in view of Theorem 1.10 we can also replace statement (i) about *discrete* universality by joint *continuous* universality: for any $\epsilon > 0$,

$$\liminf_{T \to \infty} \frac{1}{T} \operatorname{meas} \left\{ \tau \in [0, T] : \max_{1 \le j \le m} \max_{s \in \mathcal{K}} |\zeta_{\mathbb{K}_j}(s + i\tau) - f_j(s)| < \epsilon \right\} > 0.$$

Theorem 13.1 is best possible. We cannot have joint universality for any family of Dedekind zeta-functions to different number fields. If $m > \ell$, then there exist restrictive dependencies for the Dedekind zeta-functions. For instance,

$$\zeta_{\mathbb{Q}(\sqrt{-3})} \zeta_{\mathbb{Q}(2^{1/3})} = \zeta_{\mathbb{Q}} \zeta_{\mathbb{Q}(2^{1/3}, \exp(2\pi i/3))} \qquad (13.11)$$

(see Sect. 13.6). Indeed, if the Galois group has more normal subgroups than conjugacy classes, then such algebraic relations for the corresponding Dedekind zeta-functions exist and they cannot be jointly universal. Recently, Bauer [16] proved that if there is no algebraic relation for a collection of Dedekind zeta-function to finite normal extensions of the rationals, then they are functionally independent; his proof relies on the joint universality for Artin L-functions to non-equivalent characters of the Galois group of a Galois extension $\mathbb{K}/\mathbb{Q}$ (see Sect. 13.8).

It is an interesting question to which extent the Dedekind zeta-function determines the field. One can show that the Dedekind zeta-function $\zeta_{\mathbb{K}}(s)$ determines the minimal normal extension $\mathbb{L}$ of $\mathbb{Q}$ containing $\mathbb{K}$ and thus we have to ask whether there exist non-conjugate subgroups of $\mathsf{Gal}(\mathbb{L}/\mathbb{Q})$ giving the same induced trivial character. This is indeed possible! Two number fields $\mathbb{K}_1$ and $\mathbb{K}_2$ are said to be arithmetically equivalent if their Dedekind zeta-functions are the same. The first non-trivial example was given by Gassmann [93]. Perlis [291] proved that arithmetically equivalent non-isomorphic fields have at least degree 7 and that this bound cannot be improved. An explicit example of degree 8 is for instance $\mathbb{Q}((-3)^{1/8})$ and $\mathbb{Q}((-48)^{1/8})$ which is due to Perlis and Schinzel [292]. By the rank condition from Marszałek [238] it follows that two distinct abelian fields $\mathbb{K}_1$ and $\mathbb{K}_2$ cannot be arithmetically equivalent since otherwise their Dedekind zeta-functions have the same factorization into Dirichlet *L*-functions and $\mathbb{K}_1 = \mathbb{K}_2$ if and only if their character groups are equal: $\hat{G}_1 = \hat{G}_2$ (see Washington [368]). A stronger statement was shown by Nicolae [282] who proved that Dedekind zeta-functions to distinct finite Galois extensions are linearly independent over $\mathbb{C}$. This also shows that linear independence of Dedekind zeta-functions does not imply their joint universality.

13.2 Grössencharacters

In 1920, Hecke [125] introduced a new class of *L*-functions which generalize the concepts of Dedekind zeta-functions and Dirichlet *L*-functions. However, first of all we have to introduce Hecke grössencharacters which represent the most general extension of Dirichlet characters to number fields.

Given a number field $\mathbb{K}$ of degree n over $\mathbb{Q}$, there are exactly n embeddings $\mathbb{K}^{(j)}$ of $\mathbb{K}$ into $\mathbb{C}$, for $1 \leq j \leq n$ given by

$$\mathbb{K} \ni \alpha \mapsto \alpha^{(j)} \in \mathbb{K}^{(j)},$$

where the $\alpha^{(j)}$ denote the conjugates of α; we assume that among these there are r_1 real and $2r_2$ complex embeddings (that makes $n = r_1 + 2r_2$). We denote the real embeddings by

$$\mathbb{K}^{(1)}, \ldots, \mathbb{K}^{(r_1)}$$

and the complex embeddings which are pairwise complex conjugate by

$$\mathbb{K}^{(r_1+1)}, \ldots, \mathbb{K}^{(r_1+r_2+1)} = \overline{\mathbb{K}^{(r_1+1)}}, \ldots, \mathbb{K}^{(n)} = \overline{\mathbb{K}^{(r_1+r_2)}}.$$

Let $\mathfrak{f}$ be a non-zero integral ideal of $\mathbb{K}$. The unit group modulo $\mathfrak{f}$ is defined to be the set of all units $\epsilon \equiv 1 \bmod \mathfrak{f}$ which are totally positive (i.e., all its real conjugates are positive) and we denote it by $\mathsf{U}(\mathfrak{f})$. It is easily seen that $\mathsf{U}(\mathfrak{f})$ is indeed a group. By Dirichlet's unit theorem there exist $r = r_1 + r_2 - 1$ units $\eta_1, \ldots, \eta_r$ and a root of unity ζ in $\mathbb{K}$ such that any $\epsilon \in \mathsf{U}(\mathfrak{f})$ has a unique representation

$$\epsilon = \zeta^m \eta_1^{n_1} \cdots \eta_r^{n_r}$$

with integers m, n_k. The units $\eta_1, \ldots, \eta_r$ are said to be fundamental units of $U(\mathfrak{f})$ although they are not uniquely determined. Define the matrix

$$(e_j \log |\eta_k^{(j)}|)_{1 \le j,k \le r}, \qquad \text{where} \quad e_j := \begin{cases} 1 & \text{if } 1 \le j \le r_1, \\ 2 & \text{if } r_1 < j \le r_1 + r_2. \end{cases}$$

Then the regulator $R(\mathfrak{f})$ is defined to be the absolute value of the determinant of this matrix:

$$R(\mathfrak{f}) = |\det(e_j \log |\eta_k^{(j)}|)|;$$

it should be noted that the regulator does not depend on the choice of the fundamental units η_k.

We further denote by $I(\mathfrak{f})$ the multiplicative group generated by all ideals coprime with $\mathfrak{f}$. The principal ray class $P(\mathfrak{f})$ is the subgroup of $I(\mathfrak{f})$ consisting of all principal ideals of the form (α/β) satisfying

- $0 \ne \alpha, \beta \in \mathcal{O}_K$ (the ring of integers);
- $\alpha \equiv \beta \bmod \mathfrak{f}$;
- α/β is totally positive.

The factor group

$$G(\mathfrak{f}) := I(\mathfrak{f})/P(\mathfrak{f})$$

is called the ray class group mod $\mathfrak{f}$, and its elements, the ray classes, might be regarded as the analogues of the residue classes in the rational number field case. One can show that $G(\mathfrak{f})$ is a finite abelian group and we denote its order by $h(\mathfrak{f})$.

We shall give a brief example. For the sake of simplicity we shall consider the number field $\mathbb{Q}(\sqrt{-5})$ and choose $\mathfrak{f} = (1)$ in which case $G = G((1))$ is the class group. (In general, the narrow divisor class group $G((1))$ is not equal to G, however, in complex quadratic fields there are no sign conditions.) There is no unique prime factorization, and so the class number h is greater than one. One can deduce from Minkowski's theorem on linear forms that every class of G contains an integral ideal $\mathfrak{a}$ with norm

$$N(\mathfrak{a}) \le \sqrt{|d_K|},$$

where d_K is the discriminant of the number field, that is in our case $d_K = -20$. Obviously, this observation proves the finiteness of the class number. However, we can also use it to get information about the structure of the class group. For this purpose we observe that the only prime ideals $\mathfrak{p}$ with $N(\mathfrak{p}) \le 4$ can be among the prime ideal divisors of (2) and (3). By the splitting of primes in quadratic number fields, we find

$$(2) = \mathfrak{p}_1^2 \quad \text{with} \quad \mathfrak{p}_1 = (2, 1 + \sqrt{-5}) = (2, 1 - \sqrt{-5}),$$
$$(3) = \mathfrak{p}_2\mathfrak{p}_2' \quad \text{with} \quad \mathfrak{p}_2 = (3, 1 + \sqrt{-5}) \ne \mathfrak{p}_2' = (3, 1 - \sqrt{-5}).$$

Hence, the ideals with norms less than or equal to 4 are $\mathfrak{p}_1, \mathfrak{p}_2, \mathfrak{p}_2'$, and $(2) = \mathfrak{p}_1^2$. It is easy to see that $\mathfrak{p}_1$ is not principal and represents a class of order two.

Furthermore, it turns out that all other ideals lie in this class or are principal. Hence the class number of $\mathbb{Q}(\sqrt{-5})$ equals two and we have a description of the associated class group.

Now we are in the position to define Hecke characters. Suppose we are given numbers a_j and ν_k satisfying

- $a_j \in \{0, 1\}$ for $1 \leq j \leq r_1$ and $a_j \in \mathbb{Z}$ for $r_1 < j \leq r_1 + r_2$;
- $\nu_k \in \mathbb{R}$ for $1 \leq k \leq r_1 + r_2$ such that $\nu_1 + \cdots + \nu_{r_1+r_2} = 0$.

Then we define a function $\chi_\infty : \mathbb{K}^* \to \mathbb{C}^*$ by

$$\chi_\infty(\alpha) = \prod_{k=1}^{r_1+r_2} |\alpha^{(k)}|^{\mathrm{i}\nu_k} \prod_{j=1}^{r_1+r_2} \left(\frac{\alpha^{(j)}}{|\alpha^{(j)}|} \right)^{a_j}.$$

Obviously, χ_∞ is unimodular. Since the sum of the ν_ks vanishes, it follows that χ_∞ is trivial on $\mathbb{Q}^*$. We suppose that the kernel of χ_∞ contains the unit group modulo $\mathfrak{f}$, i.e., $\chi_\infty(\epsilon) = 1$ for any $\epsilon \in \mathsf{U}(\mathfrak{f})$. Then χ_∞ induces a character on $\mathsf{P}(\mathfrak{f})$.

If a non-trivial homomorphism $\chi : \mathsf{I}(\mathfrak{f}) \to \mathbb{C}^*$ is identified with χ_∞ on $\mathsf{P}(\mathfrak{f})$, that is

$$\chi(\mathfrak{a}) = \chi_\infty(\alpha) \qquad \text{for} \quad \mathfrak{a} = (\alpha) \in \mathsf{P}(\mathfrak{f}),$$

then χ is said to be a grössencharacter modulo $\mathfrak{f}$ (resp. Hecke character in some literature). If all numbers a_j, ν_k are equal to zero, then χ is said to be a ray class character, and if additionally $\mathfrak{f} = (1)$, then χ is an ideal class group character (that is one of the finitely many characters of the class group of $\mathbb{K}$). However, there are infinitely many grössencharacters since, for example, the function

$$\mathfrak{a} = (\alpha) \mapsto \chi_m(\mathfrak{a}) = \left(\frac{\alpha}{|\alpha|} \right)^{4m} = \exp(4\mathrm{i}m \arg(\alpha)) \tag{13.12}$$

for $\mathfrak{a} \neq 0$ and any integer m is a primitive grössencharacter. If there exists an ideal $\mathfrak{f}^* \subset \mathfrak{f}$ and a grössencharacter χ^* mod $\mathfrak{f}^*$ such that $\chi = \chi^*$ on $\mathsf{I}(\mathfrak{f})$, then χ is said to be induced by χ^*; otherwise χ is called primitive and $\mathfrak{f}$ is said to be the conductor of χ.

13.3 Hecke *L*-Functions

Let $\mathbb{K}$ be a number field, $\mathfrak{f}$ be an ideal of $\mathbb{K}$, and let χ modulo $\mathfrak{f}$ be a grössencharacter. We extend χ to the group I of all fractional ideals of $\mathbb{K}$ by setting $\chi(\mathfrak{a}) = 0$ if $\mathfrak{a}$ is not coprime with $\mathfrak{f}$. Then the Hecke *L*-function associated to χ is (formally) given by

$$L(s, \chi) = \sum_\mathfrak{a} \frac{\chi(\mathfrak{a})}{\mathrm{N}(\mathfrak{a})^s} = \prod_\mathfrak{p} \left(1 - \frac{\chi(\mathfrak{p})}{\mathrm{N}(\mathfrak{p})^s} \right)^{-1}, \tag{13.13}$$

where the sum is taken over all non-zero integral ideals $\mathfrak{a}$ of $\mathbb{K}$, the product is taken over all prime ideals $\mathfrak{p}$, and $N(\mathfrak{a})$ denotes the norm of the ideal $\mathfrak{a}$ (as in the case of the Dedekind zeta-function).

Hecke L-functions to grössencharacters are the analogues of Dirichlet L-functions: if $\mathbb{K} = \mathbb{Q}$, $\mathfrak{f} = (q)$ with $q \in \mathbb{Z}$, and $\chi_\infty \equiv 1$, then the construction above leads without the totally positive condition to

$$G(\mathfrak{f}) = (\mathbb{Z}/q\mathbb{Z})^*/\{\pm 1\}.$$

If χ is the trivial (principal) character, then the Hecke L-function for a number field $\mathbb{K}$ is nothing else but the Dedekind zeta-function. Note that what we call Hecke L-functions are in some literature called (generalized) Dirichlet L-functions.

Both the series and the product (13.13) defining $L(s,\chi)$ are absolutely convergent for $\sigma > 1$ and uniformly convergent in any compact subset. To see this we proceed as we did for Dedekind zeta-functions via the splitting of primes (13.1). Indeed, we can rewrite (13.13) as an ordinary Euler product

$$L(s,\chi) = \prod_{\mathfrak{p}} \left(1 - \frac{\chi(\mathfrak{p})}{N(\mathfrak{p})^s} \right)^{-1} = \prod_p \prod_{\substack{j=1 \\ \mathfrak{p}_j|(p)}}^{r} \left(1 - \frac{\chi(\mathfrak{p}_j)}{p^{sf_j}} \right)^{-1}.$$

Thus, $L(s,\chi)$ has a representation as a polynomial Euler product. Hence we may also rewrite this as an ordinary Dirichlet series:

$$L(s,\chi) = \sum_{n=1}^{\infty} \frac{a(n)}{n^s},$$

where

$$a(n) = \prod_{p|n} \sum_{\substack{k_1,\ldots,k_r \geq 0 \\ k_1 f_1 + \cdots + k_r f_r = \nu(n;p)}} \prod_{j=1}^{r} \chi(\mathfrak{p}_j)^{k_j}.$$

Since the degree of the local Euler factors is bounded by the degree of the field extension $\mathbb{K}/\mathbb{Q}$, it follows (as in the case of the Dedekind zeta-function) that the Dirichlet series and so the Euler product converge for $\sigma > 1$ absolutely and that the Ramanujan hypothesis holds.

Hecke [125] proved that $L(s,\chi)$ extends to an entire function and satisfies a functional equation of Riemann-type provided χ is primitive. Let $d_\mathbb{K}$ denote the discriminant of $\mathbb{K}$. We define

$$\gamma(\chi) = \prod_{k=r_1+1}^{r_1+r_2} 2^{i\nu_k/2}, \quad A(\mathfrak{f}) = \left(\frac{|d_\mathbb{K}|N(\mathfrak{f})}{\pi^n} \right)^{1/2} 2^{-r_2},$$

and

$$\Gamma(s,\chi) = \prod_{j=1}^{r_1} \Gamma\left(\frac{s + a_j - i\nu_j}{2} \right) \prod_{j=r_1+1}^{r_1+r_2} \Gamma\left(s + \frac{|a_j| - i\nu_j}{2} \right).$$

Then

$$\Lambda(1-s,\chi) = \omega(\chi)\Lambda(s,\overline{\chi}),$$

where $\omega(\chi)$ is a complex number with $|\omega(\chi)| = 1$, depending only on χ, and

$$\Lambda(s,\chi) := \gamma(\chi)A(\mathfrak{f})^s \Gamma(s,\chi)L(s,\chi).$$

In view of all the properties mentioned, it follows that Hecke *L*-functions $L(s,\chi)$ to primitive grössencharacters are elements of the Selberg class of degree $n = [\mathbb{K} : \mathbb{Q}]$. The Hecke *L*-function to the trivial character is, as already mentioned, identical with the Dedekind zeta-function and so it is an element of the Selberg class too.

We sketch some arithmetic consequences of the analytic properties of Hecke *L*-functions. First of all, we note that $L(s,\chi)$ does not vanish on the edge of the critical strip; the proof follows the proof for Dirichlet *L*-functions (see Heilbronn [129]). We have already mentioned that $L(s,\chi)$ is entire and so it is regular at $s = 1$ unless χ is trivial. If χ is not trivial, then

$$\sum_{\mathrm{N}(\mathfrak{p})\leq x} \chi(\mathfrak{p}) = o\left(\frac{x}{\log x}\right)$$

as $x \to \infty$; for trivial χ we obtain the prime ideal theorem (13.4).

13.4 Universality for Hecke *L*-Functions

Universality for Hecke *L*-functions was first proved by Mishou. In [249], he obtained universality for Hecke *L*-functions to ray class characters. The proof proceeds along the lines of Voronin's original proof; besides, arithmetical properties of those rational primes which split completely play a central role. In Mishou [250], Bagchi's approach and, in particular, the positive density method are applied to tackle the general case of grössencharacters.

Theorem 13.2. *Let* $\mathbb{K}$ *be a finite extension of* $\mathbb{Q}$ *of degree* n *and* χ *be a grössencharacter modulo* $\tilde{\mathfrak{f}}$. *Let* $\mathcal{K}$ *be a compact subset of the strip* $1-\frac{1}{n} < \sigma < 1$ *with connected complement and* $g(s)$ *be a non-vanishing and continuous on* $\mathcal{K}$ *which is analytic in the interior of* $\mathcal{K}$. *Then, for any* $\epsilon > 0$,

$$\liminf_{T\to\infty} \frac{1}{T} \operatorname{meas}\left\{\tau \in [0,T] : \max_{s\in\mathcal{K}} |L(s+i\tau,\chi) - g(s)| < \epsilon\right\} > 0.$$

The main part of the proof relies on a discussion of the various ways how rational primes can split. This is used to show that there exist infinitely many rational primes p for which the character sum

$$a(p) = \sum_{\mathfrak{p}\mid(p)} \chi(\mathfrak{p}),$$

where the sum is taken over all prime ideals of degree one which divide p, is bounded away from zero. In the case of ray class characters one can deduce this quite easily by applying class field theory; in fact, the set of all rational primes which split completely in the class field for the ideal class group has infinitely many elements. The general case is more delicate (for the details see [249, 250]).

For the special type of grössencharacters χ_m of the form (13.12) for $\mathbb{Q}(i)$, Koyama and Mishou [174] succeeded to prove a mixed version of universality for Hecke L-functions $L(s + i\tau, \chi_m)$ in the τ- and in the χ-aspect:

Theorem 13.3. *Let $\mathcal{K}$ be a compact set in the open right half of the critical strip. For any function $g(s)$ which is non-vanishing and continuous on $\mathcal{K}$ and which is analytic in the interior of $\mathcal{K}$, and for any $\epsilon > 0$,*

$$\liminf_{T \to \infty} \frac{1}{T^2} \, \widetilde{\text{meas}} \left\{ (\tau, m) \in \tilde{T} : \max_{s \in \mathcal{K}} |L(s + i\tau, \chi_m) - g(s)| < \epsilon \right\} > 0,$$

where $\tilde{T} := [0, T] \times \{0, \ldots, T\}$ and the measure $\widetilde{\text{meas}}$ is defined by

$$\widetilde{\text{meas}}(X) = \sum_{m=0}^{T} \text{meas} \left\{ \tau \in [0, T] : (\tau, m) \in X \right\}$$

for $X \subset \tilde{T}$.

As Koyama and Mishou mention, it is unfortunate that the range of τ and m has to be the same. They remark that a conjecture of Duke [71] on mean-values for certain Dirichlet series over $\mathbb{Z}(i)$ would imply the universality in the character aspect alone.

13.5 Artin's Reciprocity Law

Now we want to study a further class of L-functions which play a central role in algebraic number theory ever since Artin introduced them in order to find higher reciprocity laws. However, first of all, we briefly motivate their definition.

In algebraic number theory, a fundamental problem is to describe how a rational prime factors into primes in the ring of integers $\mathcal{O}_{\mathbb{K}}$ of a given number field $\mathbb{K}$. Now assume that $\mathbb{K}$ is a Galois extension over $\mathbb{Q}$ with Galois group $G := \text{Gal}(\mathbb{K}/\mathbb{Q})$ (i.e., $\mathbb{Q}$ is fixed with respect to automorphisms from G). Then $\mathbb{K}$ is the splitting field of some monic polynomial with rational coefficients, and G is the group of field automorphisms of $\mathbb{K}$ fixing $\mathbb{Q}$ pointwise. The splitting type of a prime p in $\mathcal{O}_{\mathbb{K}}$ is completely determined by the size of the subgroup of G which fixes any $\mathfrak{p}_j$ above p. For simplicity, assume that the rational prime p is unramified in $\mathbb{K}$, i.e., the primes $\mathfrak{p}_j$ in (13.1) are all distinct, then these subgroups are all cyclic. Information about the factorization of such p is encoded in the so-called Frobenius automorphism $\sigma_{\mathfrak{p}_j}$ of G, the canonical generator of the subgroup of G which maps any $\mathfrak{p}_j$ into itself. The Frobenius is determined only up to conjugacy in G; nevertheless, the resulting conjugacy class, which we denote by σ_p, completely determines the splitting type of (13.1).

If, for example, $\mathbb{K} = \mathbb{Q}(i) = \{a + bi : a, b \in \mathbb{Q}\}$, then $\mathcal{O}_{\mathbb{K}} = \mathbb{Z}[i]$, the set of Gaussian integers $m + ni$ with $m, n \in \mathbb{Z}$. In this case, σ_p is the identity if -1 is a quadratic residue mod p, and the complex conjugation otherwise. Hence, we may identify G with the subgroup $\{\pm 1\}$ of $\mathbb{C}^* := \mathbb{C} \setminus \{0\}$ via the homomorphism $\varrho : \mathsf{G} \to \{\pm 1\}$:

$$\varrho(\sigma_p) = \left(\frac{-1}{p}\right).$$

By a part of the quadratic reciprocity law, the Legendre symbol can be expressed in terms of a congruence condition on p which states for unramified (odd) primes p

$$\left(\frac{-1}{p}\right) = (-1)^{\frac{p-1}{4}} = \begin{cases} +1 & \text{if } p \equiv 1 \bmod 4, \\ -1 & \text{otherwise.} \end{cases}$$

Thus, the factorization of p in $\mathbb{Z}[i]$ depends only on its residue mod 4.

One goal of class field theory is to find a similar description of σ_p for arbitrary Galois extensions $\mathbb{K}$. In general, one cannot expect that there exists a modulus q such that σ_p is the identity if and only if p lies in some arithmetic progression mod q. However, if $\mathbb{K}$ is abelian, i.e., $\mathsf{G} = \mathsf{Gal}(\mathbb{K}/\mathbb{Q})$ is abelian, and $\varrho : \mathsf{G} \to \mathbb{C}^*$ is a homomorphism, then it is known that there exists a Dirichlet character ψ mod q such that

$$\psi : (\mathbb{Z}/q\mathbb{Z})^* \to \mathbb{C}^* \qquad \text{with} \quad \varrho(\sigma_p) = \psi(p) \tag{13.14}$$

for all primes p, unramified in $\mathbb{K}$. This is the theorem of Kronecker–Weber. It follows that the splitting properties of p in $\mathbb{K}$ depend only on its residue modulo some fixed number q depending on $\mathbb{K}$. In particular, this implies the general quadratic reciprocity law of Gauss. As a matter of fact, the factorization of Dedekind zeta-functions

$$\zeta_{\mathbb{K}}(s) = \zeta(s)L(s, \chi), \qquad \text{where} \quad \mathbb{K} = \mathbb{Q}(\sqrt{\chi(-1)q}),$$

for all quadratic fields $\mathbb{K}$ is equivalent with quadratic reciprocity.

What can be said for *non-abelian* Galois extensions? Recognizing the utility of studying groups in terms of their matrix representations, Artin focused attention on homomorphisms

$$\varrho : \mathsf{G} = \mathsf{Gal}(\mathbb{K}/\mathbb{Q}) \to \mathsf{GL}_m(\mathbb{C}),$$

i.e., m-dimensional representations of the Galois group G; note that one-dimensional representations are simply characters. Artin transferred the problem of analysing conjugacy classes in G to the analogous problem in $GL_m(\mathbb{C})$, where the corresponding classes are completely determined by their characteristic polynomials

$$\det \left(1 - \frac{\varrho(\sigma_p)}{p^s} \right),$$

where 1 denotes the unitary matrix. Introducing the so-called Artin L-function

$$L(s, \varrho) = \prod_p \det \left(1 - \frac{\varrho(\sigma_p)}{p^s} \right)^{-1}$$

(we give a precise definition in Sect. 13.6), Artin was able to reduce the problem to one involving these analytic objects: *is it possible to define $L(s, \varrho)$ in terms of the arithmetic of $\mathbb{Q}$ alone?* It was in this context that Artin proved his reciprocity law. Indeed, for abelian fields $\mathbb{K}$ and one-dimensional ϱ, Artin showed that $L(s, \varrho)$ is identical to a Dirichlet L-function $L(s, \psi)$ with an appropriate character ψ mod q. Since an identity between two Euler products implies an identity between the local Euler factors, this implies Artin's reciprocity law (13.14). Actually, Artin proved a stronger result for abelian extensions $\mathbb{L}/\mathbb{K}$ (see Theorems 13.4 and (13.16)).

13.6 Artin L-Functions

Let $\mathbb{L}/\mathbb{K}$ be a Galois extension of number fields with Galois group G. Further, let $\varrho : G \to GL_m(V)$ be a representation (group homomorphism) of G on a finite dimensional complex vector space V. In order to give the definition of the Artin L-function attached to these data, we recall some facts on prime ideals in number fields and their ramification in Galois extensions.

For each prime $\mathfrak{p}$ of $\mathbb{K}$, and a prime $\mathfrak{P}$ of $\mathbb{L}$ with $\mathfrak{P}|\mathfrak{p}$, we define the decomposition group by

$$D_{\mathfrak{P}} = \{ \varrho \in G : \mathfrak{P}^{\varrho} = \mathfrak{P} \} = Gal(\mathbb{L}_{\mathfrak{P}}/\mathbb{K}_{\mathfrak{p}}),$$

where $\mathbb{L}_{\mathfrak{P}}$ and $\mathbb{K}_{\mathfrak{p}}$ are the completions of $\mathbb{L}$ at $\mathfrak{P}$ and $\mathbb{K}$ at $\mathfrak{p}$, respectively. Denote by $k_{\mathfrak{P}}/k_{\mathfrak{p}}$ the residue field extension. By Hensel's lemma, we have a surjective map from $D_{\mathfrak{P}}$ to $Gal(k_{\mathfrak{P}}/k_{\mathfrak{p}})$; its kernel $I_{\mathfrak{P}}$ is the inertia group at $\mathfrak{P}$, defined by

$$I_{\mathfrak{P}} = \{ \varrho \in G : \varrho(\alpha) \equiv \alpha \bmod \mathfrak{P} \text{ for all } \alpha \in \mathcal{O}_{\mathbb{L}} \}.$$

We thus have an exact sequence

$$1 \to I_{\mathfrak{P}} \to D_{\mathfrak{P}} \to Gal(k_{\mathfrak{P}}/k_{\mathfrak{p}}) \to 1.$$

Hence, there is an isomorphism

$$D_{\mathfrak{P}}/I_{\mathfrak{P}} \simeq \mathsf{Gal}(k_{\mathfrak{P}}/k_{\mathfrak{p}}).$$

Now $k_{\mathfrak{P}}/k_{\mathfrak{p}}$ is a Galois extension of finite fields, and hence the group $\mathsf{Gal}(k_{\mathfrak{P}}/k_{\mathfrak{p}})$ is cyclic, generated by the map $\alpha \mapsto \alpha^{\mathrm{N}(\mathfrak{p})}$, where $\mathrm{N}(\mathfrak{p})$, the absolute norm of $\mathfrak{p}$, is the cardinality of $k_{\mathfrak{p}}$. We can choose an element $\sigma_{\mathfrak{P}} \in D_{\mathfrak{P}}$ whose image in $\mathsf{Gal}(k_{\mathfrak{P}}/k_{\mathfrak{p}})$ is this generator; this $\sigma_{\mathfrak{P}}$ is called Frobenius element at $\mathfrak{P}$, i.e.,

$$\sigma_{\mathfrak{P}}(\alpha) \equiv \alpha^{\mathrm{N}(\mathfrak{p})} \bmod \mathfrak{P}$$

for all $\alpha \in \mathcal{O}_{\mathbb{L}}$. Note that the Frobenius element is only defined mod $I_{\mathfrak{P}}$. The Frobenius describes how prime ideals split when lifted from a smaller ring of integers to a larger ring of integers. For unramified $\mathfrak{p}$ (and in particular, these are all but finitely many $\mathfrak{p}$), the Frobenius is well defined since $I_{\mathfrak{P}} = \{1\}$. The action of the Galois group on the set of primes in $\mathbb{L}$ above $\mathfrak{p}$ is transitive, and thus for any pair of primes $\mathfrak{P}_1$ and $\mathfrak{P}_2$ lying above $\mathfrak{p}$, there exists an automorphism in G which simultaneously conjugates $D_{\mathfrak{P}_1}$ into $D_{\mathfrak{P}_2}$, $I_{\mathfrak{P}_1}$ into $I_{\mathfrak{P}_2}$, and $\sigma_{\mathfrak{P}_1}$ into $\sigma_{\mathfrak{P}_2}$. This implies an identity for the characteristic polynomials of $\sigma_{\mathfrak{P}_j}$ on the subspace $V_{\mathfrak{P}_j}$ of V on which $I_{\mathfrak{P}_j}$ acts trivially:

$$\det\left(1 - \frac{\varrho(\sigma_{\mathfrak{P}_1})}{\mathrm{N}(\mathfrak{p})^s}\Big|V_{\mathfrak{P}_1}\right) = \det\left(1 - \frac{\varrho(\sigma_{\mathfrak{P}_2})}{\mathrm{N}(\mathfrak{p})^s}\Big|V_{\mathfrak{P}_2}\right).$$

Thus, these characteristic polynomials are independent of the choice of $\sigma_{\mathfrak{P}}$. Denote by $\sigma_{\mathfrak{p}}$ the conjugacy class of Frobenius elements at primes $\mathfrak{P}$ above $\mathfrak{p}$; in case of unramified $\mathfrak{p}$ the inertia group is trivial, and $\sigma_{\mathfrak{p}}$ is called Artin symbol.

Following Artin [4], we define the Artin *L*-function attached to ϱ by

$$L(s, \varrho, \mathbb{L}/\mathbb{K}) = \prod_{\mathfrak{p}} \det\left(1 - \frac{\varrho(\sigma_{\mathfrak{p}})}{\mathrm{N}(\mathfrak{p})^s}\Big|V_{\mathfrak{P}}\right)^{-1}, \tag{13.15}$$

where $\mathfrak{p}$ runs through the prime ideals of the ring of integers in $\mathbb{K}$; this Euler product converges for $\sigma > 1$ by the same reasoning as for previous *L*-functions of number fields.

We shall illustrate this by an example (see also Heilbronn [129] and Stark [337]). Assume that $\mathbb{L}/\mathbb{Q}$ is normal with Galois group equal to the symmetric group S_3 on three letters:

$$\mathsf{G} := \{1, (\alpha\beta\gamma), (\alpha\gamma\beta), (\alpha\beta), (\alpha\gamma), (\beta\gamma)\}$$

say. For instance, one may consider the cubic field $\mathbb{K} = \mathbb{Q}(2^{1/3})$ and its normal closure $\mathbb{L} = \mathbb{Q}(\alpha, \mathrm{e}^{2\pi\mathrm{i}/3}) = \mathbb{Q}(\alpha, \beta, \gamma)$, where

$$\alpha = 2^{1/3}, \quad \beta = \mathrm{e}^{2\pi\mathrm{i}/3}2^{1/3}, \quad \gamma = \mathrm{e}^{4\pi\mathrm{i}/3}2^{1/3}.$$

Since automorphisms of $\mathbb{L}$ are determined by their action on α, β and γ, we find $\mathsf{G} = \mathsf{Gal}(\mathbb{L}/\mathbb{Q})$ for the Galois group of $\mathbb{L}$. The splitting of primes from $\mathbb{Q}$ to $\mathbb{K}$, and likewise from $\mathbb{K}$ to $\mathbb{L}$, is ruled by the Frobenius automorphisms. Suppose that $\mathfrak{P}$ is an unramified prime of $\mathbb{L}$ which lies above $\mathfrak{p}$ of $\mathbb{K}$ which in turn lies above the rational prime p of $\mathbb{Q}$. Then the Frobenius automorphism of $\mathfrak{P}$ relative to $\mathbb{Q}$ is given by one of the following conjugacy classes:

- $\sigma_{\mathfrak{P}} = 1$. Since the Frobenius has order one, by (13.1), there are six primes in $\mathbb{L}$ above p. Obviously, $\sigma_{\mathfrak{P}} \in \mathsf{Gal}(\mathbb{L}/\mathbb{K}) = \{1, (\beta\gamma)\}$. In this case p splits in $\mathbb{K}$ into three different primes $\mathfrak{p}_j$ $(1 \leq j \leq 3)$ each of which splits into two prime ideals $\mathfrak{P}_k$ $(1 \leq k \leq 6)$ of $\mathbb{L}$.

- $\sigma_{\mathfrak{P}}$ is in the conjugacy class $\{(\alpha\beta), (\alpha\gamma), (\beta\gamma)\}$ of elements of order two. We may choose $\mathfrak{P}$ such that $\sigma_{\mathfrak{P}} = (\beta\gamma) \in \mathsf{Gal}(\mathbb{L}/\mathbb{K})$. Then $f = 2$ in (13.1) and so there are three second degree primes $\mathfrak{P}_k$ $(1 \leq k \leq 3)$ above p; we may assume that $\mathfrak{P} = \mathfrak{P}_1$. We observe that the Frobenius automorphism of $\mathfrak{P}$ relative to $\mathbb{K}$ is equal to $\sigma_{\mathfrak{P}}$. Hence, we find $\mathrm{N}(\mathfrak{P}) = \mathrm{N}(\mathfrak{p})^2$ and $\mathrm{N}(\mathfrak{p}) = p$ for some prime $\mathfrak{p} = \mathfrak{p}_1$ of $\mathbb{K}$. For the other two primes $\mathfrak{P}_2$ and $\mathfrak{P}_3$ the Frobenius $\sigma_{\mathfrak{P}}$ is equal to $(\alpha\beta)$ and $(\alpha\gamma)$, respectively. In these cases we find $\sigma_{\mathfrak{P}}^2 = 1 \in \mathsf{Gal}(\mathbb{L}/\mathbb{K})$ and $\mathrm{N}(\mathfrak{P}) = \mathrm{N}(\mathfrak{p})$ and $\mathrm{N}(\mathfrak{p}) = p^2$ for some prime $\mathfrak{p} = \mathfrak{p}_2$ of $\mathbb{K}$. Thus, the primes $\mathfrak{P}_2$ and $\mathfrak{P}_3$ have relative degree one over a single prime $\mathfrak{p}_2$ of $\mathbb{K}$ (which is of degree two).

- $\sigma_{\mathfrak{P}}$ is in the conjugacy class $\{(\alpha\beta\gamma), (\alpha\gamma\beta)\}$ of elements of order three. In this case we have $f = 3$ in (13.1) and there are two third degree primes $\mathfrak{P}_1$ and $\mathfrak{P}_2$ of $\mathbb{L}$ above p, for one of them $\sigma_{\mathfrak{P}} = (\alpha\beta\gamma)$ and for the other $\sigma_{\mathfrak{P}} = (\alpha\gamma\beta)$. In both cases neither $\sigma_{\mathfrak{P}}$ nor $\sigma_{\mathfrak{P}}^2$ lie in $\mathsf{Gal}(\mathbb{L}/\mathbb{K}) = \{1, (\beta\gamma)\}$ and so both $\mathfrak{P}_1$ and $\mathfrak{P}_2$ lie above a single prime $\mathfrak{p}$ of $\mathbb{K}$ (which must be of degree three).

Before we continue we remark that the splitting of primes can be computed by use of the following statement (analogous to (13.3)): suppose that $F(X)$ is the minimal polynomial of $\alpha \in \mathbb{K}$ over $\mathbb{Q}$ and that it splits factors mod p into irreducible pieces as

$$F(X) \equiv F_1(X)^{e_1} \cdot \ldots \cdot F_r(X)^{e_r} \bmod \mathfrak{p}.$$

If the power of $\mathfrak{p}$ in the polynomial discriminant of $F(X)$ is the same as the power of $\mathfrak{p}$ in the relative discriminant $D_{\mathbb{L}/\mathbb{K}}$ of $\mathbb{L}/\mathbb{K}$, then $\mathfrak{p}$ splits in $\mathbb{L}$ as

$$\mathfrak{p} = \mathfrak{P}_1^{e_1} \cdot \ldots \cdot \mathfrak{P}_r^{e_r},$$

where $\mathfrak{P}_j = (\mathfrak{p}, F_j(\alpha))$ is of relative degree $\deg F_j$. This together with Eisenstein's irreducibility criterion gives the basic tools to do arithmetic computations in number fields.

We may represent the Galois group G by matrices as follows. For $g \in \mathsf{G}$ we write

$$\begin{pmatrix} \alpha \\ \beta \\ \gamma \end{pmatrix} g = \mathsf{M}(g) \begin{pmatrix} \alpha \\ \beta \\ \gamma \end{pmatrix},$$

where $\mathsf{M}(g)$ is the permutation matrix corresponding to g. Thus, in our example, we can represent the six elements of G by

$$1 \mapsto \begin{pmatrix} 1\,0\,0 \\ 0\,1\,0 \\ 0\,0\,1 \end{pmatrix}, \quad (\alpha\beta\gamma) \mapsto \begin{pmatrix} 0\,1\,0 \\ 0\,0\,1 \\ 1\,0\,0 \end{pmatrix}, \quad (\alpha\gamma\beta) \mapsto \begin{pmatrix} 0\,0\,1 \\ 1\,0\,0 \\ 0\,1\,0 \end{pmatrix},$$

$$(\alpha\beta) \mapsto \begin{pmatrix} 0\,1\,0 \\ 1\,0\,0 \\ 0\,0\,1 \end{pmatrix}, \quad (\alpha\gamma) \mapsto \begin{pmatrix} 0\,0\,1 \\ 0\,1\,0 \\ 1\,0\,0 \end{pmatrix}, \quad (\beta\gamma) \mapsto \begin{pmatrix} 1\,0\,0 \\ 0\,0\,1 \\ 0\,1\,0 \end{pmatrix}.$$

The map $\varrho : g \mapsto \mathsf{M}(g)$ defines a homomorphism: $\mathsf{M}(gh) = \mathsf{M}(g)\mathsf{M}(h)$; it is an example of a three-dimensional permutation representation of the group G. The conjugacy classes of the symmetric group on α, β, γ are precisely the conjugacy classes of Frobenius automorphisms arising from prime numbers which split in the indicated form; that are

$$\mathsf{C}_1 : \{1\}, \qquad \mathsf{C}_2 : \{(\alpha\beta\gamma), (\alpha\gamma\beta)\}, \qquad \mathsf{C}_3 : \{(\alpha\beta), (\alpha\gamma), (\beta\gamma)\}.$$

For each of them we observe that the associated Euler factors are of the form as predicted by (13.15). To see this we have a look on every individual Euler factor. Since the field extension $\mathbb{K}/\mathbb{Q}$ has degree 3, there are the following possibilities to consider.

- The prime p splits completely into three different prime divisors; e.g., $(31) = \mathfrak{p}_1\mathfrak{p}_2\mathfrak{p}_3$ with

$$\mathfrak{p}_1 = (31, \alpha - 4), \quad \mathfrak{p}_2 = (31, \alpha - 7), \quad \mathfrak{p}_3 = (31, \alpha - 20).$$

In this case the local Euler factor at p is of the form

$$\left(1 - \frac{1}{p^s}\right)^{-3} = \det\left(1 - \begin{pmatrix} 1\,0\,0 \\ 0\,1\,0 \\ 0\,0\,1 \end{pmatrix} \frac{1}{p^s}\right)^{-1}.$$

Obviously, the appearing matrix has the eigenvalue $+1$ with multiplicity three. This Euler factor corresponds to the class C_1.

- The prime p can be factored into a product of two factors, one of first degree and one of second degree; for example, $(5) = \mathfrak{p}_1\mathfrak{p}_2$ with

$$\mathfrak{p}_1 = (5, \alpha - 3), \quad \mathfrak{p}_2 = (5, \alpha^2 + 3\alpha + 9).$$

Here we have

$$\left(1 - \frac{1}{p^s}\right)^{-1}\left(1 - \frac{1}{p^{2s}}\right)^{-1} = \det\left(1 - \mathsf{M}\frac{1}{p^s}\right)^{-1},$$

for any of the matrices

$$M = \begin{pmatrix} 0\,1\,0 \\ 1\,0\,0 \\ 0\,0\,1 \end{pmatrix}, \quad \begin{pmatrix} 0\,0\,1 \\ 0\,1\,0 \\ 1\,0\,0 \end{pmatrix}, \quad \text{and} \quad \begin{pmatrix} 1\,0\,0 \\ 0\,0\,1 \\ 0\,1\,0 \end{pmatrix},$$

corresponding to C_3. The eigenvalues of the (similar) matrices are -1 and $+1$ with multiplicities one and two, respectively.

- The prime p is a prime ideal of third degree; e.g., $(7) = \mathfrak{p}$. In this case we have

$$\left(1 - \frac{1}{p^{3s}}\right)^{-1} = \det\left(1 - M\frac{1}{p^s}\right)^{-1}$$

for the matrices associated with C_2:

$$M = \begin{pmatrix} 0\,1\,0 \\ 0\,0\,1 \\ 1\,0\,0 \end{pmatrix} \quad \text{and} \quad \begin{pmatrix} 0\,0\,1 \\ 1\,0\,0 \\ 0\,1\,0 \end{pmatrix}.$$

Here the eigenvalues of the (similar) matrices are the third roots of unity.

Now we introduce a more convenient notation of Artin L-functions. To any representation ϱ of G, we attach a character χ of G by setting

$$\chi(g) = \text{trace}(\varrho(g))$$

for $g \in \mathsf{G}$. The degree of a character is defined by $\deg \chi = \chi(1)$. If h is another element of G, then

$$\varrho(h^{-1}gh) = \varrho(h)^{-1}\varrho(g)\varrho(h),$$

so that $\varrho(h^{-1}gh)$ and $\varrho(g)$ are similar matrices and thus have the same trace. This shows that characters χ of G are constant on the conjugacy classes. Two representations are said to be equivalent if they have the same character. If ϱ_1 and ϱ_2 are representations of G with characters χ_1 and χ_2, then

$$\varrho(g) = \begin{pmatrix} \varrho_1(g) & 0 \\ 0 & \varrho_2(g) \end{pmatrix}$$

also defines a representation of G with character $\chi_1 + \chi_2$, and in this case ϱ is said to be reducible; any representation which is not reducible is called irreducible. We shall use the same attributes for the associated character.

It turns out that any conjugacy class of G corresponds to an irreducible representation and one can show that there are not more; of course, distinct irreducible representations are non-equivalent (these observations are analogous to the case of Dirichlet characters and the group of residue classes of $\mathbb{Z}$). In our example we find for the three conjugacy classes of G:

	C_1	C_2	C_3
χ_1	$+1$	$+1$	$+1$
χ_2	$+1$	$+1$	-1
χ_3	$+2$	-1	0

Hence, for $G = S_3$, there are three irreducible characters (in some literature "simple characters").

It is easily seen that our characters satisfy the orthogonality relations, that are

$$\frac{1}{\sharp G} \sum_{\chi \in \hat{G}} \overline{\chi}(C)\chi(D) = \begin{cases} 1/\sharp C & \text{if} \quad C = D, \\ 0 & \text{otherwise,} \end{cases}$$

where C and D are conjugacy classes. Since the Euler factors in (13.15) depend only on the conjugacy class σ_p, in the sequel we will talk sometimes in terms of characters and denote the Artin *L*-function (13.15) by $L(s, \chi, \mathbb{L}/\mathbb{K})$ (and sometimes we shall even write $L(s, \chi)$ for short).

For further illustration we continue with our example. We can construct more characters from the irreducible characters listed above, for example, a third degree character χ related to the permutation representation $(\alpha\beta)$. Taking the character relations into account we find $\chi = \chi_1 + \chi_3$. For the related Artin *L*-functions we note that

$$L(s, \chi, \mathbb{L}/\mathbb{K}) = L(s, \chi_1 + \chi_3, \mathbb{L}/\mathbb{K}) = L(s, \chi_1, \mathbb{L}/\mathbb{K})L(s, \chi_3, \mathbb{L}/\mathbb{K}).$$

This additivity property holds in general.

For the field $\mathbb{L} = \mathbb{Q}(\alpha, \beta, \gamma)$ there are four subfields up to conjugacy: firstly, the field $\mathbb{Q}$ itself, fixed by all of G, secondly, $\mathbb{Q}(\sqrt{-3})$ fixed by $G_2 := \{1, (\alpha\beta\gamma), (\alpha\gamma\beta)\}$ (corresponding to the conjugacy class C_2), thirdly, $\mathbb{K} = \mathbb{Q}(2^{1/3})$ fixed by $G_3 := \{1, (\beta\gamma)\}$ (corresponding to the conjugacy class C_3), and finally $\mathbb{L}$ fixed just by $\{1\}$.

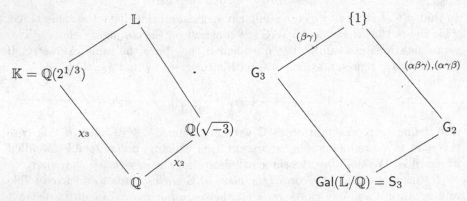

We obtain the following factorizations of the associated Dedekind zeta-functions into products of Artin *L*-functions to $\mathbb{L}/\mathbb{Q}$:

$$\zeta(s) = \zeta_{\mathbb{Q}}(s) = L(s, \chi_1),$$
$$\zeta_{\mathbb{Q}(\sqrt{-3})}(s) = L(s, \chi_1)\,L(s, \chi_2),$$
$$\zeta_{\mathbb{Q}(2^{1/3})}(s) = L(s, \chi_1)\,L(s, \chi_3),$$
$$\zeta_{\mathbb{L}}(s) = L(s, \chi_1)\,L(s, \chi_2)\,L(s, \chi_3).$$

We observe that any of the Dedekind zeta-functions on the left-hand side is divisible by the Riemann zeta-function. It follows from these factorizations and the analytic behaviour of Dedekind zeta-functions that each of the involved Artin L-functions with $\chi \neq \chi_0$ possesses a meromorphic continuation to the whole complex plane; the only possible poles can occur at zeros of other Artin L-functions. Furthermore, one can deduce functional equations of the Riemann-type. This is a remarkable way to deduce analytic properties for L-functions!

13.7 The Artin Conjecture

One of the most fundamental conjectures in algebraic number theory is

Artin's Conjecture. *Let* $\mathbb{L}/\mathbb{K}$ *be a finite Galois extension with Galois group* G. *For any irreducible character* $\chi \neq 1$ *of* G *the Artin L-function* $L(s, \chi, \mathbb{L}/\mathbb{K})$ *extends to an entire function.*

We discuss briefly one of its important consequences. Dedekind's conjecture claims that *the quotient* $\zeta_{\mathbb{L}}(s)/\zeta_{\mathbb{K}}(s)$ *is entire provided* $\mathbb{L}/\mathbb{K}$ *is an extension of number fields*, not necessarily Galois. Here one may recall the factorizations of the Dedekind zeta-functions in our example from Sect.13.6. If $\mathbb{L}/\mathbb{K}$ is a Galois extension, then the so-called Artin–Takagi factorization gives a factorization of the Dedekind zeta-function of a number field relative to a subfield (see Heilbronn's survey article [129]); more precisely,

$$L(s, 1, \mathbb{L}/\mathbb{K}) = \zeta_{\mathbb{K}}(s) \quad \text{and} \quad L(s, \mathsf{R}_{\mathsf{G}}, \mathbb{L}/\mathbb{K}) = \zeta_{\mathbb{L}}(s),$$

where R_{G} is the regular character of G (the character defined by $\sum_{\chi} \chi(1)\chi$), and

$$\zeta_{\mathbb{L}}(s) = \prod_{\chi \in \tilde{\mathsf{G}}} L(s, \chi, \mathbb{L}/\mathbb{K})^{\chi(1)},$$

where $\tilde{\mathsf{G}}$ denotes the set of irreducible characters of G. In the case of Galois extensions $\mathbb{L}/\mathbb{K}$, a theorem of Aramata [2] yields the truth of Dedekind's conjecture; its proof relies mainly on the Artin–Takagi factorization (an easier proof was given by Brauer [42] by his induction theorem; see also Heilbronn [129] or Murty and Murty [270], Sect. 2.3). In the general case, if $\mathbb{L}/\mathbb{K}$ is a finite (not necessarily Galois) extension, then Dedekind's conjecture follows from Artin's conjecture by studying the normal closure of $\mathbb{L}/\mathbb{K}$.

As indicated in a previous section, Artin [5] proved his conjecture if χ is one-dimensional and $\mathbb{L}/\mathbb{K}$ is abelian. In this case, the related Artin L-function coincides with a Hecke L-function attached to some ray class group. The Frobenius $\sigma_{\mathfrak{p}}$ induces a map from the fractional ideals of $\mathbb{K}$ coprime with $\mathfrak{f}$ to the Galois group:

$$\Phi(\mathfrak{f}) : \mathsf{I}(\mathfrak{f}) \to \mathsf{Gal}(\mathbb{L}/\mathbb{K}).$$

By class field theory it is surjective and if the finite primes dividing $\mathfrak{f}$ are sufficiently large, then its kernel is a congruence subgroup for $\mathfrak{f}$. This is Artin's reciprocity law:

Theorem 13.4. *Let* $\mathbb{L}/\mathbb{K}$ *be abelian and let* χ *be a one-dimensional character of* $\mathsf{G} = \mathsf{Gal}(\mathbb{L}/\mathbb{K})$. *Then there exists a modulus* $\mathfrak{f} = \mathfrak{f}(\mathbb{L}/\mathbb{K})$ *divisible by all ramified primes (finite and infinite) such that*

$$\mathsf{I}(\mathfrak{f})/\ker(\varPhi(\mathfrak{f})) \simeq \mathsf{G}.$$

Using the isomorphism $\varPhi(\mathfrak{f})$, we define a character ψ of the class group by setting $\psi(\mathfrak{p}) = \chi(\sigma_\mathfrak{p})$ for $\mathfrak{p}$ coprime with $\mathfrak{f}$. Then

$$L(s, \chi, \mathbb{L}/\mathbb{K}) = L(s, \psi). \qquad (13.16)$$

Artin proved this theorem by means of class field theory and, in particular, using Chebotarev's density theorem. We shall briefly explain the latter result. Let $\mathbb{L}/\mathbb{K}$ be a *finite* Galois extension with Galois group G and let C be a subset of G, closed under conjugation. Further, denote by $\pi_\mathsf{C}(x)$ the number of prime ideals $\mathfrak{p}$ of $\mathbb{K}$, unramified in $\mathbb{L}$, for which $\sigma_\mathfrak{p} \subset \mathsf{C}$ and which have norm $\mathrm{N}(\mathfrak{p}) \leq x$ in $\mathbb{K}$. Then, Chebotarev's density theorem [53] states

$$\pi_\mathsf{C}(x) \sim \frac{\#\mathsf{C}}{\#\mathsf{G}} \pi(x). \qquad (13.17)$$

This rather deep theorem can be seen as a higher analogue of the prime number theorem in arithmetic progressions. A modern proof can be found, for example, in Narkiewicz [277]. The Chebotarev density theorem can be used to determine the Galois group of a given irreducible polynomial $P(X)$ of degree n by counting the number of unramified primes up to a certain bound for which P factors in a certain way and comparing the results with the fractions of elements of each of the transitive subgroups of the symmetric group S_n with the same cyclic structure.

We illustrate this by returning to our example from Sect.13.6. By means of class field theory one can show that $\mathbb{K} = \mathbb{Q}(2^{1/3})$ is the class field for a ray class group $\mathsf{G}(\mathfrak{f})$ over $\mathsf{Gal}(\mathbb{L}/\mathbb{K})$ which is cyclic of order three. Besides the trivial character, there are two cubic characters χ_2 and χ_3 each of which is the square of the other one. Their common conductor is $\mathfrak{f} = (6)$ and, by Chebotarev's density theorem it follows that:

- 1/6 of the rational primes split completely in $\mathbb{K}$ (and they have a representation as $p = x^2 + 27y^2$ with $x, y \in \mathbb{Z}$),
- 1/2 of the rational primes split as the product of first and second degree prime factors (namely $p \equiv 2 \bmod 3$), and
- 1/3 of the rational primes generate third degree primes of $\mathbb{K}$.

Brauer [42] obtained a significant extension of Artin's result. Let H be a subgroup of G. Then any character χ on H induces a character $\mathsf{Ind}_\mathsf{H}^\mathsf{G}(\chi)$ on G and

$$L(s, \chi, \mathbb{L}/\mathbb{L}^\mathsf{H}) = L(s, \mathsf{Ind}_\mathsf{H}^\mathsf{G}(\chi), \mathbb{L}/\mathbb{K}), \qquad (13.18)$$

where $\mathbb{L}^\mathsf{H}$ is the corresponding fixed field. Brauer [42] proved that any character of a finite group G can be written as a linear combination

$$\chi = \sum_j n_j \mathsf{Ind}_{\mathsf{H}_j}^{\mathsf{G}}(\chi_j),$$

where the H_j are nilpotent subgroups of G, the χ_j are one-dimensional, and the n_js are integers. Hence, it follows that

$$L(s,\chi,\mathbb{L}/\mathbb{K}) = \prod_j L(s,\mathsf{Ind}_{\mathsf{H}_j}^{\mathsf{G}}(\chi_j),\mathbb{L}/\mathbb{K})^{n_j}.$$

However, by (13.18),

$$L(s,\chi,\mathbb{L}/\mathbb{K}) = \prod_j L(s,\chi_j,\mathbb{L}/\mathbb{L}^{\mathsf{H}_j})^{n_j}.$$

Now applying Artin's reciprocity law to the one-dimensional characters χ_j, we obtain

$$L(s,\chi,\mathbb{L}/\mathbb{K}) = \prod_j L(s,\psi_j)^{n_j} \tag{13.19}$$

for certain ray class characters ψ_j related to χ_j. Using the analytic properties of Hecke L-functions, Brauer [42] deduced a functional equation of the Riemann-type for Artin L-functions which gives a meromorphic continuation throughout the complex plane (see also Neukirch [279]), Sect. VII.12).

Brauer's theorem implies the Artin conjecture if G is nilpotent or supersolvable since then every character of G is monomial (see Serre [325]). Murty [267] observed that

Theorem 13.5. *Selberg's Conjecture B implies Artin's conjecture.*

Murty and Perelli [271] replaced Selberg's conjecture by the pair correlation conjecture (as already mentioned above).

The proof uses some easy properties of Artin L-functions which we did not prove or even did not mention above. The reader may have a look into the literature, e.g., Heilbronn [129], and may consult the examples from Sect.13.6.

Proof. Let $\tilde{\mathbb{L}}$ be the normal closure of $\mathbb{L}$ over $\mathbb{Q}$. Then, $\tilde{\mathbb{L}}/\mathbb{K}$ and $\tilde{\mathbb{L}}/\mathbb{Q}$ are Galois. Thus, χ can be considered as a character $\tilde{\chi}$ of $\mathsf{Gal}(\tilde{\mathbb{L}}/\mathbb{K})$, and by the properties of Artin L-functions it turns out that

$$L(s,\tilde{\chi},\tilde{\mathbb{L}}/\mathbb{K}) = L(s,\chi,\mathbb{L}/\mathbb{K}).$$

By the invariance induction of $\tilde{\chi}$ from $\mathsf{Gal}(\tilde{\mathbb{L}}/\mathbb{K})$ to $\mathsf{Gal}(\tilde{\mathbb{L}}/\mathbb{Q})$, it follows that

$$L(s,\chi,\mathbb{L}/\mathbb{K}) = \prod_\psi L(s,\psi,\tilde{\mathbb{L}}/\mathbb{Q})^{m(\psi)},$$

where the product is taken over all irreducible characters ψ of $\mathsf{Gal}(\tilde{\mathbb{L}}/\mathbb{Q})$ and $m(\psi)$ are non-negative integers. To prove Artin's conjecture, it suffices to show that all appearing $L(s, \psi, \tilde{\mathbb{L}}/\mathbb{Q})$ are entire. By Brauer's induction theorem (13.18) and Artin's reciprocity law, Theorem 13.4, respectively, formula (13.19), we have

$$L(s, \psi, \tilde{\mathbb{L}}/\mathbb{Q}) = \frac{L(s, \chi_1)}{L(s, \chi_2)},$$

where χ_1, χ_2 are characters of $\mathsf{Gal}(\tilde{\mathbb{L}}/\mathbb{Q})$ and $L(s, \chi_1)$, $L(s, \chi_2)$ are products of Hecke *L*-functions (13.13). Since Hecke *L*-functions belong to the Selberg class $\mathcal{S}$, and $\mathcal{S}$ is multiplicatively closed, the functions $L(s, \chi_1)$ and $L(s, \chi_2)$ belong to $\mathcal{S}$ too. Now, by Theorem 6.2, there exist primitive functions $\mathcal{L}_j \in \mathcal{S}$ such that

$$L(s, \psi, \tilde{\mathbb{L}}/\mathbb{Q}) = \prod_{j=1}^{f} \mathcal{L}_j(s)^{e_j}, \tag{13.20}$$

where $e_j \in \mathbb{Z}$. By comparing the pth coefficient in the Dirichlet series expansions of both sides, we get

$$\psi(p) = \sum_{j=1}^{f} e_j a_{\mathcal{L}_j}(p).$$

Thus,

$$\sum_{p \leq x} \frac{|\psi(p)|^2}{p} = \sum_{p \leq x} \frac{1}{p} \left| \sum_{j=1}^{f} e_j a_{\mathcal{L}_j}(p) \right|^2. \tag{13.21}$$

Selberg's Conjecture *B* yields the asymptotic formula

$$\sum_{p \leq x} \frac{1}{p} \left| \sum_{j=1}^{f} e_j a_{\mathcal{L}_j}(p) \right|^2 = \left(\sum_{j=1}^{f} e_j^2 \right) \log\log x + O(1). \tag{13.22}$$

Next, we decompose the sum on the left-hand side of (13.21) according to the conjugacy classes C of $\mathsf{G} := \mathsf{Gal}(\tilde{\mathbb{L}}/\mathbb{Q})$ to which the Frobenius element σ_p belongs. If g_{C} denotes any element of C, this leads to

$$\sum_{p \leq x} \frac{|\psi(p)|^2}{p} = \sum_{\mathsf{C}} |\psi(g_{\mathsf{C}})|^2 \sum_{\substack{p \leq x \\ \sigma_p \subset \mathsf{C}}} \frac{1}{p}.$$

By partial summation, we deduce from Chebotarev's density theorem (13.17)

$$\sum_{\substack{p \leq x \\ \sigma_p \subset \mathsf{C}}} \frac{1}{p} = \frac{\sharp \mathsf{C}}{\sharp \mathsf{G}} \log\log x + O(1).$$

This gives

$$\sum_{p\leq x}\frac{|\psi(p)|^2}{p}=\sum_{C}|\psi(g_C)|^2\frac{\#C}{\#G}\log\log x+O(1).$$

Since ψ is irreducible, we have

$$\sum_{C}|\psi(g_C)|^2\frac{\#C}{\#G}=\frac{1}{\#G}\sum_{C}\#C=1,$$

which implies via (13.22) and (13.21)

$$\sum_{j=1}^{f}e_j^2=1.$$

Thus, $f=1$ and $e_1=\pm1$. The case $e_1=-1$ implies

$$L(s,\psi,\tilde{\mathbb{L}}/\mathbb{Q})=\frac{1}{\mathcal{L}_1(s)},$$

which is impossible since $L(s,\psi,\tilde{\mathbb{L}}/\mathbb{Q})$ has trivial zeros (their existence follows from the functional equation). Hence, $e_1=+1$, and we conclude that $L(s,\psi,\tilde{\mathbb{L}}/\mathbb{Q})=\mathcal{L}_1(s)$ is entire. $\qquad\square$

The proof shows that if χ is an irreducible non-trivial character of $\mathsf{Gal}(\mathbb{K}/\mathbb{Q})$, then the Artin L-function $L(s,\chi,\mathbb{K}/\mathbb{Q})$ is a primitive element in the Selberg class $\mathcal{S}$ if Selberg's Conjecture B is true; the case of Artin L-functions to reducible characters follows from the fact that $\mathcal{S}$ is multiplicatively closed. Furthermore, the splitting of primes implies axiom (iv) on the polynomial Euler product from the definition of $\tilde{\mathcal{S}}$. Decomposing as in Murty's proof, we find

$$\frac{1}{\pi(x)}\sum_{p\leq x}|\chi(p)|^2=\frac{1}{\pi(x)}\sum_{C}|\chi(g_C)|^2\sum_{\substack{p\leq x\\\sigma_p\subset C}}1.$$

By Chebotarev's density theorem (13.17), this is asymptotically equal to

$$\frac{1}{\#G}\sum_{C}\#C=1,$$

which implies axiom (v) on the prime mean-square. So, $L(s,\chi,\mathbb{K}/\mathbb{Q})\in\tilde{\mathcal{S}}$ provided that Artin's conjecture is true; note that in the case of one-dimensional characters this is unconditional since these L-functions are known to be entire. Thus, on behalf of Theorem 5.14 we obtain universality for Artin L-functions $L(s,\chi,\mathbb{K}/\mathbb{Q})$ to Galois extensions $\mathbb{K}/\mathbb{Q}$ subject to Selberg's Conjecture B. However, following the lines of Voronin's proof, Bauer [16] proved universality for any Artin L-function without assumption of any unproved hypothesis. In Sect.13.8 we shall prove and extend this result by use of Bagchi's method.

13.8 Joint Universality for Artin *L*-Functions

We shall use joint universality for Artin *L*-functions to one-dimensional representations in combination with Brauer's induction theorem (13.19) in order to deduce joint universality for arbitrary Artin *L*-functions of $\mathbb{K}/\mathbb{Q}$.

Theorem 13.6. *Let $\mathbb{K}$ be a finite Galois extension of $\mathbb{Q}$ and let $\chi_1, \ldots, \chi_\ell$ be $\mathbb{C}$-linearly independent characters of $\mathsf{Gal}(\mathbb{K}/\mathbb{Q})$. Let $\mathcal{K}$ be a compact subset of $\max\{\frac{1}{2}, 1 - [\mathbb{K} : \mathbb{Q}]^{-1}\} < \sigma < 1$ with connected complement. Further, for each $1 \le k \le \ell$ let $g_k(s)$ be a continuous non-vanishing function on $\mathcal{K}$ which is analytic in the interior of $\mathcal{K}$. Then, for any $\epsilon > 0$,*

$$\liminf_{T \to \infty} \frac{1}{T} \operatorname{meas} \left\{ \tau \in [0, T] : \max_{1 \le k \le \ell} \max_{s \in \mathcal{K}} |L(s + i\tau, \chi_k, \mathbb{K}/\mathbb{Q}) - g_k(s)| < \varepsilon \right\} > 0.$$

Proof. Without loss of generality we may assume that the characters $\chi_1, \ldots, \chi_\ell$ form a basis of the set of class functions on $\mathsf{Gal}(\mathbb{K}/\mathbb{Q})$. In view of Brauer's induction theorem (13.19) any of the Artin *L*-functions has a representation

$$L(s, \chi_k, \mathbb{K}/\mathbb{Q}) = \prod_j L(s, \psi_{jk})^{n_{jk}}, \tag{13.23}$$

where the $L(s, \psi_{jk})$ are Hecke *L*-functions to number fields $\mathbb{K}_{jk}$ contained in $\mathbb{K}$, the ψ_{jk} are ray class characters of $\mathbb{K}_{jk}$, and the numbers n_{jk} are non-zero integers.

It suffices to prove that the family of Artin *L*-functions to all pairwise non-equivalent one-dimensional characters ψ_k to subgroups of $\mathsf{Gal}(\mathbb{K}/\mathbb{Q})$ is jointly universal. Then the assertion of the theorem follows from (13.23) in just the same way as in the final step of the proof of Theorem 13.1.

Now let $\{L(s, \psi_k)\}$ be such a family of distinct Artin *L*-functions to pairwise non-equivalent one-dimensional characters ψ_k to subgroups of $\mathsf{Gal}(\mathbb{K}/\mathbb{Q})$. We want to prove their joint universality. In view of Artin's reciprocity law, Theorem 13.4, each $L(s, \psi_k)$ is a Hecke *L*-function to a ray class group character for some subfield of $\mathbb{K}$; in particular, they belong to $\tilde{\mathcal{S}}$ (as we have seen in Sect. 13.7). According to (13.1),

$$p = \prod_{j=1}^{r_p} \mathfrak{p}_j^{e_j} \quad \text{with} \quad \mathsf{N}(\mathfrak{p}_j) = p^{f_j}, \quad \sum_{j=1}^{r_p} e_j f_j = [\mathbb{K} : \mathbb{Q}],$$

we write

$$L(s, \psi_k) = \prod_p \det \left(1 - \frac{\varrho_k(\sigma_p)}{p^s} \right)^{-1} = \prod_{\mathfrak{p}} \left(1 - \frac{\psi_k(\mathfrak{p})}{p^s} \right)^{-1}$$

$$= \prod_p \prod_{j=1}^{r_p} \left(1 - \frac{\psi_k(\mathfrak{p}_j)}{p^{f_j s}} \right)^{-1} = \sum_{n=1}^{\infty} \frac{a_k(n)}{n^s}.$$

with

$$a_k(p) = \sum_{\substack{j=1 \\ f_j=1}}^{r_p} \psi_k(\mathfrak{p}_j).$$

Clearly, $m := \max r_p \leq [\mathbb{K} : \mathbb{Q}]$; this defines the local roots α_{jk}. In order to prove joint universality for $\{L(s, \psi_k)\}$ we shall apply Theorem 12.5, i.e., we have to show that the set of all convergent series

$$\sum_p \underline{g}_p(s, b(p)) \qquad \text{with} \quad b(p) \in \gamma$$

is dense in $\mathcal{H}(\mathcal{D})^\ell$, where the $\underline{g}_p$ are defined by (12.11) and (12.12), that are

$$g_{pk}(s, b(p)) = -\sum_{j=1}^m \log\left(1 - \frac{\alpha_{jk}(p)b(p)}{p^s}\right),$$

$$\underline{g}_p(s, b(p)) = (g_{p1}(s, b(p)), \ldots, g_{p\ell}(s, b(p))),$$

and ℓ is the cardinality of $\{L(s, \psi_k)\}$. Most of the reasoning is similar as in the proof of Theorem 12.8; however, the verification of the first assumption of Theorem 12.9 is slightly different. We sketch the main idea.

Let $\mu_1, \ldots, \mu_\ell$ be complex Borel measures on $(\mathbb{C}, \mathcal{B}(\mathbb{C}))$ with compact support contained in $\mathcal{D}$ such that

$$\sum_p \left| \sum_{k=1}^\ell \int_{\mathbb{C}} \hat{b}(p)\hat{g}_{pk}(s) \, d\mu_k(s) \right| < \infty;$$

here, as in Sect.13.7,

$$\hat{g}_{pk}(s) = \begin{cases} \tilde{g}_{pk}(s) & \text{if} \quad p > N, \\ 0 & \text{if} \quad p \leq N, \end{cases}$$

$$\underline{\hat{g}}_p(s) = (\hat{g}_{p1}(s), \ldots, \hat{g}_{p\ell}(s)),$$

and the $\hat{b}(p)$ form a sequence such that $\sum_p \hat{b}(p)\hat{g}_{pk}(s)$ converges in $\mathcal{H}(\mathcal{D})$. Following the proof of Theorem 12.8, we find

$$\tilde{g}_{pk}(s) = \frac{a_k(p)}{p^s} + r_{pk}(s)$$

with a sufficiently small remainder $r_{pk}(s)$. It is easily seen that we can replace (12.25) by

$$\sum_p \left| \sum_{k=1}^\ell \int_{\mathbb{C}} a_k(p)p^{-s} \, d\mu_k(s) \right| < \infty. \tag{13.24}$$

Denote by $\mathsf{C}_1, \ldots, \mathsf{C}_J$ the different conjugacy classes of $\mathsf{G} = \mathrm{Gal}(\mathbb{K}/\mathbb{Q})$. Then J is the dimension of the vector space of class functions on G, and so J is greater than or equal to the number of $\mathbb{C}$-linear independent characters. Define

$$\mathbb{P}_j = \{p \in \mathbb{P} : \sigma_p \in C_j\} \qquad \text{for} \quad 1 \le j \le J;$$

recall that the Frobenius σ_p completely determines the splitting of the primes. Clearly, $\mathbb{P} = \cup_{1 \le j \le J} \mathbb{P}_j$, and

$$a_k(p) = \mathsf{trace}(\varrho_k(\sigma_p)) = \mathsf{trace}(\varrho_k(C_j)) \qquad \text{for} \quad p \in \mathbb{P}_j.$$

Hence, we can replace (13.24) by

$$\sum_{p \in \mathbb{P}_j} |\eta_j(\log p)| < \infty,$$

where

$$\eta_j(z) := \int_{\mathbb{C}} \exp(-sz)\, d\nu_j(s) \qquad \text{and} \qquad \nu_j(s) := \sum_{k=1}^{\ell} \mathsf{trace}(\varrho_k(C_j)) \mu_k(s).$$

Next we need Chebotarev's prime number theorem (13.17) with remainder term due to Artin [4]

$$\pi(x; C_j) := \sum_{\substack{p \le x \\ p \in \mathbb{P}_j}} 1 = \frac{\#C_j}{\#G} \mathrm{li}(x) + O(x \exp(-c(\log x)^{1/2}))$$

(see also [129]). This implies

$$\sum_{\substack{p \le x \\ p \in \mathbb{P}_j}} \frac{1}{p} = \pi(x; C_j)\frac{1}{x} + \int_2^x \pi(u; C_j)\frac{du}{u^2}$$

$$= \frac{\#C_j}{\#G} \log\log x + c_j + O((\log x)^{-2}),$$

where the c_j certain constants depending only on C_j. Now we can proceed as in the proof of Theorem 12.8 and obtain that $\eta_j(z)$ vanishes identically, respectively,

$$\int_{\mathbb{C}} s^r\, d\nu_j(s) \equiv 0 \qquad \text{for} \quad r = 0, 1, 2, \ldots$$

That means

$$\sum_{k=1}^{\ell} \mathsf{trace}(\varrho_k(C_j)) \int_{\mathbb{C}} s^r \mu_k(s) \equiv 0$$

for $r = 0, 1, 2, \ldots$ and all $1 \le j \le J$. Hence

$$\sum_{k=1}^{\ell} \psi_k(g_j) \int_{\mathbb{C}} s^r \nu_k(s) \equiv 0$$

for any g_j in the conjugacy class C_j. Since the characters ψ_k are linear independent over $\mathbb{C}$, it follows that all

$$\varphi(q) \int_{\mathbb{C}} s^r \mu_k(s) \equiv 0.$$

Thus, any $\mu_k(s)$ vanishes identically and so we are done. Now we continue in the standard way. This yields the assertion of the theorem.   □

We conclude our studies on Artin L-functions by some interesting consequences of Theorem 13.6 due to Bauer [16]. First of all, any single Artin L-function to an arbitrary normal extension $\mathbb{K}/\mathbb{L}$ is universal.

Corollary 13.7. *Let $\mathbb{L}/\mathbb{K}$ be a normal extension and χ an arbitrary character of $\mathsf{Gal}(\mathbb{L}/\mathbb{K})$. Further, let $\mathcal{K}$ be a compact subset of $\max\{\frac{1}{2}, 1 - \frac{1}{d}\} < \sigma < 1$ with connected complement, where d is the degree of a normal extension of $\mathbb{Q}$ containing $\mathbb{L}$, and let $g(s)$ be a continuous non-vanishing function on $\mathcal{K}$ which is analytic in the interior. Then, for any $\epsilon > 0$,*

$$\liminf_{T \to \infty} \frac{1}{T} \operatorname{meas} \left\{ \tau \in [0, T] \, : \, \max_{s \in \mathcal{K}} |L(s + i\tau, \chi, \mathbb{L}/\mathbb{K}) - g(s)| < \varepsilon \right\} > 0.$$

The proof follows from Theorem 13.6 by observing that a normal extension $\tilde{\mathbb{L}}$ of $\mathbb{Q}$ containing $\mathbb{L}$ is also a normal extension of $\mathbb{K}$ and that $L(s, \chi, \mathbb{L}/\mathbb{K}) = L(s, \chi, \tilde{\mathbb{L}}/\mathbb{K})$ by well-known properties of Artin L-functions. Moreover, $\mathsf{Gal}(\tilde{\mathbb{L}}/\mathbb{K})$ is a subgroup of $\mathsf{Gal}(\mathbb{L}/\mathbb{K})$ and so $L(s, \chi, \tilde{\mathbb{L}}/\mathbb{K}) = L(s, \chi^*, \tilde{\mathbb{L}}/\mathbb{Q})$ for the induced character χ^*. Application of Theorem 13.6 to $L(s, \chi^*, \tilde{\mathbb{L}}/\mathbb{Q})$ yields the assertion.

As a second application we note that there is no non-trivial continuous relation between Artin L-functions to irreducible characters of any fixed Galois extension $\mathbb{K}/\mathbb{Q}$. This remarkable result follows in a similar way as the theorem on the functional independence; for a precise formulation and the details we refer to Bauer [16]. Similarly, it follows that there is no non-trivial continuous relation between Dedekind zeta-functions to finite normal extensions $\mathbb{K}_j$ of $\mathbb{Q}$ satisfying $\mathbb{K}_i \cap \mathbb{K}_j = \mathbb{Q}$ for $i \neq j$. The latter condition cannot be removed as the example (13.11) shows. Bauer also showed that the only non-trivial relations are monomial algebraic.

Davenport and Heilbronn [65] proved that the zeta-function of an ideal class of an imaginary quadratic field has an infinity of zeros in the half-plane of convergence $\sigma > 1$ provided that the class number is greater than one. Voronin [364] extended this result to the right half of the critical strip and gave lower bounds for the number of these zeros. Following Voronin's reasoning in addition with some basics from class field theory, Bauer [16] deduced from the joint universality of Artin L-functions that any partial zeta-function $\zeta(s, \mathsf{A})$ associated with an arbitrary class A of the ray class of any algebraic number field has more than cT many zeros in $\frac{1}{2} < \sigma < 1$, $|t| < T$ for T sufficiently large, where c is a positive constant, provided the ray class group has cardinality greater than one (in particular for class groups of number fields with class number greater than one).

Finally, we shall remark that Hecke *L*-functions and Artin *L*-functions form overlapping classes of Euler products, however, neither is contained in the other. For example, a Hecke *L*-functions associated with a grössencharacter which is not a ray class group character is certainly not an Artin *L*-function. Conversely, if $\mathbb{L}/\mathbb{K}$ is a non-abelian Galois extension of $\mathbb{Q}$ and ϱ is an irreducible representation of $\mathsf{Gal}(\mathbb{L}/\mathbb{K})$, then the attached Artin *L*-function is a Hecke *L*-function if and only if ϱ is one-dimensional. Weil [369] gave generalizations of both Hecke and Artin *L*-functions which can be unified in a single theory; the corresponding objects are so-called Artin–Weil *L*-functions. Their universality is still open.

13.9 *L*-Functions to Automorphic Representations

The Langlands program has emerged in the late 1960s of the last century in a series of far-reaching conjectures tying together seemingly unrelated objects in number theory, algebraic geometry, and the theory of automorphic forms. These disciplines are linked by Langlands' *L*-functions associated with automorphic representations, and by the relations between the analytic properties and the underlying algebraic structures. There are two kinds of *L*-functions: *motivic L*-functions which generalize Artin *L*-functions and are defined purely arithmetically, and *automorphic L*-functions, defined by transcendental data. In its comprehensive form, an identity between a motivic *L*-function and an automorphic *L*-function is called a reciprocity law. Langlands' reciprocity conjecture claims, roughly, that every *L*-function, motivic or automorphic, is equal to a product of *L*-functions attached to automorphic representations. For an introduction to the Langlands program we refer to the excellent surveys of Gelbart [97], Murty [266] and Langlands' lecture [178] at the International Congress in Helsinki. We shall see that the axioms defining $\tilde{\mathcal{S}}$, resp. $\mathcal{S}$ are intimately related to fundamental conjectures concerning general *L*-functions; the paper [145] by Iwaniec and Sarnak is a nice survey on these conjectures and the progress made in the last years. Besides, we give some applications of our universality Theorem 5.14.

At the heart of Langlands's program is the notion of an *automorphic representation* π and its *L*-function $L(s, \pi)$. These objects, both defined via group theory and the theory of harmonic analysis on adèle groups, will be briefly explained below, omitting the technical details; for a more detailed description we refer to the monographs of Bump [46], Gelbart [96], Langlands [177], and the paper [41] of Borel.

Let $\mathbb{K}$ be a number field (one loses not too much by restricting to $\mathbb{Q}$). For each absolute value ν on $\mathbb{K}$, there is a completion $\mathbb{K}_\nu$ of $\mathbb{K}$ which is $\mathbb{R}$, $\mathbb{C}$, or a $\mathfrak{p}$-adic field, where $\mathfrak{p}$ is a prime ideal in $\mathbb{K}$. Denote by $\mathcal{O}_\nu$ the ring of integers in $\mathbb{K}_\nu$. In discussing local–global problems it is often necessary to consider several places simultaneously. At first sight it seems natural to form the product of all the $\mathbb{K}_\nu$ which is a topological ring, but it does not have satisfactory compactness properties. Since any $\alpha \in \mathbb{K}$ is a $\mathfrak{p}$-adic integer for almost all $\mathfrak{p}$, we restrict to elements

$$\alpha = \prod_\nu \alpha_\nu,$$

where α_ν lies in $\mathcal{O}_\nu$ for all but finitely many places ν; such elements are called adèles. The adèles form a set-theoretic (restricted) product. This product is a topological ring, the adèle ring $\mathbb{A}_\mathbb{K}$ of $\mathbb{K}$. One can think of $\mathbb{K}$ as embedded in $\mathbb{A}_\mathbb{K}$ via the map $\alpha \mapsto (\alpha, \alpha, \ldots)$.

For $m \geq 1$ let $\mathsf{GL}_m(\mathbb{A}_\mathbb{K})$ be the group of $m \times m$ matrices over $\mathbb{A}_\mathbb{K}$ whose determinant is a unit in $\mathbb{A}_\mathbb{K}$. By the product topology of the adèle ring, $\mathsf{GL}_m(\mathbb{A}_\mathbb{K})$ becomes a locally compact group in which $\mathsf{GL}_m(\mathbb{K})$, embedded diagonally, is a discrete subgroup of $\mathsf{GL}_m(\mathbb{A}_\mathbb{K})$. A character ψ of $\mathbb{K}^* \setminus \mathsf{GL}_1(\mathbb{A}_\mathbb{K})$ is called grössencharacter, where $\mathbb{K}^* := \mathbb{K} \setminus \{0\}$. For a fixed grössencharacter ψ we consider the Hilbert space

$$L^2 := L^2(\mathsf{GL}_m(\mathbb{K}) \setminus \mathsf{GL}_m(\mathbb{A}_\mathbb{K}), \psi)$$

of measurable functions f on $\mathsf{GL}_m(\mathbb{K}) \setminus \mathsf{GL}_m(\mathbb{A}_\mathbb{K})$ satisfying the conditions

- $f(zg) = \psi(z)f(g)$ for any $z \in \mathbb{Z}$, $g \in \mathsf{GL}_m(\mathbb{K}) \setminus \mathsf{GL}_m(\mathbb{A}_\mathbb{K})$;
- the integral

$$\int_{\mathbb{Z}\mathsf{GL}_m(\mathbb{K}) \setminus \mathsf{GL}_m(\mathbb{A}_\mathbb{K})} |f(g)|^2 \, \mathrm{d}g$$

  is bounded.

Elements $f \in L^2$ generalize the concept of twisted modular forms to discrete subgroups of the full modular group. In order to introduce a subspace of cusp forms we have to consider appropriate subgroups. Any parabolic subgroup P of $\mathsf{GL}_m(\mathcal{R})$, where $\mathcal{R}$ is a commutative ring with identity, has a decomposition, called the Levi decomposition, of the form $P = MN$, where N is the unipotent radical of P; M is called the Levi component of P. We denote the unipotent radical of P in the Levi decomposition of a parabolic subgroup P in $\mathsf{GL}_m(\mathcal{R})$ by $N_P(\mathcal{R})$.

The subspace of cusp forms

$$L_0^2 := L^2(\mathsf{GL}_m(\mathbb{K}) \setminus \mathsf{GL}_m(\mathbb{A}_\mathbb{K}), \psi)$$

of L^2 is defined by the additional vanishing condition

- for all parabolic subgroups P of $\mathsf{GL}_m(\mathbb{A}_\mathbb{K})$ and every $g \in \mathsf{GL}_m(\mathbb{A}_\mathbb{K})$,

$$\int_{N_P(\mathbb{K}) \setminus N_P(\mathbb{A}_\mathbb{K})} f(ng) \, \mathrm{d}n = 0.$$

The right regular representation R of $\mathsf{GL}_m(\mathbb{K})$ on L^2 is given by

$$(R(g)f)(\alpha) = f(\alpha g)$$

for each $f \in L^2$ and any $\alpha, g \in \mathsf{GL}_m(\mathbb{A}_\mathbb{K})$. An automorphic representation is a subquotient of the right regular representation of $\mathsf{GL}_m(\mathbb{A}_\mathbb{K})$ on L^2, and a cuspidal automorphic representation is a subrepresentation of the right regular representation of $\mathsf{GL}_m(\mathbb{A}_\mathbb{K})$ on L_0^2.

A representation of $\mathsf{GL}_m(\mathbb{A}_\mathbb{K})$ is called admissible if its restriction to the maximal subgroup

$$K := \prod_{\nu \text{ complex}} \mathsf{U}_m(\mathbb{C}) \times \prod_{\nu \text{ real}} \mathsf{O}_m(\mathbb{R}) \times \prod_{\nu \text{ finite}} \mathsf{GL}_m(\mathcal{O}_\nu)$$

contains each irreducible representation of K with finite multiplicity; here U_m and O_m denote the groups of unitary and orthogonal $m \times m$ matrices, respectively.

Now let π be an irreducible, admissible, cuspidal automorphic representation of $\mathsf{GL}_m(\mathbb{K})$. Then π can be factored into a direct product

$$\pi = \otimes_\nu \pi_\nu,$$

where ν ranges over all (finite and infinite) places of $\mathbb{K}$, and each π_ν is an irreducible representation of $\mathsf{GL}_m(\mathbb{K}_\nu)$; see Flath [79]. For all but a finite number of places ν the representation π_ν is unramified (that means the quotient obtained by inducing a quasi-character from the Borel subgroup of $\mathsf{GL}_m(\mathbb{K}_\nu)$ to $\mathsf{GL}_m(\mathbb{K}_\nu)$ is unique).

In order to define the L-function attached to an automorphic representation π we define the local Euler factors for non-archimedean (finite) unramified places ν by

$$L_\nu(s, \pi) = \det\left(1 - \frac{A_\nu}{\mathrm{N}(\mathfrak{p})^s}\right)^{-1},$$

where A_ν is the semi-simple conjugacy class corresponding to π_ν and $\mathfrak{p}$ is the prime ideal of $\mathbb{K}$ belonging to the place ν. We do not explain here the rather technical definition of the Euler factors $L_\nu(s, \pi)$ for ramified places ν. However, any Euler factor $L_\nu(s, \pi)$ for a non-archimedean place ν associated with the prime ideal $\mathfrak{p}$, unramified or not, can be rewritten as

$$L_\nu(s, \pi) = \prod_{j=1}^m \left(1 - \frac{\alpha_j(\mathfrak{p})}{\mathrm{N}(\mathfrak{p})^s}\right)^{-1}, \tag{13.25}$$

where the numbers $\alpha_j(\mathfrak{p})$ for $1 \leq j \leq m$ are so-called Satake, resp. Langlands parameters, determined from the local representations π_ν. At the archimedean (infinite) places ν we put for certain numbers $\alpha_j(\nu)$

$$L_\nu(s, \pi) = \prod_{j=1}^m \Gamma_\nu(s - \alpha_j(\nu)).$$

with

$$\Gamma_\nu(s) := \begin{cases} \pi^{-\frac{s}{2}} \Gamma(s/2) & \text{if } \mathbb{K}_\nu \simeq \mathbb{R}, \\ (2\pi)^{-s} \Gamma(s) & \text{if } \mathbb{K}_\nu \simeq \mathbb{C}, \end{cases} \qquad (13.26)$$

where, again, the appearing numbers $\alpha_j(\nu)$ for $1 \le j \le m$ are determined from the local representations π_ν. Then the global L-function associated with π is given by

$$L(s,\pi) = \prod_{\nu \text{ non-archimedean}} L_\nu(s,\pi),$$

and the completed L-function is defined by

$$\Lambda(s,\pi) = L(s,\pi) \prod_{\nu \text{ archimedean}} L_\nu(s,\pi).$$

By the work of Hecke [126], Jacquet and Langlands [147], and Godement and Jacquet [101] we have

Theorem 13.8. *Let* $\mathbb{K}$ *be a number field and* π *be an irreducible, admissible, cuspidal automorphic representation of* $\mathsf{GL}_m(\mathbb{A}_\mathbb{K})$. *Then* $\Lambda(s,\pi)$ *has a meromorphic continuation throughout the complex plane and satisfies the functional equation*

$$\Lambda(s,\pi) = \epsilon_\pi N_\pi^{s-1/2} \Lambda(1-s,\tilde{\pi}),$$

where $\tilde{\pi}$ *is the contragradient representation of* π, $N_\pi \in \mathbb{N}$ *is the conductor of* π *and* ϵ_π *is the root number (these quantities are completely determined by the local representations).* $\Lambda(s,\pi)$ *is entire unless* $m = 1$ *and* π *is trivial, in which case it has a pole at* $s = 1$.

For $m = 1$ one simply obtains the Riemann zeta-function, Dirichlet L-functions and Hecke L-functions attached to grössencharacters, whereas for $m = 2$ one gets L-functions associated with newforms; see Jacquet's paper [146] for the details. The similarities between these general L-functions and those of the Selberg class are obvious. On one hand we have the Selberg class defined by axioms which are known to be the most common pattern of many L-functions in number theory, on the other hand we have Langlands' construction of general L-functions out of group representations.

Langlands visionary program can be regarded as a continuation of the famous Artin conjecture. One of his central conjectures claims that all zeta-functions arising in number theory are special realizations of L-functions to automorphic representations constructed above.

Langlands' Reciprocity Conjecture. *Suppose* $\mathbb{L}$ *is a finite Galois extension of a number field* $\mathbb{K}$ *with Galois group* G, *and* $\varrho : \mathsf{G} \to V$ *is an irreducible representation of* G, *where* V *is an* m-*dimensional vector space. Then there exists an automorphic cuspidal representation* π *of* $\mathsf{GL}_m(\mathbb{A}_\mathbb{K})$ *such that*

$$L(s, \varrho, \mathbb{L}/\mathbb{K}) = L(s,\pi).$$

This means that there are identities between certain L-functions, which are a priori of different type! Since Hecke grössencharacters are automorphic representations of $\mathsf{GL}_1(\mathbb{A})$, Artin's conjecture is a special case of the Langlands reciprocity conjecture. By Artin's work, if $m = 1$ and $\mathbb{L}/\mathbb{K}$ is abelian, Langlands' reciprocity law is settled by means of class field theory. In the case of function fields, the Langlands conjecture has been proved by Drinfeld [70] in dimension 2, and recently by Lafforgue [175] for arbitrary dimension (for which both of them were awarded with a Fields medal).

Now we consider the local Euler factors of L-functions attached to automorphic representations. Petersson [293] extended Ramanujan's conjecture on the values of the τ-function to modular forms. Deligne's estimate (1.34) proved the desired bound for newforms but it is expected that an analogue should hold for all L-functions of arithmetical nature.

Ramanujan–Petersson Conjecture. *Let π be a cuspidal automorphic representation of $\mathsf{GL}_m(\mathbb{A}_{\mathbb{K}})$ which is unramified at a place ν. If ν is non-archimedean, then*

$$|\alpha_j(\mathfrak{p})| = 1 \qquad for \quad 1 \leq j \leq m,$$

where $\mathfrak{p}$ is the prime ideal associated with the place ν. If ν is archimedean, then $\mathrm{Re}\ \alpha_j(\nu) = 0$ *for* $1 \leq j \leq m$.

Note that, for $\mathbb{K} = \mathbb{Q}$ and $\mathbb{Q}_\nu = \mathbb{R}$, this conjecture includes Selberg's eigenvalue conjecture [322] on the smallest eigenvalue of the Laplacian as a special case. The Ramanujan–Petersson conjecture might look very restrictive on the first view, but it is nothing else than the local analogue of the Grand Riemann hypothesis. We refer to Iwaniec and Sarnak [145] for details and the current knowledge concerning this conjecture.

We shall speculate a little bit about all these widely believed conjectures and the axioms defining the Selberg class and the class $\tilde{S}$, in particular. It is expected that all functions in the Selberg class are automorphic L-functions. If $\mathcal{L} \in \mathcal{S}$ is primitive and automorphic, then it is also attached to an irreducible automorphic representation. Conversely, every irreducible automorphic representation should give a primitive function in $\mathcal{S}$. This is not known in general, but it has been proved by Murty [267, 269] for GL_1 and GL_2. The axioms on the analytic continuation and on the functional equation follow immediately from Theorem 13.8. The polynomial Euler product in the definition of $\tilde{S}$ fits (by the splitting of primes in $\mathbb{K}$) perfectly to the Euler product of Langlands' L-functions attached to automorphic representations (13.25) and the Ramanujan–Petersson conjecture. The axiom on the mean-square was already discussed in the context of the Selberg conjectures. Finally, let us note that the Euler factor at the infinite places (13.26) is of the form, predicted by the strong λ-conjecture (from Sect. 6.1). Of course, all these axioms and the hypotheses too, are guessed from known examples of L-functions in number theory, and so they have to share certain patterns. Anyway, we are led to see a close connection between Langlands' general *L*-functions and the elements of the Selberg class.

Assuming Selberg's Conjecture *B*, Murty [267] proved that

- if π is an irreducible cuspidal automorphic representation of $\mathsf{GL}_m(\mathbb{A}_\mathbb{Q})$ which satisfies the Ramanujan–Petersson conjecture, then $L(s, \pi)$ is a primitive function in $\mathcal{S}$,
- if $\mathbb{K}$ is a Galois extension of $\mathbb{Q}$ with solvable Galois group G, and if χ is an irreducible character of G of degree m, then there exists an irreducible cuspidal automorphic representation π of $\mathsf{GL}_m(\mathbb{A}_\mathbb{Q})$ such that

$$L(s, \chi)' = L(s, \pi).$$

The first assertion identifies certain *L*-functions to automorphic representations as primitive functions in the Selberg class subject to the truth of Selberg's Conjecture *B* and the Ramanujan–Petersson conjecture; if we assume additionally axiom (v) on the mean-square, $L(s, \pi)$ is an element of $\tilde{\mathcal{S}}$ too.

The second assertion is Langland's reciprocity conjecture if $\mathbb{K}/\mathbb{Q}$ is solvable. The constraint of solvability arises from a theorem of Arthur and Clozel [3], who showed that the maps of base change and automorphic inductions for automorphic representations exist if the extension is cyclic of prime degree. We do not explain here what that means, but these maps are conjectured to exist in general, which would yield Langlands' reciprocity conjecture in its full generality. However, Murty's proof shows that if the Dedekind zeta-function of $\mathbb{K}$ is the *L*-function of an automorphic representation over $\mathbb{Q}$, then Selberg's Conjecture *B* implies Langlands' reciprocity conjecture.

Concerning universality, Theorem 5.14, in addition to the work of Liu, Wang, and Ye [225], implies that if n is a positive integer, π a cuspidal automorphic representation of $\mathsf{GL}_n(\mathbb{Q})$ satisfying the Ramanujan–Petersson conjecture (including the infinite place), then the Godement–Jacquet *L*-function $L(s, \pi)$ is universal.

A

Appendix: A Short History of Universality

> Thus it appears that universality is a generic phenomenon in analysis.
> K.-G. Grosse-Erdmann

We conclude with a brief historical overview on the phenomenon of universality. We refer to Grosse-Erdmann [109] for a detailed and interesting survey on more or less all known types of universalities and an approach to unify them all.

The first *universal* object appearing in the mathematical literature was discovered by Fekete in 1914/15 (see [287]); he proved that there exists a real power series $\sum_{n=1}^{\infty} a_n x^n$ with the property that for any continuous function $f(x)$ on the interval $[-1, 1]$ satisfying $f(0) = 0$ there is a sequence of positive integers m_k such that

$$\sum_{n=1}^{m_k} a_n x^n \xrightarrow[k \to \infty]{} f(x) \quad \text{uniformly on } [-1, 1].$$

The proof relies essentially on Weierstrass' approximation theorem which states that every continuous function on a compact interval is the limit of a uniformly convergent sequence of polynomials (see [367]).

We illustrate Fekete's theorem and its proof by a p-adic version. The p-adic numbers were introduced by Hensel [131] in 1897. Given a prime number p, we define the p-adic absolute value $|\,.\,|_p$ on $\mathbb{Z}$ by

$$|\alpha|_p = \begin{cases} p^{-\nu(\alpha;p)} & \text{if } \alpha \neq 0, \\ 0 & \text{otherwise}; \end{cases}$$

here $\nu(\alpha; p)$ is the exponent of p in the prime factorization of $\alpha \in \mathbb{Z}$. It is an easy task to extend $|\cdot|_p$ from $\mathbb{Z}$ to $\mathbb{Q}$. The p-adic absolute values satisfies the *strong* triangle inequality: for any $\alpha, \beta \in \mathbb{Q}$,

$$|\alpha + \beta|_p \leq \max\{|\alpha|_p, |\beta|_p\}$$

in contrast to the standard absolute value. The p-adic absolute value is non-archimedean (i.e., $\sup\{|n|_p : n \in \mathbb{N}\}$ is bounded). The integers that are p-adically close to zero are precisely the ones that are *highly* divisible by p. The set of p-adic numbers is the completion of $\mathbb{Q}$ with respect to the p-adic absolute value, and is denoted by $\mathbb{Q}_p$. The p-adic absolute value extends to a non-archimedean absolute value on $\mathbb{Q}_p$, and $\mathbb{Q}_p$ becomes a complete, locally compact, totally disconnected Hausdorff space. Similarly to the completion $\mathbb{R}$ of $\mathbb{Q}$ with respect to the standard (archimedean) absolute value, $\mathbb{Q}_p$ is a field, and its elements have a p-adically convergent series representation: for any $\alpha \in \mathbb{Q}_p$, there exist integers ν and a_k such that

$$\alpha = \sum_{k \geq \nu(\alpha;p)} a_k p^k \quad \text{with} \quad 0 \leq a_k < p.$$

The set of p-adic numbers α with $|\alpha|_p \leq 1$ forms a ring, the ring $\mathbb{Z}_p$ of p-adic integers. Another construction of p-adic numbers uses the representation of $\mathbb{Z}_p$ as a projective limit of the ring of residue classes mod p^k. It is customary to write $|\cdot|_\infty$ for the standard absolute value on $\mathbb{Q}$, $\mathbb{Q}_\infty$ for $\mathbb{R}$, and then call $\mathbb{R}$ the completion of $\mathbb{Q}$ at *the infinite prime* $p = \infty$. Two absolute values are called equivalent if they induce the same topology. Ostrowski (see [279, Sect. II.3]) showed that every non-trivial absolute value on $\mathbb{Q}$ is equivalent to one of the absolute values $|\cdot|_p$, where p is a prime number or $p = \infty$. Thus, completion of $\mathbb{Q}$ with respect to its non-equivalent absolute values yields a family of complete, locally compact topological fields $\mathbb{Q}_p$ which contain $\mathbb{Q}$, one for each place $p \leq \infty$:

$$\mathbb{Q} \hookrightarrow \mathbb{Q}_2, \mathbb{Q}_3, \mathbb{Q}_5, \ldots, \quad \text{and} \quad \mathbb{Q}_\infty = \mathbb{R}.$$

p-adic analysis is quite different from real analysis. A non-archimedean absolute value induces a curious (ultrametric) topology. In p-adic analysis, the role of the intervals in $\mathbb{R}$ are played by the balls

$$a + p^\nu \mathbb{Z}_p := \{\alpha \in \mathbb{Q}_p : |\alpha - a|_p \leq p^{-\nu}\},$$

where $a \in \mathbb{Q}_p$ and $\nu \in \mathbb{Z}$. These balls are *clopen* sets, i.e., they are *clo*sed and *open*. It follows that each point inside a ball is center of the ball, and that any two balls are either disjoint or contained one in another.

The algebraic closure $\overline{\mathbb{Q}}_p$ of $\mathbb{Q}_p$ has infinite degree over $\mathbb{Q}_p$, but is not complete, and so it is not the right field for doing analysis. However, the completion $\mathbb{C}_p$ of $\overline{\mathbb{Q}}_p$ is algebraically closed. If

$$n = \alpha_0 + \alpha_1 p + \alpha_2 p^2 + \cdots \quad \text{and} \quad k = \beta_0 + \beta_1 p + \beta_2 p^2 + \cdots$$

are the p-adic expansions of the integers n and k, then one can show that

$$\binom{n}{k} \equiv \binom{\alpha_0}{\beta_0}\binom{\alpha_1}{\beta_1} \cdots \mod p.$$

In view of the binomic inversion formula this leads for any continuous functions $f : \mathbb{Z}_p \to \mathbb{C}_p$ to the representation

$$f(n) = \sum_{k=0}^{n} a_k(f) \binom{n}{k},$$

where $n \in \mathbb{N}$ and

$$a_k(f) := \sum_{j=0}^{k} (-1)^{k-j} \binom{k}{j} f(j).$$

It is remarkable that this yields a series representation, the so-called Mahler series of f. For $k \in \mathbb{N}_0$, we define the finite difference operator $\bigtriangledown$ by

$$\bigtriangledown^k f(x) = \sum_{j=0}^{k} (-1)^{k-j} \binom{k}{j} f(x+j).$$

Then $a_k = \bigtriangledown^k f(0)$ for $k \in \mathbb{N}$. Mahler [233] proved

Theorem A.1. Let $f : \mathbb{Z}_p \to \mathbb{C}_p$ be a continuous function. Then $a_k \to 0$ as $k \to \infty$, and

$$\sum_{k=0}^{n} a_k \binom{x}{k} \xrightarrow[n \to \infty]{} f(x) \quad \text{uniformly on } \mathbb{Z}_p.$$

A proof can be found in Robert [311, Sect. 4.2.4]. Notice that this theorem does not hold over $\mathbb{R}$.

In particular, Mahler's Theorem A.1 implies a p-adic version of Weierstrass' approximation theorem: for any continuous function $f : \mathbb{Z}_p \to \mathbb{C}_p$, there exists a sequence of polynomials f_n with coefficients in $\mathbb{C}_p$ that converges uniformly to f. Further applications of Mahler's theorem are the construction of p-adic analogues of classical functions like the exponential function, the logarithm or the Gamma-function; for details we refer again to Robert's monograph [311].

Now, it is not too surprising that an analogue of Fekete's universality theorem holds in the p-adic case too. Steuding [341] obtained

Theorem A.2. There exists a p-adic power series $\sum_{n=1}^{\infty} a_n x^n$ with the property that for any ball $a + p^\nu \mathbb{Z}_p$, where $a, \nu \in \mathbb{Z}$, $0 < a < p$, and any continuous function f, defined on $a + p^\nu \mathbb{Z}_p$ and satisfying $f(a) = 0$, there exists a sequence of positive integers m_k such that

$$\sum_{n=1}^{m_k} a_n x^n \xrightarrow[k \to \infty]{} f(x) \quad \text{uniformly on } a + p^\nu \mathbb{Z}_p.$$

We follow Luh [227] in his proof of a version of Fekete's theorem.

Proof. Let $\{Q_n\}$ be a sequence of all polynomials with integral coefficients. We construct a sequence of polynomials $\{P_n\}$ as follows: let $P_0 = Q_0$ and assume that $P_0, \ldots, P_{n-1}$ are known. Denote by d_n the degree of P_{n-1}. Now let ϕ_n be a continuous function on $\mathbb{Z}_p$ such that

$$\phi_n(x) = \left(Q_n(x) - \sum_{k=0}^{n-1} P_k(x) \right) x^{-d_n-1}$$

for $x \in a + p^{-n}\mathbb{Z}_p$. In view of Mahler's Theorem A.1 there exists a polynomial f_n so that

$$\max_{x \in a+p^{-n}\mathbb{Z}_p} |f_n(x) - \phi_n(x)|_p \le p^{-d_n}.$$

Setting $P_n(x) = f_n(x)x^{d_n+1}$, the sequence $\{P_n\}$ is constructed. We note

$$\max_{x \in a+p^{-n}\mathbb{Z}_p} \left| \sum_{k=0}^{n} P_k(x) - Q_n(x) \right|_p = \max_{x \in a+p^{-n}\mathbb{Z}_p} |(f_n(x) - \phi_n(x))x^{d_n+1}|_p$$

$$\le p^{-d_n}.$$

By construction, distinct P_n have no powers in common. Thus we can rearrange formally the polynomial series into a power series:

$$\sum_{n=1}^{\infty} a_n x^n := \sum_{n=0}^{\infty} P_n(x).$$

Again by Mahler's Theorem A.1, for any continuous function f on $a + p^\nu \mathbb{Z}_p$ there exists a sequence of positive integers n_k, tending to infinity with k, such that

$$\max_{x \in a+p^{-n_k}\mathbb{Z}_p} |f(x) - Q_{n_k}(x)|_p \le p^{-n_k}.$$

For sufficiently large k the ball $a + p^\nu \mathbb{Z}_p$ is contained in $a + p^{-n_k}\mathbb{Z}_p$. In view of the above estimates we obtain

$$\max_{x \in a+p^{-n_k}\mathbb{Z}_p} \left| f(x) - \sum_{n=1}^{d_{n_k}+1} a_n x^n \right|_p = \max_{x \in a+p^{-n_k}\mathbb{Z}_p} \left| f(x) - \sum_{n=0}^{n_k} P_n(x) \right|_p$$

$$\le \max_{x \in a+p^{-n_k}\mathbb{Z}_p} \left\{ |f(x) - Q_{n_k}(x)|_p, \left| Q_{n_k}(x) - \sum_{n=0}^{n_k} P_n(x) \right|_p \right\}$$

$$\le \max\{p^{-n_k}, p^{-d_{n_k}}\},$$

which tends to zero as $k \to \infty$. Thus, putting $m_k = d_{n_k} + 1$, the assertion of the theorem follows. $\qquad\square$

In the years after Fekete's discovery many universal objects were found. For instance, Birkhoff [23] proved in 1929 the existence of an entire function $f(z)$ with the property that to any given entire function $g(z)$ there exists a sequence of complex numbers a_n such that

$$f(z + a_n) \xrightarrow[n \to \infty]{} g(z) \qquad \text{uniformly on compacta in } \mathbb{C}.$$

The proof relies in the main part on Runge's approximation theorem. This type of universality is very similar to the one of the Riemann zeta-function and other Dirichlet series. Birkhoff's theorem states the existence of an entire function with wild behaviour near infinity. Luh [228] constructed *holomorphic monsters*, that are holomorphic functions with an extraordinary wild boundary behaviour in arbitrary simply connected open sets. More precisely: let $\mathcal{G}$ be a proper open subset of $\mathbb{C}$ with simply connected components. Then there exists a function f holomorphic on $\mathcal{G}$ such that for every boundary point z of $\mathcal{G}$, every compact subset $\mathcal{K}$ with connected complement and every continuous function g on $\mathcal{K}$ which is holomorphic in the interior of $\mathcal{K}$, there exist linear transformations $\tau_n(z) = a_n z + b_n$ with $\tau_n(\mathcal{K}) \subset \mathcal{G}$ and $\text{dist}(\tau_n(\mathcal{K}), z) \to 0$ as $n \to \infty$ for which

$$f(\tau_n(z)) \xrightarrow[n \to \infty]{} g(z) \qquad \text{uniformly on } \mathcal{K};$$

in addition, each derivative of f and each antiderivative of f of arbitrary order has the boundary behaviour described above. We shall explain the notion antiderivative of a holomorphic function f defined on a simply connected open subset of $\mathbb{C}$; in fact, this notion is not unique. Here, for a *negative* integer j, the jth antiderivative $f^{(j)}$ of f with order $|j|$ is defined by

$$\frac{\mathrm{d}^{|j|}}{\mathrm{d}^{|j|}z} f^{(j)}(z) = f(z).$$

Any other antiderivative of f with the same order $|j|$ differs from $f^{(j)}$ on each component of $\mathcal{G}$ by some polynomial of degree less than $|j|$. Moreover, in [230], Luh proved the existence of multiply universal functions, that are holomorphic functions that satisfy, along with their derivatives and antiderivatives, six universal properties at the *same time*.

Marcinkiewicz [237] was in 1935 the first to use the notion *universality* when he proved the existence of a continuous function whose difference quotients approximate any measurable function in the sense of convergence almost everywhere. This should be compared with the result of Blair and Rubel [25] who proved that there exists an entire function f such that the set $\{f^{(n)} : n \in \mathbb{N}_0\}$ of all derivatives of f is dense in the space of all entire functions in the topology of uniform convergence on compact subsets of the complex plane. Other universal objects are, for example, conformal mappings composed with universal functions, discovered by Luh [229]. However, for a long time no *explicit* example of a universal object was found until Voronin discovered in 1975 that the Riemann zeta-function is universal!

In all these examples of universality there are two characteristic aspects of universality, namely the existence of a single object which

- is *maximal* divergent
- (via a countable process) allows to approximate a *maximal* class of objects.

This observation led to understand universality as a phenomenon which occurs quite naturally in certain limiting processes. Meanwhile it turned out that the phenomenon of universality is anything but a rare event in analysis! Many analytical processes which diverge or behave irregularly in some cases produce universal objects.

Grosse-Erdmann gave a rather general description of universality as follows. There is a topological space $\mathcal{X}$ of objects, a topological space $\mathcal{Y}$ of elements to be approximated, and a family of continuous mappings $T_j : \mathcal{X} \to \mathcal{Y}$ for $j \in J$. Then an object $x \in \mathcal{X}$ is called universal if every element $y \in \mathcal{Y}$ can be approximated by certain $T_j(x)$, i.e., the set $\{T_j(x) : j \in J\}$ lies dense in $\mathcal{Y}$. In the special case when the mappings T_j form a group of homeomorphisms, the concept of universality is well-known in topological dynamics under the name of topological transitivity (this reminds us of Bagchi's reformulation of Riemann's hypothesis in Sect. 8.2). In operator theory, where iterates T^j of an operator T are studied, universal elements are said to be hypercyclic. The general setting of Grosse-Erdmann covers quite many universality results. In [108, 109] he proved the following universality criterion.

Theorem A.3. *Suppose that $\mathcal{X}$ is a Baire space and $\mathcal{Y}$ is of the second category. Then the following assertions are equivalent:*

- *The set $\mathcal{U}$ of universal elements is residual in $\mathcal{X}$.*
- *The set $\mathcal{U}$ of universal elements is dense in $\mathcal{X}$.*
- *To every pair of non-empty sets $V \subset \mathcal{X}$ and $W \subset \mathcal{Y}$ there exists some $j \in J$ with*

$$T_j(V) \cap W \neq \emptyset.$$

If one of these conditions holds, then $\mathcal{U}$ is a dense G_δ-subset of $\mathcal{X}$.

We briefly explain the topological notions before we discuss a special case of this theorem; for details we refer to Grosse-Erdmann [108], Kelley's monograph [169] and Rudin's monograph [312]. In 1899, Baire introduced the notion of category to measure the size of subsets of topological spaces. A subset $\mathcal{E}$ of a topological space $\mathcal{X}$ is called nowhere dense if its closure $\overline{\mathcal{E}}$ contains no non-empty open subset of $\mathcal{X}$. Any countable union of nowhere dense sets is called a set of the first category (meager); all other subsets of $\mathcal{X}$ are said to be of the second category (non-meager). The complement of a set of the first category is called residual (co-meager). A topological space $\mathcal{X}$ is said to be a Baire space if the intersection of any countable family of open and dense subsets of $\mathcal{X}$ is dense in $\mathcal{X}$. Countable intersections of open sets are called G_δ-sets. The theorem of Baire states that any complete metric space is a Baire space.

In many applications of Theorem A.3, both $\mathcal{X}$ and $\mathcal{Y}$ are metric spaces. Then the first two assumptions of the theorem are fulfilled if $\mathcal{X}$ is complete and $\mathcal{Y}$ is separable. Furthermore, the third assertion can be rewritten as follows:

- For every $x \in \mathcal{X}$ and $y \in \mathcal{Y}$ there exist sequences $\{x_n\}$ in $\mathcal{X}$ and $\{j_n\}$ in J such that $T_{j_n}(x_n)$ tends to y as $x_n \to x$.

For a verification of this statement one needs a suitable approximation theorem; for example, Weierstrass' approximation theorem for Fekete's universality theorem, respectively, Runge's approximation theorem for Birkhoff's universality result. With regard to universality for Dirichlet series, this fits to the denseness Theorem 5.10 for the space of analytic functions (which is a complete separable metric space by Theorem 3.15) and the use of Mergelyan's approximation Theorem 5.15, in the proof of the general universality Theorem 5.14, for $\tilde{S}$. There is another remarkable aspect of Theorem A.3: if we observe the phenomenon of universality in some space $\mathcal{X}$, then the set of universal elements is dense. This observation supports the Linnik–Ibragimov conjecture (see Sect. 1.6)! This can be stated in a more explicit way: Nestoridis and Papadimitropoulos [278] proved the existence of a Dirichlet series $\sum_{n=1}^{\infty} \frac{a_n}{n^s}$, absolutely convergent for any $\sigma > 0$, with the property that for every admissible set $K \subset \{s \in \mathbb{C} : \sigma \le 0\}$ and every entire function g, there exists a sequence of positive integers $\{\ell_j\}_j$ such that for any $\ell = 0, 1, 2, \dots$

$$\left(\sum_{n=1}^{\ell_j} \frac{a_n}{n^s} \right)^{(\ell)} \to g(s)^{(\ell)} \qquad \text{uniformly on } K,$$

as $j \to \infty$. Furthermore, the set of such Dirichlet series is dense and G_δ (in the Baire sense) in the space of absolutely convergent Dirichlet series in the right half-plane. Here a set K is called admissible if K is compact with connected complement, and K is the finite union of sets K_r each of which contained in a vertical strip of width less than $\frac{1}{2}$. Their approach does not produce any explicit example of a universal Dirichlet series.

Universality is far away from being completely understood. In particular, the discovery of explicit examples of universal objects (zeta- and L-functions) has led to many new and interesting questions. It seems that universality of general Dirichlet series is not an arithmetic phenomenon at all, but it is much easier to find universal Dirichlet series explicitly among those associated with arithmetic objects.

References

1. J. Andersson, Disproof of some conjectures of K. Ramachandra, *Hardy–Ramanujan J.* **22** (1999), 2–7
2. H. Aramata, Über die Teilbarkeit der Dedekindschen Zetafunktionen, *Proc. Imp. Acad. Jpn.* **9** (1933), 31–34
3. J. Arthur, L. Clozel, *Simple Algebras, Base Change and the Advanced Theory of the Trace Formula*, Ann. Math. Stud., vol. 120, Princeton University Press (1990)
4. E. Artin, Über eine neue Art von *L*-Reihen, *Abh. Math. Sem. Univ. Hamburg* **3** (1923), 89–108
5. E. Artin, Beweis des allgemeinen Reziprozitätsgesetzes, *Abh. Math. Sem. Univ. Hamburg* **5** (1927), 353–363
6. A.O.L. Atkin, J. Lehner, Hecke operators on $\Gamma_0(m)$, *Math. Ann.* **185** (1970), 134–160
7. R. Ayoub, Euler and the zeta-function, *Am. Math. Mon.* **81** (1974), 1067–1086
8. R. Backlund, Über die Beziehung zwischen Anwachsen und Nullstellen der Zetafunktion, *Öfversigt Finska Vetensk. Soc.* **61**(9) (1918–1919)
9. B. Bagchi, *The statistical Behaviour and Universality Properties of the Riemann Zeta-Function and Other Allied Dirichlet Series*, Ph.D. Thesis, Calcutta, Indian Statistical Institute, (1981)
10. B. Bagchi, A joint universality theorem for Dirichlet *L*-functions, *Math. Z.* **181** (1982), 319–334
11. B. Bagchi, Recurrence in topological dynamics and the Riemann hypothesis, *Acta Math. Hung.* **50** (1987), 227–240
12. A. Baker, Linear Forms in the Logarithms of Algebraic Numbers. I, *Mathematika* **13** (1966), 204–216
13. A. Baker, *Transcendental Number Theory*, Cambridge University Press (1975)
14. E.P. Balanzario, Remark on Dirichlet series satisfying functional equations, *Divulg. Mat.* **8** (2000), 169–175
15. R. Balasubramanian, K. Ramachandra, On the frequency of Titchmarsh's phenomenon for $\zeta(s)$. III. *Proc. Indian Acad. Sci.* **86** (1977), 341–351
16. H. Bauer, The value distribution of Artin *L*-series and zeros of zeta-functions, *J. Number Theory* **98** (2003), 254–279
17. P. Bauer, *Über den Anteil der Nullstellen der Riemannschen Zeta-Funktion auf der Kritischen Geraden*, Diploma Thesis, Frankfurt University (1992), available at www.math.uni-frankfurt.de/~pbauer/diplom.ps

18. P. Bauer, Zeros of Dirichlet L-series on the critical line, *Acta Arith.* **93** (2000), 37–52

19. S.N. Bernstein, The extension of properties of trigonometric polynomials to entire functions of finite degree, *Izvestiya Akad. Nauk SSSR* **12** (1948), 421–444 (Russian)

20. A. Beurling, Analyse de la loi asymptotique de la distribution des nombres premiers généralisés, I, *Acta. Math.* **68** (1937), 255–291

21. P. Billingsley, *Convergence of Probability Measures*, Wiley, New York (1968)

22. P. Billingsley, *Probability and Measure* 3rd Edition, Wiley, New York (1995)

23. G.D. Birkhoff, Démonstration d'un théorème élémentaire sur les fonctions entières, *C. R. Acad. Sci. Paris* **189** (1929), 473–475

24. K.M. Bitar, N.N. Khuri, H.C. Ren, Path integrals and Voronin's theorem on the universality of the Riemann zeta-function, *Ann. Phys.* **211** (1991), 172–196

25. C. Blair, L.A. Rubel, A universal entire function, *Am. Math. Mon.* **90** (1983), 331–332

26. R.P. Boas, *Entire Functions*, Academic, New York (1954)

27. S. Bochner, On Riemann's functional equation with multiple gamma factors, *Ann. Math.* **67** (1958), 29–41

28. H. Bohr, Über das Verhalten von $\zeta(s)$ in der Halbebene $\sigma > 1$, *Nachr. Akad. Wiss. Göttingen II Math. Phys. Kl.* (1911), 409–428

29. H. Bohr, Zur Theorie der Riemannschen Zetafunktion im kritischen Streifen, *Acta Math.* **40** (1915), 67–100

30. H. Bohr, Über eine quasi-periodische Eigenschaft Dirichletscher Reihen mit Anwendung auf die Dirichletschen L-Funktionen, *Math. Ann.* **85** (1922), 115–122

31. H. Bohr, B. Jessen, Über die Werteverteilung der Riemannschen Zetafunktion, erste Mitteilung, *Acta Math.* **54** (1930), 1–35

32. H. Bohr, B. Jessen, Über die Werteverteilung der Riemannschen Zetafunktion, zweite Mitteilung, *Acta Math.* **58** (1932), 1–55

33. H. Bohr, E. Landau, Über das Verhalten von $\zeta(s)$ und $\zeta^{(k)}(s)$ in der Nähe der Geraden $\sigma = 1$, *Nachr. Ges. Wiss. Göttingen Math. Phys. Kl.* (1910), 303–330

34. H. Bohr, E. Landau, Ein Satz über Dirichletsche Reihen mit Anwendung auf die ζ-Funktion und die L-Funktionen, *Rend. di Palermo* **37** (1914), 269–272

35. H. Bohr, E. Landau, Sur les zéros de la fonction $\zeta(s)$ de Riemann, *Comptes Rendus Acad. Sci. Paris* **158** (1914), 106–110

36. H. Bohr, E. Landau, Nachtrag zu unseren Abhandlungen aus den Jahren 1910 und *Nachr. Ges. Wiss. Göttingen Math. Phys. Kl.* (1924), 168–172

37. E. Bombieri, On the large sieve, *Mathematika* **12** (1965), 201–225

38. E. Bombieri, D.A. Hejhal, Sur les zéros des fonctions zêta d'Epstein, *Comptes Rendus Acad. Sci. Paris* **304** (1987), 213–217

39. E. Bombieri, D.A. Hejhal, On the distribution of zeros of linear combinations of Euler products, *Duke Math. J.* **80** (1995), 821–862

40. E. Bombieri, A. Perelli, Distinct zeros of L-functions, *Acta Arith.* **83** (1998), 271–281

41. A. Borel, Automorphic L-functions, in: *Automorphic Forms, Representations and L-Functions*, Proceedings of the Symposium in Pure Mathematics, vol. 33, part 2, 27–61, Am. Math. Soc., Providence (1979)

42. R. Brauer, On Artin's L-series with general group characters, *Ann. Math.* **48** (1947), 502–514

43. R. Brauer, Beziehungen zwischen Klassenzahlen von Teilkörpern eines galoisschen Körpers, *Math. Nachr.* **4** (1951), 158–174

44. C. Breuil, B. Conrad, F. Diamond, R. Taylor, On the modularity of elliptic curves over ℚ: Wild 3-adic exercises, *J. Am. Math. Soc.* **14** (2001), 843–939

45. V.V. Buldygin, *Convergence of Random Series in Topological Spaces*, Naukova Dumka, Kiev (1985) (in Russian)

46. D. Bump, *Automorphic Forms and Representations*, Cambridge University Press (1997)

47. D. Bump, J.W. Cogdell, D. Gaitsgory, E. de Shallit, S.S. Kudla, *An Introduction to the Langlands Program*, J. Bernstein, S. Gelbart (Eds.), Birkhäuser, Boston (2003)

48. R.B. Burckel, *Introduction to Classical Complex Analysis, vol. I*, Birkhäuser, Boston (1979)

49. F. Carlson, Über die Nullstellen der Dirichletschen Reihen und der Riemannschen ζ-Funktion, *Arkiv för Mat. Astr. och Fysik* **15**(20) (1920)

50. F. Carlson, Contributions a la theorie des series de Dirichlet. I, *Arkiv för Mat. Astr. och Fysik* **16**(18) (1922)

51. F. Carlson, Contributions a la theorie des series de Dirichlet. II, *Arkiv för Mat. Astr. och Fysik* **19**(26) (1926)

52. K. Chandrasekharan, R. Narasimhan, The approximate functional equation for a class of zeta-functions, *Math. Ann.* **152** (1963), 30–64

53. N. Chebotarev, Determination of the density of the set of prime numbers, belonging to a given substitution class, *Izv. Ross. Akad. Nauk* **17** (1924), 205–250 (in Russian)

54. P.L. Chebyshev, Sur la fonction qui détermine la totalité des nombres premiers inférieurs à une limite donnée, *Mémoires des savants étrangers de l'Acad. Sci. St. Pétersbourg* **5** (1848), 1–19

55. P.L. Chebyshev, Mémoire sur nombres premiers, *Mémoires des savants étrangers de l'Acad. Sci. St. Pétersbourg* **7** (1850), 17–33

56. J. Cogdell, P. Michel, On the complex moments of symmetric power L-functions at $s = 1$, *Int. Math. Res. Not.* **31** (2004), 1561–1617

57. J.B. Conrey, More than two fifths of the zeros of the Riemann zeta-function are on the critical line, *J. Reine Angew. Math.* **399** (1989), 1–26

58. J.B. Conrey, The Riemann hypothesis, *Notices Am. Math. Soc.* **50** (2003), 341–353

59. J.B. Conrey, A. Ghosh, On the Selberg class of Dirichlet series: Small degrees, *Duke Math. J.* **72** (1993), 673–693

60. J.B. Conrey, A. Ghosh, Remarks on the generalized Lindelöf hypothesis, preprint, available at www.math.okstate.edu/~conrey/papers.html

61. J.B. Conrey, S.M. Gonek, High moments of the Riemann zeta-function, *Duke Math. J.* **107** (2001), 577–604

62. J.B. Conway, *Functions of One Complex Variable I* 2nd Edition, Springer, Berlin Heidelberg New York (1978)

63. H. Cramér, Ein Mittelwertsatz in der Primzahltheorie, *Math. Z.* **12** (1922), 147–153

64. H. Cramér, M. Leadbetter, *Stationary and Related Stochastic Processes*, Wiley, New York (1967)

65. H. Davenport, H. Heilbronn, On the zeros of certain Dirichlet series I, II, *J. Lond. Math. Soc.* **11** (1936), 181–185; 307–312

66. R. Dedekind, Über die Anzahl der Ideal-Klassen in den verschiedenen Ordnungen eines endlichen Körpers, in: *Festschrift der Technischen Hochschule in Braunschweig zur Säkularfeier des Geburtstages von C.F. Gauß*, Braunschweig (1877), 1–55; *Gesammelte Werke*, vol. 1, Vieweg, Braunschweig (1930), 104–158

67. P. Deligne, La Conjecture de Weil I, II, *Publ. I.H.E.S.* **43** (1974), 273–307; **52** (1981), 313–428

68. A. Denjoy, L'Hypothèse de Riemann sur la distribution des zéros de $\zeta(s)$, reliée à la théorie des probabilités, *Comptes Rendus Acad. Sci. Paris* **192** (1931), 656–658

69. P.G.L. Dirichlet, Beweis des Satzes, dass jede unbegrenzte arithmetische Progression, deren erstes Glied und Differenz ganze Zahlen ohne gemeinschaftlichen Factor sind, unendlich viele Primzahlen enthält, *Abhdl. Königl. Preuss. Akad. Wiss.* (1837), 45–81

70. V.G. Drinfeld, *Langlands Conjecture for* $GL(2)$ *Over Function Fields*, Proceedings of the International Congress of Mathematicians, Helsinki 1978, 565–574, Acad. Sci. Fennica, Helsinki (1980)

71. W. Duke, Some problems in multidimensional analytic number theory, *Acta Arith.* **52** (1989), 203–228

72. N. Dunford, J.T. Schwartz, *Linear Operators*, vol. 1, Interscience, New York (1958)

73. P.L. Duren, *Theory of H^p-Spaces*, Academic, New York (1970)

74. H.M. Edwards, *Riemann's Zeta-Function*, Academic, New York (1974)

75. K.M. Eminyan, χ-universality of the Dirichlet L-function, *Mat. Zametki* **47** (1990), 132–137 (Russian); *Math. Notes* **47** (1990), 618–622

76. L. Euler, Variae observationes circa series infinitas, *Comment. Acad. Sci. Petropol* **9** (1737), 160–188

77. D.W. Farmer, Long mollifiers of the Riemann zeta-function, *Matematika* **40** (1993), 71–87

78. D.W. Farmer, Counting distinct zeros of the Riemann zeta-function, *Electron. J. Combin.* **2** (1995), 5 (electronic)

79. D. Flath, Decomposition of representations into tensor products, in: *Automorphic Forms, Representations and L-Functions*, vol. 33, Am. Math. Soc., Providence (1979), 179–184

80. M. van Frankenhuijsen, Arithmetic progressions of zeros of the Riemann zeta-function, *J. Number Theor.* **115** (2005), 360–370

81. M.Z. Garaev, On a series with simple zeros of $\zeta(s)$, *Math. Notes* **73** (2003), 585–587

82. M.Z. Garaev, One inequality involving simple zeros of $\zeta(s)$, *Hardy–Ramanujan J.* **26** (2003), 18–22

83. M.Z. Garaev, On vertical zeros of $\Re\zeta(s)$ and $\Im\zeta(s)$, *Acta Arith.* **108** (2003), 245–251

84. V. Garbaliauskienė, A. Laurinčikas, Discrete value-distribution of L-functions of elliptic curves, *Publ. Inst. Math. (Beograd)* **76** (2004), 65–71

85. R. Garunkštis, On zeros of the Lerch zeta-function II, in: *Probability Theory and Mathematical Statistics*, Proceedings of the Seventh Vilnius Conference, Grigelionis et al. (Eds.), TEV Vilnius (1998), 267–276

86. R. Garunkštis, The effective universality theorem for the Riemann zeta-function, in: *Special Activity in Analytic Number Theory and Diophantine Equations*, Proceedings of a Workshop at the Max Planck-Institut Bonn 2002, R.B. Heath-Brown and B. Moroz (Eds.), Bonner Math. Schriften **360** (2003)

87. R. Garunkštis, On Voronin's universality theorem for the Riemann zeta-function, *Fiz. Mat. Fak. Moksl. Semin. Darb.* **6** (2003), 29–33

88. R. Garunkštis, A. Laurinčikas, On zeros of the Lerch zeta-function, in: *Number Theory and its Applications*, S. Kanemitsu and K. Györy (Eds.), Kluwer Academicdelete Publishers", Dordrecht (1999), 129–143

89. R. Garunkštis, A. Laurinčikas, *The Lerch Zeta-Function*, Kluwer, Dordrecht (2002)

90. R. Garunkštis, A. Laurinčikas, R. Šleževičienė, J. Steuding, On the universality of the Estermann zeta-function, *Analysis* **22** (2002), 285–296

91. R. Garunkštis, J. Steuding, On the zero distribution of Lerch zeta-functions, *Analysis* **22** (2002), 1–12

92. R. Garunkštis, J. Steuding, Does the Lerch zeta-function satisfy the Lindelöf hypothesis?, in: *Analytic and Probabilistic Methods in Number Theory*, Proceedings of the Third International Conference in Honour of J. Kubilius, Palanga 2001, A. Dubickas et al. (Eds.), TEV, Vilnius (2002), 61–74

93. F. Gaßmann, Über Beziehungen zwischen den Primidealen eines algebraischen Körpers und den Substitutionen seiner Gruppen, *Math. Z.* **25** (1926), 661–675

94. C.F. Gauss, *Asymptotische Gesetze der Zahlentheorie*, *Werke*, vol. 10.1 (1791), 11–16, Teubner 1917

95. P.M. Gauthier, N. Tarkhanov, Approximation by the Riemann zeta-function, *Complex Variables* **50** (2005), 211–215

96. S.S. Gelbart, *Automorphic Forms on Adele Groups*, Princeton University Press and University of Tokyo Press, Princeton (1975)

97. S.S. Gelbart, An elementary introduction to the Langlands program, *Bull. Am. Math. Soc.* **10** (1984), 177–219

98. J. Genys, A. Laurinčikas, Value distribution of general Dirichlet series. V, *Lith. Math. J.* **44** (2004), 145–156

99. P. Gérardin, W. Li, Functional equations and periodic sequences, in: *Théorie des nombres*, Quebec 1987, J.-M. De Koninck and C. Levesque (Eds.), de Gruyter, Berlin (1989), 267–279

100. E. Girondo, J. Steuding, Effective estimates for the distribution of values of Euler products, *Monatshefte Math.* **145** (2005), 97–106

101. R. Godement, H. Jacquet, *Zeta-Functions of Simple Algebras*, Lecture Notes 260, Springer, Berlin Heidelberg New York (1972)

102. D. Goldfeld, Gauss' class number problem for imaginary quadratic number fields, *Bull. Am. Math. Soc.* **13** (1985), 23–37

103. D.A. Goldston, S.M. Gonek, H.L. Montgomery, Mean values of the logarithmic derivative of the Riemann zeta-function with aplications to primes in short intervals, *J. Reine Angew. Math.* **537** (2001), 105–126

104. S.M. Gonek, *Analytic Properties of Zeta and L-Functions*, Ph.D. Thesis, University of Michigan, (1979)

105. A. Good, On the distribution of the values of Riemann's zeta-function, *Acta Arith.* **38** (1981), 347–388

106. S.A. Gritsenko, On zeros of linear combinations of analogues of the Riemann function, *Tr. Mat. Inst. Steklova* **218** (1997), 134–150 (Russian); translation in *Proc. Steklov Inst. Math.* **218** (1997), 129–145

107. B. Gross, D.B. Zagier, Heegner points and derivatives of L-series, *Invent. Math.* **84** (1986), 225–320

108. K.-G. Grosse-Erdmann, Holomorphe Monster und universelle Funktionen, *Mitt. Math. Sem. Giessen* **176** (1987), 1–81

109. K.-G. Grosse-Erdmann, Universal families and hypercyclic operators, *Bull. Am. Math. Soc.* **36** (1999), 345–381

110. E. Grosswald, F.J. Schnitzer, A class of modified ζ and L-functions, *Pacific J. Math.* **74**(2) (1978), 357–364

111. G.G. Gundersen, Meromorphic functions that share four values, *Trans. Am. Math. Soc.* **277** (1983), 545–567

112. M.C. Gutzwiller, Stochastic behavior in quantum scattering, *Physica D* **7** (1983), 341–355

113. J. Hadamard, Étude sur le propiétés des fonctions entières et en particulier d'une fonction considérée par Riemann, *J. Math. Pures Appl.* **9** (1893), 171–215

114. J. Hadamard, Sur les zéros de la fonction $\zeta(s)$ de Riemann, *Comptes Rendus Acad. Sci. Paris* **122** (1896), 1470–1473

115. H. Hamburger, Über die Riemannsche Funktionalgleichung der ζ-Funktion, *Math. Z.* **10** (1921), 240–254

116. G.H. Hardy, Sur les zéros de la fonction $\zeta(s)$ de Riemann, *Comptes Rendus Acad. Sci. Paris* **158** (1914), 1012–1014

117. G.H. Hardy, J.E. Littlewood, Some problems of 'partitio numerorum'; III: On the expression of a number as a sum of primes, *Acta Math.* **44** (1922), 1–70

118. G.H. Hardy, J.E. Littlewood, The approximate functional equation in the theory of the zeta-function, with applications to the divisor problems of Dirichlet and Piltz, *Proc. Lond. Math. Soc.* **21** (1922), 39–74

119. G.H. Hardy, J.E. Littlewood, On Lindelöf's hypothesis concerning the Riemann zeta-function, *Proc. Royal Soc.* **103** (1923), 403–412

120. G.H. Hardy, E.M. Wright, *An Introduction to the Theory of Numbers*, 5th Edition, Oxford Science, Oxford (1979)

121. T. Hattori, K. Matsumoto, A limit theorem for Bohr–Jessen's probability measures for the Riemann zeta-function, *J. Reine Angew. Math.* **507** (1999), 219–232

122. D.R. Heath-Brown, Simple zeros of the Riemann zeta-function on the critical line, *Bull. Lond. Math. Soc.* **11** (1979), 17–18

123. D.R. Heath-Brown, Gaps between primes and the pair correlation between zeros of the Riemann zeta-function, *Acta Math.* **41** (1982), 85–99

124. E. Hecke, Über die Zetafunktion beliebiger algebraischer Zahlkörper, *Nachr. Ges. Wiss. Göttingen* (1917), 77–89

125. E. Hecke, Über eine neue Art von Zetafunktionen, *Math. Z.* **6** (1920), 11–51

126. E. Hecke, Über die Bestimmung Dirichletscher L-Reihen durch ihre Funktionalgleichung, *Math. Ann.* **112** (1936), 664–699

127. K. Heegner, Diophantische Analysis und Modulfunktionen, *Math. Z.* **56** (1952), 227–253

128. H. Heilbronn, On the class number in imaginary quadratic fields, *Quarterly J. Math.* **5** (1934), 150–160

129. H. Heilbronn, Zeta-functions and L-functions, in: *Algebraic Number Theory*, J.W.S. Cassels and A. Fröhlich (Eds.), Academic, New York, (1967), 204–230

130. D.A. Hejhal, On a result of Selberg concerning zeros of linear combinations of L-functions, *Internat. Math. Res. Notices* **11** (2000), 551–577

131. K. Hensel, Über eine neue Begründung der Theorie der algebraischen Zahlen, *Jahresberichte DMV* **6** (1897), 83–88

132. E. Hewitt, K.A. Ross, *Abstract Harmonic Analysis, vol. I*, Springer, Berlin Heidelberg New York (1979)

133. H. Heyer, *Probability Measures on Locally Compact Groups*, Springer, Berlin Heidelberg New York (1977)
134. D. Hilbert, Mathematische Probleme, *Archiv Math. Physik* **1** (1901), 44–63, 213–317
135. O. Hölder, Über die Eigenschaft der Gammafunktion keiner algebraischen Differentialgleichung zu genügen, *Math. Ann.* **28** (1887), 1–13
136. C.P. Hughes, *On the Characteristic Polynomial of a Random Unitary Matrix and the Riemann Zeta-Function*, Ph.D. Thesis, University of Bristol (2001)
137. A. Hurwitz, Einige Eigenschaften der Dirichletschen Funktionen $F(s) = \sum \left(\frac{D}{n}\right)\frac{1}{n^s}$, die bei der Bestimmung der Klassenzahlen binärer quadratischer Formen auftreten, *Zeitschrift Math. Physik* **27** (1887), 86–101
138. M.N. Huxley, Exponential sums and the Riemann zeta-function, V, *Proc. Lond. Math. Soc.* **90** (2005), 1–41
139. A.E. Ingham, Mean-value theorems in the theory of the Riemann zeta-function, *Proc. Lond. Math. Soc.* **27** (1926), 273–300
140. A.E. Ingham, On two conjectures in the theory of numbers, *Am. J. Math.* **64** (1942), 313–319
141. A. Ivić, *The Theory of the Riemann Zeta-Function with Applications*, Wiley, New York (1985)
142. A. Ivić, On the multiplicity of zeros of the zeta-function, *Bull. Acad. Serbe Sci. Arts, Classe Sci. Math.* **24** (1999), 119–132
143. A. Ivić, On small values of the Riemann zeta-function on the critical line and gaps between zeros, *Lith. Math. J.* **42** (2002), 25–36
144. H. Iwaniec, *Topics in Classical Automorphic Forms*, Am. Math. Soc., Providence (1997)
145. H. Iwaniec, P. Sarnak, Perspectives on the analytic theory of *L*-functions, *Geom. Funct. Anal.*, special volume - GAFA (2000), 705–741
146. H. Jacquet, Principal *L*-functions for GL(n), *Proc. Symp. Pure Math.* **61** (1997), 321–329
147. H. Jacquet, R.P. Langlands, *Automorphic Forms on* GL(2), Lecture Notes 114, Springer, Berlin Heidelberg New York (1970)
148. H. Jacquet, J.A. Shalika, A non-vanishing theorem for zeta-functions on GL$_n$, *Invent. Math.* **38** (1976), 1–16
149. B. Jessen, A. Wintner, Distribution functions and the Riemann zeta-function, *Trans. Am. Math. Soc.* **38** (1935), 48–88
150. D. Joyner, *Distribution Theorems of L-Functions*, Pitman Research Notes in Mathematics, (1986)
151. G. Julia, *Lecons sur les fonctions uniformes à point singulier essentiel isolè*, Gauthier-Villars, Paris (1924)
152. A. Kačėnas, A. Laurinčikas, On the periodic zeta-function, *Liet. Mat. Rink.* **38** (1998), 82–97 (Russian); *Lith. Math. J.* **38** (1998), 64–76
153. R. Kačinskaitė, A discrete limit theorem for the Matsumoto zeta-function in the space of analytic functions, *Liet. Mat. Rink.* **41** (2001), 441–448 (in Russian); *Lith. Math. J.* **41** (2001), 344–350
154. R. Kačinskaitė, *Discrete Limit Theorems for the Matsumoto Zeta-Function*, Ph.D. Thesis, Vilnius University (2002)
155. R. Kačinskaitė, A. Laurinčikas, On the value distribution of the Matsumoto zeta-function, *Acta Math. Hung.* **105** (2004), 339–359

156. J. Kaczorowski, M. Kulas, On the non-trivial zeros off the critical line for
     L-functions from the extended Selberg class, *Monatshefte Math.* **150** (2007),
     217–232
157. J. Kaczorowski, A. Laurinčikas, J. Steuding, On the value distribution of shifts
     of universal Dirichlet series, *Monatshefte Math.* **147** (2006), 309–317
158. J. Kaczorowski, G. Molteni, A. Perelli, J. Steuding, J. Wolfart, Hecke's theory
     and the Selberg class, *Funct. Approx. Comment. Math.* **35** (2006), 183–194
159. J. Kaczorowski, A. Perelli, Functional independence of the singularities of a
     class of Dirichlet series, *Am. J. Math.* **120** (1998), 289–303
160. J. Kaczorowski, A. Perelli, The Selberg class: A survey, in: *Number Theory
     in Progress.* Proceedings of the International Conference in Honor of the 60th
     Birthday of Andrej Schinzel, Zakopane 1997. vol. 2: Elementary and analytic
     number theory. De Gruyter (1999), 953–992
161. J. Kaczorowski, A. Perelli, On the structure of the Selberg class, I: $0 \leq d \leq 1$,
     Acta Math. **182** (1999), 207–241
162. J. Kaczorowski, A. Perelli, On the structure of the Selberg class, V: $1 < d <$
     $5/3$, *Invent. Math.* **150** (2002), 485–516
163. J. Kaczorowski, A. Perelli, On the prime number theorem in the Selberg class,
     Arch. Math. **80** (2003), 255–263
164. A.A. Karatsuba, On the zeros of the function $\zeta(s)$ on short intervals of the
     critical line, *Izv. Akad. Nauk. SSSR Ser. Mat.* **48** (1984) (Russian); *Math.
     USSR-Izv.* **24** (1985), 523–537
165. A.A. Karatsuba, *Complex Analysis in Number Theory*, CRC, Boca Raton, FL
     (1995)
166. A.A. Karatsuba, S.M. Voronin, *The Riemann Zeta-Function*, de Gruyter, New
     York (1992)
167. N.M. Katz, P. Sarnak, *Random Matrices, Frobenius Eigenvalues, and
     Monodromy*, Am. Math. Soc., Providence (1999)
168. J.P. Keating, N.C. Snaith, Random matrix theory and $\zeta(\frac{1}{2}+it)$, *Comm. Math.
     Phys.* **214** (2000), 57–89
169. J.L. Kelley, *General Topology*, Van Nostrand, New Jersey (1955)
170. H.H. Kim, F. Shahidi, Symmetric cube L-functions for GL_2 are entire, *Ann.
     Math.* **150** (1999), 645–662
171. N. Koblitz, *Introduction to Elliptic Curves and Modular Forms*, Springer,
     Berlin Heidelberg New York (1984)
172. H. von Koch, Sur la distribution des nombres premiers, *Comptes Rendus Acad.
     Sci. Paris* **130** (1900), 1243–1246
173. N.M. Korobov, Estimates of trigonometric sums and their applications, *Uspehi
     Mat. Nauk* **13** (1958), 185–192 (Russian)
174. S. Koyama, H. Mishou, Universality of Hecke L-functions in the
     grossencharacter-aspect, *Proc. Jpn. Acad.* **78** (2002), 63–67
175. L. Lafforgue, Chtoucas de Drinfeld et correspondance des Langlands, *Invent.
     Math.* **147** (2002), 1–241
176. E. Landau, Neuer Beweis des Primzahlsatzes und Beweis des Primidealsatzes,
     Math. Ann. **56** (1903), 645–670
177. R.P. Langlands, *Problems in the Theory of Automorphic Forms*, Lecture Notes
     170, Springer, Berlin Heidelberg New York (1970), 18–86
178. R.P. Langlands, L-functions and automorphic representations, Proceedings of
     the International Congress of Mathematicians, Helsinki 1978, 165–175, *Acad.
     Sci. Fennica*, Helsinki (1980)

179. M.L. Lapidus, M. van Frankenhuijsen, *Fractal Geometry and Number Theory*, Birkhäuser, Boston (2000)

180. A. Laurinčikas, A limit theorem for Dirichlet L-functions, *Mat. Zametki* **25** (1979), 481–485

181. A. Laurinčikas, Distributions des valeurs de certaines séries de Dirichlet, *Comptes Rendus Acad. Sci. Paris* **289** (1979), 43–45

182. A. Laurinčikas, Zeros of the derivative of the Riemann zeta-function, *Litovsk. Mat. Sb.* **25** (1985), 111–118 (Russian)

183. A. Laurinčikas, Limit theorems for the Riemann zeta-function in the complex space, in: *Probability Theory and Mathematical Statistics*, Proceedings Sixth Vilnius Conference, B. Grigelionis et al. (Eds.), VSP/Mokslas, Vilnius 1990, 59–69

184. A. Laurinčikas, A limit theorem for the Riemann zeta-function near the critical line in the complex space, *Acta Arith.* **59** (1991), 1–9

185. A. Laurinčikas, On limit theorems for the Riemann zeta-function in some spaces, in: *Probability Theory and Mathematical Statistics*, Proceedings of the Sixth Vilnius Conference 1993, Grigelionis et al. (Eds.) VSP Utrecht/Vilnius (1994), 457–483

186. A. Laurinčikas, *Limit Theorems for the Riemann Zeta-Function*, Kluwer, Dordrecht 1996

187. A. Laurinčikas, On the limit distribution of the Matsumoto zeta-function II, *Liet. Mat. Rink.* **36** (1996), 464–485 (Russian); *Lith. Math. J.* **36** (1996), 371–387

188. A. Laurinčikas, On the limit distribution of the Matsumoto zeta-function I, *Acta Arith.* **79** (1997), 31–39

189. A. Laurinčikas, The universality of the Lerch zeta-functions, *Liet. Mat. Rink.* **37** (1997), 367–375 (Russian); *Lith. Math. J.* **37** (1997), 275–280

190. A. Laurinčikas, On the Matsumoto zeta-function, *Acta Arith.* **84** (1998), 1–16

191. A. Laurinčiaks, On the zeros of linear combinations of Matsumoto zeta-functions, *Liet. Mat. Rink.* **38** (1998) (Russian), 185–204; *Lith. Math. J.* **38** (1998), 144–159

192. A. Laurinčikas, A limit theorem in the theory of finite Abelian groups, *Publ. Math. Debrecen* **52** (1998), 517–533

193. A. Laurinčikas, On the effectivization of the universality theorem for the Lerch zeta-function, *Liet. Mat. Rink.* **40** (2000), 172–178 (Russian); *Lith. Math. J.* **40** (2000), 135–139

194. A. Laurinčikas, The universality of Dirichlet series attached to finite abelian groups, in: *Number Theory*, Proceedings of the Turku Symposium on Number theory in Memory of Kustaa Inkeri (1999), M. Jutila, T. Metsänkylä (Eds.), Walter de Gruyter, Berlin, New York (2001), 179–192

195. A. Laurinčikas, A probabilistic equivalent of the Lindelöf hypothesis, in: *Analytic and Probabilistic Methods in Number Theory*, Proceedings of the Third International Conference in Honour of J. Kubilius, Palanga 2001, A. Dubickas et al. (Eds.), TEV, Vilnius (2002), 157–161

196. A. Laurinčikas, The universality of zeta-functions, Proceedings of the Eighth Vilnius Conference on Probability Theory and Mathematical Statistics, Part I (2002), *Acta Appl. Math.* **78**(1–3) (2003), 251–271

197. A. Laurinčikas, On the derivatives of zeta-functions of certain cusp forms, *Glasgow Math. J.* **47** (2005), 87–96

198. A. Laurinčikas, Prehistory of the Voronin universality theorem, *Šiauliai Math. J.* **1** (9) (2006), 41–53

199. A. Laurinčikas, K. Matsumoto, On estimation of the number of zeros of linear combinations of certain zeta-functions, *LMD Mokslo Darbai* (1998), 43–48

200. A. Laurinčikas, K. Matsumoto, The joint universality and the functional independence for Lerch zeta-functions, *Nagoya Math. J.* **157** (2000), 211–227

201. A. Laurinčikas, K. Matsumoto, The universality of zeta-functions attached to certain cusp forms, *Acta Arith.* **98** (2001), 345–359

202. A. Laurinčikas, K. Matsumoto, The joint universality of twisted automorphic *L*-functions, *J. Math. Soc. Jpn.* **56** (2004), 923–939

203. A. Laurinčikas, K. Matsumoto, Joint value-distribution theorems on Lerch zeta-functions II, *Liet. Mat. Rink.* **46** (2006), 332–350 (Russian); *Lith. Math. J.* **46** (2006), 271–286

204. A. Laurinčikas, K. Matsumoto, J. Steuding, The universality of *L*-functions associated with newforms, *Izvestija Math.* **67** (2003), 77–90; *Izvestija Ross. Akad. Nauk Ser. Mat.* **67** (2003), 83–96 (Russian)

205. A. Laurinčikas, K. Matsumoto, J. Steuding, Discrete universality of *L*-functions for newforms, *Math. Notes* **78** (2005), 551–558; *Mat. Zametki* **78** (2005), 595–603 (Russian)

206. A. Laurinčikas, W. Schwarz, J. Steuding, The universality of general Dirichlet series, *Analysis* **23** (2003), 13–26

207. A. Laurinčikas, R. Šleževičienė, The universality of zeta-functions with multiplicative coefficients, *Integral Transforms and Special Functions* **13** (2002), 243–257

208. A. Laurinčikas, J. Steuding, A short note on the Lindelöf hypothesis, *Lith. Math. J.* **43** (2003), 51–55; *Liet. Mat. Rink.* **43** (2003), 59–64 (Russian)

209. A. Laurinčikas, J. Steuding, Joint universality for *L*-functions attached to a family of elliptic curves, Proceedings of the ELAZ 2004, Conference on elementary and analytic number theory, Mainz 2004, W. Schwarz, J. Steuding (Eds.), Steiner, Stuttgart (2006), 155–163

210. A. Laurinčikas, J. Steuding, A limit theorem for the Hurwitz zeta-function with an algebraic irrational parameter, *Arch. Math.* **85** (2005), 419–432

211. A. Laurinčikas, J. Steuding, Limit theorems in the space of analytic functions for the Hurwitz zeta-function with an algebraic irrational parameter, (submitted)

212. D.H. Lehmer, The Vanishing of Ramanujan's Function, *Duke Math. J.* **14** (1947), 429–433

213. D. Leitmann, D. Wolke, Periodische und multiplikative zahlentheoretische Funktionen, *Monatsh. Math.* **81** (1976), 279–289

214. C.G. Lekkerkerker, *On the Zeros of a Class of Dirichlet Series*, Proefschrift, van Gorcum, NV (1955)

215. M. Lerch, Note sur la fonction $\mathcal{K}(w,x,s) = \sum_{k=0}^{\infty} e^{2k\pi i x}(w+k)^{-s}$, *Acta Math.* **11** (1887), 19–24

216. N. Levinson, More than one third of Riemann's zeta-function are on $\sigma = \frac{1}{2}$, *Adv. Math.* **13** (1974), 383–436

217. N. Levinson, Almost all roots of $\zeta(s) = a$ are arbitrarily close to $\sigma = 1/2$, *Proc. Nat. Acad. Sci. USA* **72** (1975), 1322–1324

218. N. Levinson, H.L. Montgomery, Zeros of the derivative of the Riemann zeta-function, *Acta Math.* **133** (1974), 49–65

219. H.-Z. Li, J. Wu, The universality of symmetric power L-functions and their Rankin–Selberg L-functions, *J. Math. Soc. Jpn.* (in print)
220. E. Lindelöf, Quelques remarques sur la croissance de la fonction $\zeta(s)$, *Bull. Sci. Math.* **32** (1908), 341–356
221. E. Lindelöf, E. Phragmén, Sur une extension d'un principe classique d'analyse et sur quelques propriétés de fonctions monogènes dans le voisinage d'un point singulier, *Acta Math.* **31** (1908), 381–406
222. R. Lipschitz, Untersuchung einer aus vier Elementen gebildeten Reihe, *J. Reine und Angew. Math.* **54** (1857), 313–328
223. J.E. Littlewood, Quelques conséquences de l'hypothése que la fonction $\zeta(s)$ de Riemann n'a pas de zéros dans de demi-plan $R(s) > \frac{1}{2}$, *Compt. Rend. Acad. Sci. Paris* **154** (1912), 263–266
224. J.E. Littlewood, On the zeros of the Riemann zeta-function, *Proc. Camb. Philol. Soc.* **22** (1924), 295–318
225. J. Liu, Y. Wang, Y. Ye, A proof of Selberg's orthogonality for automorphic L-functions, *Manuscripta Math.* **118** (2005), 135–149
226. M. Loève, *Probability Theory*, Van Nostrand, Toronto (1955)
227. W. Luh, Universalfunktionen in einfach zusammenhängenden Gebieten, *Aequationes Mathematicae* **19** (1979), 183–193
228. W. Luh, Holomorphic monsters, *J. Approx. Theory* **53** (1988), 128–144
229. W. Luh, Universal functions and conformal mappings, *Serdica* **19** (1993), 161–166
230. W. Luh, Multiply universal holomorphic functions, *J. Approx. Theory* **89** (1997), 135–155
231. E. Lukacs, *Characteristic Functions*, 2nd Edition, Griffin, London (1970)
232. J. van de Lune, H.J.J. te Riele, D.T. Winter, On the zeros of the Riemann zeta-function in the critical strip, IV, *Math. Comp.* **46** (1986), 667–681
233. K. Mahler, *p-adic Numbers and their Functions* 2nd Edition, Cambridge University Press (1981)
234. S. Mandelbrojt, *Dirichlet Series*, Reidel, Dordrecht (1972)
235. H. von Mangoldt, Zu Riemanns' Abhandlung "Über die Anzahl der Primzahlen unter einer gegebenen Grösse", *J. Reine Angew. Math.* **114** (1895), 255–305
236. H. von Mangoldt, Zur Verteilung der Nullstellen der Riemannschen Funktion $\xi(t)$, *Math. Ann.* **60** (1905), 1–19
237. J. Marcinkiewicz, Sur les nombres dérivés, *Fund. Math.* **24** (1935), 305–308
238. R. Marszałek, The multiplicative and functional independence of Dedekind zeta-functions of abelian fields, *Coll. Math.* **103** (2005), 11–16
239. K. Matsumoto, *Value-Distribution of Zeta-Functions*, in *Analytic Number Theory*, Proceedings of the Japanese-French Symposium, Tokyo 1988, K. Nagasaka and É. Fouvry (Eds.), Lecture Notes in Maths 1434, Springer, Berlin Heidelberg New York (1990), 178–187
240. K. Matsumoto, Recent developments in the mean square theory of the Riemann zeta and other zeta-functions, in: *Number Theory*, Trends Math., Birkhäuser, Boston (2000), 241–286
241. K. Matsumoto, The mean values and the universality of Rankin-Selberg L-functions, in: *Number Theory*, Proceedings of the Turku Symposium on Number Theory in Memory of Kustaa Inkeri (1999), M. Jutila, T. Metsänkylä (Eds.), Walter de Gruyter, Berlin, New York (2001), 201–221

242. K. Matsumoto, Probabilistic value-distribution theory of zeta-functions, *Sugaku* **53** (2001), 279–296 (in Japanese); English translation in *Sugaku Expositions* **17** (2004), 51–71

243. K. Matsumoto, Some problems on mean values and the universality of zeta and multiple zeta-functions, in: *Analytic and Probabilistic Methods in Number Theory*, Proceedings of the Third International Conference in Honour of J. Kubilius, Palanga 2001, Λ. Dubickas et al. (Eds.), TEV, Vilnius (2002), 195–199

244. K. Matsumoto, An introduction to the value-distribution theory of zeta-functions, *Šiauliai Math. J.* **1** (9) (2006), 61–83

245. J.-L. Mauclaire, Almost periodicity and Dirichlet series, preprint

246. S.N. Mergelyan, Uniform approximations to functions of a complex variable, *Usp. Mat. Nauk* **7** (1952), 31–122 (Russian); *Am. Math. Soc.* (Translated) **101** (1954), 99

247. F. Mertens, Ein Beitrag zur analytischen Zahlentheorie, *J. Reine Angew. Math.* **78** (1874), 46–62

248. F. Mertens, Über eine zahlentheoretische Funktion, *Sem. Ber. Kais. Akad. Wiss. Wien* **106** (1897), 761–830

249. H. Mishou, The universality theorem for *L*-functions associated with ideal class characters, *Acta Arith.* **98** (2001), 395–410

250. H. Mishou, The universality theorem for Hecke *L*-functions, *Acta Arith.* **110** (2003), 45–71

251. H. Mishou, H. Nagoshi, Functional distribution of $L(s, \chi_d)$ with real characters and denseness of Quadratic Class Numbers, *Trans. Am. Math. Soc.* **358** (2006), 4343–4366

252. H. Mishou, H. Nagoshi, The universality of quadratic *L*-series for prime discriminants, *Acta Arith.* **123** (2006), 143–161

253. H. Mishou, H. Nagoshi, Equivalents of the Riemann hypothesis, *Arch. Math.* **86** (2006), 419–424

254. H. Mishou, H. Nagoshi, The universality for Rankin-Selberg L-functions in the level aspect, preprint

255. D.S. Mitrinovic, J. Sándor, B. Crstici, *Handbook of Number Theory*, Kluwer, Dordrecht 1995

256. G. Molteni, A note on a result of Bochner and Conrey-Ghosh about the Selberg class, *Arch. Math.* **72** (1999), 219–222

257. G. Molteni, J. Steuding, (Almost) primitivity of Hecke *L*-functions, *Monatshefte Math.* (in print)

258. H.L. Montgomery, *Topics in Multiplicative Number Theory*, Lecture Notes 227, Springer, Berlin Heidelberg New York (1971)

259. H.L. Montgomery, The pair correlation of zeros of the zeta-function, *Proc. Symp. Pure Math.* **24** (1973), 181–193

260. H.L. Montgomery, Extreme values of the Riemann zeta-function, *Comment. Math. Helv.* **52** (1977), 511–518

261. L.J. Mordell, On Ramanujan's empirical expansions of modular functions, *Proc. Camb. Philol. Soc.* **19** (1920), 117–124

262. C.J. Moreno, Explicit formulas in the theory of automorphic forms, in: *Number Theory Day*, Proceedings of the Conference in Rockefeller University, New York 1976, Lecture Notes in Math. **626**, Springer, Berlin Heidelberg New York (1977), 73–216

263. Y. Motohashi, *Spectral Theory of the Riemann Zeta-Function*, Cambridge University Press (1997)
264. H. Müller, Über eine Klasse modifizierter ζ- und L-Funktionen, *Arch. Math.* **36** (1981), 157–161
265. M.R. Murty, Oscillations of Fourier coefficients of modular forms, *Math. Ann.* **262** (1983), 431–446
266. M.R. Murty, A motivated introduction to the Langlands program, in: *Advances in Number theory*, F. Gouvea and N. Yui, (Eds.), Clarendon Press Oxford (1993), 37–66
267. M.R. Murty, Selberg's conjectures and Artin L-functions, *Bull. Am. Math. Soc.* **31** (1994), 1–14
268. M.R. Murty, V.K. Murty, Strong multiplicity one for Selberg's class, *C.R. Acad. Sci. Paris Ser. I Math.* **319** (1994), 315–320
269. M.R. Murty, Selberg's conjectures and Artin L-functions, II, in: *Current Trends in Mathematics and Physics*, S.D. Adhikari, (Ed.), Narosa, New Delhi (1995), 154–168
270. M.R. Murty, V.K. Murty, *Non-Vanishing of L-Functions and Applications*, Birkhäuser, Boston (1997)
271. M.R. Murty, A. Perelli, The pair correlation of zeros of functions in the Selberg class, *Int. Math. Res. Notices* **10** (1999), 531–545
272. V.K. Murty, On the Sato-Tate conjecture, in: *Number Theory Related to Fermat's Last Theorem*, N. Koblitz, (Ed.), Birkhäuser, Boston (1982), 195–205
273. H. Nagoshi, On the universality for L-functions attached to Maass forms, *Analysis* **25** (2005), 1–22
274. H. Nagoshi, The universality of L-functions attached to Maass forms, *Advanced Studies in Pure Mathematics*, Proceedings of the International Conference on Probability and Number Theory, Kanazawa (2005) (in print)
275. H. Nagoshi, The universality of families of automorphic L-functions, preprint
276. T. Nakamura, The existence and the non-existence of joint t-universality for Lerch zeta-functions, *J. Number Theory* (in print)
277. W. Narkiewicz, *The Development of Prime Number Theory*, Springer, Berlin Heidelberg New York (2000)
278. V. Nestoridis, C. Papadimitropoulos, Abstract theory of universal series and an application to Dirichlet series, *C.R. Acad. Sci. Paris* **341** (2005), 539–543
279. J. Neukirch, *Algebraische Zahlentheorie*, Springer 1992; English Translation: *Algebraic Number Theory*, Springer, Berlin Heidelberg New York (1999)
280. R. Nevanlinna, Einige Eindeutigkeitssätze in der Theorie der meromorphen Funktionen, *Acta Math.* **48** (1926), 367–391
281. R. Nevanlinna, *Analytic functions*, Springer, Berlin Heidelberg New York (1970), translated from the 2nd German edition (1953)
282. F. Nicolae, Über die lineare Unabhängigkeit der Dedekindschen Zetafunktionen galoisscher Zahlkörper, *Math. Nachr.* **220** (2000), 111–113
283. A.M. Odlyzko, The 10^{20}th zero of the Riemann zeta-function and 70 million of its neighbors, in: *Dynamical, spectral, and arithmetic zeta functions.* San Antonio, TX, (1999), 139–144, Contemp. Math. **290**, Am. Math. Soc., Providence (2001)
284. A.M. Odlyzko, H.J.J. te Riele, Disproof of Mertens conjecture, *J. Reine Angew. Math.* **367** (1985), 138–160
285. A.P. Ogg, A remark on the Sato-Tate conjecture, *Invent. Math.* **9** (1970), 198–200

286. A. Ostrowski, Über Dirichletsche Reihen und algebraische Differentialgleichungen, *Math. Z.* **8** (1920), 241–298

287. J. Pál, Zwei kleine Bemerkungen, *Tohoku Math. J.* **6** (1914/15), 42–43

288. D.V. Pechersky, On the permutation of the terms of functional series, *Dokl. Akad. Nauk SSSR* **209** (1973), 1285–1287 (Russian)

289. A. Perelli, General L-functions, *Ann. Mat. Pura Appl.* **130** (1982), 287–306

290. A. Perelli, A survey of the Selberg class of L-functions, Part I, *Milan J. Math.* **73** (2005), 19–52

291. R. Perlis, On the equation $\zeta_{\mathbb{K}}(s) = \zeta_{\mathbb{K}'}(s)$, *J. Number Theory* **9** (1977), 342–360

292. R. Perlis, A. Schinzel, Zeta functions and the equivalence of integral forms, *J. Reine Angew. Math.* **309** (1979), 176–182

293. H. Petersson, Konstruktion der sämtlichen Lösungen einer Riemannschen Funktionalgleichung durch Dirichletreihen mit Eulerscher Produktentwicklung II, *Math. Ann.* **117** (1940/41), 39–64

294. I. Piatetski-Shapiro, R. Rhagunathan, On Hamburger's theorem, *Am. Math. Soc. Transl.* **169** (1995), 109–120

295. J. Popken, A measure for the differential-transcendence of the zeta-function of Riemann, in: *Number Theory and Analysis*, (papers in honor of Edmund Landau), Plenum, New York 1969, 245–255

296. H.S.A. Potter, The mean values of certain Dirichlet series I, *Proc. Lond. Math. Soc.* **46** (1940), 467–468

297. C.R. Putnam, On the non-periodicity of the zeros of the Riemann zeta-function, *Am. J. Math.* **76** (1954), 97–99

298. C.R. Putnam, Remarks on periodic sequences and the Riemann zeta-function, *Am. J. Math.* **76** (1954), 828–830

299. K. Ramachandra, On Riemann zeta-function and allied questions, *Astérisque* **209** (1992), 57–72 Journées Arithmétiques, Geneva (1991)

300. S. Ramanujan, On certain arithmetical functions, *Trans. Camb. Phil. Soc.* **22** (1916), 159–184

301. S. Ramanujan, On certain trigonometrical sums and their applications in the theory of numbers, *Trans. Camb. Philol. Soc.* **22** (1918), 259–276

302. R.A. Rankin, Contributions to the theory of Ramanujan's τ-function and similar arithmetical functions. I. The zeros of the function $\sum_{n=1}^{\infty} \frac{\tau(n)}{n^s}$ on the line $\mathcal{R}s = \frac{13}{2}$, *Proc. Camb. Philol. Soc.* **35** (1939), 351–356

303. R.A. Rankin, Contributions to the theory of Ramanujan's τ-function and similar arithmetical functions. II. The order of the Fourier coefficients of integral modular forms, *Proc. Camb. Philol. Soc.* **35** (1939), 357–372

304. R.A. Rankin, An Ω-result for the coefficients of cusp forms, *Math. Ann.* **203** (1973), 239–250

305. A. Reich, Universelle Wertverteilung von Eulerprodukten, *Nach. Akad. Wiss. Göttingen, Math.-Phys. Kl.* (1977), 1–17

306. A. Reich, Wertverteilung von Zetafunktionen, *Arch. Math.* **34** (1980), 440–451

307. A. Reich, Zetafunktionen und Differenzen-Differentialgleichungen, *Arch. Math.* **38** (1982), 226–235

308. H.-E. Richert, Über Dirichletreihen mit Funktionalgleichungen, *Publ. Inst. Math. Acad. Serbe Sci.* **11** (1957), 73–124

309. G.J. Rieger, Effective simultaneous approximation of complex numbers by conjugate algebraic integers, *Acta Arith.* **63** (1993), 325–334

310. B. Riemann, Über die Anzahl der Primzahlen unterhalb einer gegebenen Grösse, *Monatsber. Preuss. Akad. Wiss. Berlin* (1859), 671–680

311. A.M. Robert, *A course in p-adic Analysis*, Springer, Berlin Heidelberg New York (2000)
312. W. Rudin, *Real and Complex Analysis*, Mc Graw-Hill, New York (1966)
313. W. Rudin, *Functional Analysis*, Mc Graw-Hill, New York (1973)
314. Z. Rudnick, P. Sarnak, Zeros of principal L-functions and Random Matrix Theory, *Duke Math. J.* **81** (1996), 269–322
315. J. Sander, J. Steuding, Joint universality for sums and products of Dirichlet L-functions, *Analysis* **26** (2006), 295–312
316. W. Schnee, Die Funktionalgleichung der Zetafunktion und der Dirichletschen Reihen mit periodischen Koeffizienten, *Math. Z.* **31** (1930), 378–390
317. W. Schwarz, Geschichte der analytischen Zahlentheorie seit 1890, in: *Ein Jahrhundert Mathematik* 1890–1990, DMV, Vieweg (1990), 741–780
318. W. Schwarz, Some remarks on the history of the prime number theorem from 1896 to 1960, in: *Development of Mathematics* 1900–1950, J.-P. Pier, (Ed.) Birkhäuser, Boston 1994 (Symposium Luxembourg 1992), 565–615
319. W. Schwarz, R. Steuding, J. Steuding, Universality for Euler products and related arithmetical functions, (submitted)
320. A. Selberg, Bemerkungen über eine Dirichletsche Reihe, die mit der Theorie der Modulformen nahe verbunden ist, *Arch. Math. Naturvid.* **43** (1940), 47–50
321. A. Selberg, On the zeros of the Riemann zeta-function, *Skr. Norske Vid. Akad.* Oslo **10** (1942), 1–59
322. A. Selberg, On the estimation of Fourier coefficients of modular forms, *Proc. Sym. Pure Math.* VIII (1965), 1–15
323. A. Selberg, Old and new conjectures and results about a class of Dirichlet series, Proceedings of the Amalfi Conference on Analytic Number Theory, Maiori 1989, E. Bombieri et al. (Eds.), Università di Salerno (1992), 367–385
324. J.-P. Serre, *Abelian l-adic representations and elliptic curves*, Research Notes in Mathematics, AK Peters (1968)
325. J.-P. Serre, *Linear representations of finite groups*, Springer, Berlin Heidelberg New York (1977)
326. F. Shahidi, Symmetric power L-functions for $GL(2)$, in: *Elliptic curves and related objects*, H. Kisilevsky and M.R. Murty, (Ed.), CRM Proceedings and Lecture Notes, vol. 4 (1994), 159–182
327. F. Shahidi, Automorphic L-functions and functoriality, Proceedings of the International Congress of Mathematicians, Bejing (2002), vol. III, 655–666
328. G. Shimura, On the holomorphy of certain Dirichlet series, *Proc. Lond. Math. Soc.* **31** (1975), 79–98
329. C.L. Siegel, Über Riemanns Nachlass zur analytischen Zahlentheorie, *Quellen u. Studien zur Geschichte der Math. Astr. Phys.* **2** (1932), 45–80
330. C.L. Siegel, Über die Classenzahl quadratischer Zahlkörper, *Acta Arith.* **1** (1935), 83–86
331. R. Šleževičienė, *Joint limit theorems and universality for the Riemann and allied zeta-functions*, Ph.D. Thesis, Vilnius University (2002)
332. R. Šleževičienė, The joint universality for twists of Dirichlet series with multiplicative coefficients, in: *Analytic and probabilistic methods in Number Theory*, Proceedings of the Third International Conference in Honour of J. Kubilius, Palanga 2001, A. Dubickas et al. (Eds.), TEV, Vilnius (2002), 303–319
333. R. Šleževičienė, J. Steuding, Short series over simple zeros of the Riemann zeta-function, *Indagationes Math.* **15** (2004), 129–132

334. R. Šleževičiené-Steuding, Universality for generalized Euler products, *Analysis* **26** (2006), 337–345

335. R. Spira, Zeros of Hurwitz zeta-functions, *Math. Comp.* **30** (1976), 863–866

336. H.M. Stark, A Complete Determination of the Complex Quadratic Fields of Class Number One, *Michigan Math. J.* **14** (1967), 1–27

337. H.M. Stark, Galois theory, algebraic numbers and zeta functions, in: *From Number Theory to Physics*, Waldschmitd et al. (Eds.), Springer, Berlin Heidelberg New York (1989), 313–393

338. J. Steuding, Grosse Werte der Riemannschen Zetafunktion auf rektifizierbaren Kurven im kritischen Streifen, *Arch. Math.* **70** (1998), 371–376

339. J. Steuding, On simple zeros of the Riemann zeta-function in short intervals on the critical line, *Acta Math. Hung.* **96** (2002), 259–308

340. J. Steuding, On Dirichlet series with periodic coefficients, *Ramanujan J.* **6** (2002), 295–306

341. J. Steuding, The world of p-adic numbers and p-adic functions, *Fiz. Mat. Fak. Moksl. Semin. Darb.* **5** (2002), 90–107

342. J. Steuding, Upper bounds for the density of universality, *Acta Arith.* **107** (2003), 195–202

343. J. Steuding, A note on the extended Selberg class (degrees 0 and 1), *Arch. Math.* **81** (2003), 22–25

344. J. Steuding, Extremal values of Dirichlet L-functions in the half-plane of absolute convergence, *J. Théo. Nombres Bordeaux* **16** (2004), 221–232

345. J. Steuding, Universality in the Selberg class, in: *Special Activity in Analytic Number Theory and Diophantine Equations*, Proceedings of a Workshop at the Max Planck-Institut Bonn 2002, R.B. Heath-Brown and B. Moroz (Eds.), Bonner math. Schriften **360** (2003)

346. J. Steuding, On the value-distribution of L-functions, *Fiz. Mat. Fak. Moksl. Semin. Darb.* **6** (2003), 87–119

347. J. Steuding, How many values can L-functions share? *Fiz. Mat. Fak. Moksl. Semin. Darb.* **7** (2004), 70–81

348. J. Steuding, On the zero-distribution of Epstein zeta-functions, *Math. Annalen* **333** (2005), 689–697

349. J. Steuding, On the value-distribution of Epstein zeta-functions, *Publicacions Mat.*, Proceedings of 'Primeras Jornadas de Teoria de Números', Vilanova i la Geltrú (Barcelona) 2005 (in print)

350. J. Steuding, Upper bounds for the density of universality II, *Acta Mathematica Universitatis Ostraviensis* **13** (2005), 73–82

351. R. Taylor, *Automorphy for some l-adic lifts of automorphic mod l representations II.* (preprint) available at http://www.math.harvard.edu/~rtaylor/

352. E.C. Titchmarsh, *The Theory of Functions*, 2nd Edition, Oxford University Press (1939)

353. E.C. Titchmarsh, *The Theory of the Riemann Zeta-Function*, 2nd Edition, revised by D.R. Heath-Brown, Oxford University Press (1986)

354. V.A. Trenogin, *Functional analysis*, Nauka, Moscow (1980) (in Russian)

355. K.M. Tsang, *Distribution of values of the Riemann zeta-function*, Ph.D. Thesis, Princeton (1984)

356. J. Tunnell, A classical diophantine problem and modular forms of weight 3/2, *Inventiones Math.* **72** (1983), 323–334

357. C.J. de la Vallée-Poussin, Recherches analytiques sur la théorie des nombres premiers, I-III, *Ann. Soc. Sci. Bruxelles* **20** (1896), 183–256, 281–362, 363–397

358. M.-F. Vignéras, Facteurs gamma et équations fonctionnelles, in: *Modular Functions of one Variable I*, Proceedings of the International Conference, University of Bonn, 1976, A. Dold and B. Eckmann (Eds.), Lecture Notes in Math. **627**, Springer, Berlin Heidelberg New York (1977), 79–103

359. I.M. Vinogradov, A new estimate for the function $\zeta(1 + it)$, *Izv. Akad. Nauk SSSR, Ser. Mat.* **22** (1958), 161–164 (Russian)

360. A.I. Vinogradov, On the density hypothesis for Dirichlet L-functions, *Izv. Akad. Nauk SSSR* **29** (1965), 903–934 (Russian)

361. S.M. Voronin, The distribution of the non-zero values of the Riemann zeta-function, *Izv. Akad. Nauk Inst. Steklov* **128** (1972), 131–150 (Russian)

362. S.M. Voronin, On the functional independence of Dirichlet L-functions, *Acta Arith.* **27** (1975), 493–503 (Russian)

363. S.M. Voronin, Theorem on the 'universality' of the Riemann zeta-function, *Izv. Akad. Nauk SSSR, Ser. Matem.*, **39** (1975), 475–486 (Russian); *Math. USSR Izv.* **9** (1975), 443–445

364. S.M. Voronin, *Analytic Properties of Dirichlet Generating Functions of Arithmetic Objects*, Ph.D. Thesis, Moscow, Steklov Math. Institute, (1977) (Russian)

365. M. Waldschmidt, A lower bound for linear forms in logarithms, *Acta Arith.* **37** (1980), 257–283

366. A. Walfisz, *Weylsche Exponentialsummen in der neueren Zahlentheorie*, VEB Deutscher Verlag der Wissenschaften (1963)

367. J.L. Walsh, Interpolation and approximation by rational functions in the complex domain, *Am. Math. Soc. Coll. Publ.* **20**, (1960)

368. L.C. Washington, *Introduction to Cyclotomic Fields*, Springer, Berlin Heidelberg New York (1980)

369. A. Weil, Sur la théorie du corps de classes, *J. Math. Soc. Jpn.* **3** (1951), 1–35

370. H. Weyl, Über ein Problem aus dem Gebiete der diophantischen Approximation, *Göttinger Nachrichten* (1914), 234–244

371. A. Wiles, Modular elliptic curves and Fermat's last theorem, *Ann. Math.* **141** (1995), 443–551

372. Z. Ye, The Nevanlinna functions of the Riemann zeta-function, *J. Math. Analysis Appl.* **233** (1999), 425–435

Notations

We indicate here some of the notations and conventions used in these notes; most of them are standard. However, this list is not complete; we omit notions which only appear in one chapter (where they are defined in situ) or which are covered by the index or which are standard.

As usual, we denote by $\mathbb{N} = \{1, 2, 3, \ldots\}$ the set of positive integers. The sets of integers, rational numbers, real numbers, and complex numbers are denoted by $\mathbb{Z}, \mathbb{Q}, \mathbb{R}$, and $\mathbb{C}$, respectively. Following an old tradition in the theory of the zeta-function, the complex variable is given by a mixture of greek and latin letters: we write $s = \sigma + it$, where $\sigma, t \in \mathbb{R}$, and 'i' is the imaginary unit $\sqrt{-1}$ in the upper half-plane.

The letter p, with and without a subscript, denotes a prime number. We write $d \mid n$ (respectively, $d \nmid n$) if the integer d divides (respectively, does not divide) the integer n. The symbol $\equiv$ stands either for some congruence or it denotes that a function is constant. The number of elements of a finite set $\mathcal{A}$ is denoted by $\sharp \mathcal{A}$. The function $\pi(x)$ counts the number of primes $p \leq x$. Besides, we use many other arithmetical functions; in our notation we follow the classic [120].

The logarithm is, as usual in number theory, always taken to the basis $e = \exp(1)$. The integer part and fractional part of a real number x are indicated by $[x]$ and $\{x\}$, respectively. Very convenient is the use of the Landau- and Vinogradov-symbols. Given two functions $f(x)$ and $g(x)$, both defined for $x \in X$, where $g(x)$ is positive for all $x \in X$, we write:

- $f(x) = \mathrm{O}(g(x))$ and $f(x) \ll g(x)$, respectively, if there exists a constant $C \geq 0$ such that

$$|f(x)| \leq Cg(x) \quad \text{for all} \quad x \in X;$$

- $f(x) \asymp g(x)$ if $f(x) \ll g(x) \ll |f(x)|$;

here X is specified either explicitly or implicitly. Usually, the set X is an interval $[\xi, \infty)$ for some real number ξ; in this case we also write:

- $f(x) \sim g(x)$ if the limit

$$\lim_{x \to \infty} \frac{|f(x)|}{g(x)}$$

exists and is equal to 1.
- $f(x) = \mathrm{o}(g(x))$ if the latter limit exists and is equal to zero.
- $f(x) = \Omega(g(x))$ if

$$\liminf_{x \to \infty} \frac{|f(x)|}{g(x)} > 0$$

(this is the negation of $f(x) = \mathrm{O}(g(x))$).

Sometimes the limit $x \to \infty$ is replaced by another limit $x \to x_0$, where x_0 is some complex number; in this case the limit x_0 is explicitly stated. In estimates, ϵ always denotes a small positive number, not necessarily the same at each appearance.

The letters $\mathbf{P}$ and $\mathbf{Q}$ always denote probability measures, often with a subscript. $\mathrm{meas}(\mathcal{A})$ is the Lebesgue measure and $\mathrm{m}(\mathcal{A})$ stands for the Haar measure of a measurable set $\mathcal{A}$. By $N_{\mathcal{L}}(\cdot)$ and $N_{\mathcal{L}}^c(.)$ we denote the numbers of zeros and of c-values of the function $\mathcal{L}(s)$ in some region, often specified by some data in the brackets or by subscripts.

Finally, we list all axioms which were used in these notes:

(i), (1) *Ramanujan hypothesis.* $a(n) \ll n^{\epsilon}$ for any $\epsilon > 0$, where the implicit constant may depend on ϵ.

(ii) *Analytic continuation.* there exists a real number $\sigma_{\mathcal{L}}$ such that $\mathcal{L}(s)$ has an analytic continuation to the half-plane $\sigma > \sigma_{\mathcal{L}}$ with $\sigma_{\mathcal{L}} < 1$ except for at most a pole at $s = 1$.

(2) *Analytic continuation.* There exists a non-negative integer k such that $(s-1)^k \mathcal{L}(s)$ is an entire function of finite order.

(iii) *Finite order.* There exists a constant $\mu_{\mathcal{L}} \geq 0$ such that, for any fixed $\sigma > \sigma_{\mathcal{L}}$ and any $\epsilon > 0$,

$$\mathcal{L}(\sigma + \mathrm{i}t) \ll |t|^{\mu_{\mathcal{L}} + \epsilon} \quad \text{as} \quad |t| \to \infty;$$

the implicit constant may depend on ϵ.

(3) *Functional equation.* $\mathcal{L}(s)$ satisfies a functional equation of type

$$\Lambda_{\mathcal{L}}(s) = \omega \overline{\Lambda_{\mathcal{L}}(1 - \bar{s})},$$

where

$$\Lambda_{\mathcal{L}}(s) := \mathcal{L}(s) Q^s \prod_{j=1}^{f} \Gamma(\lambda_j s + \mu_j)$$

with positive real numbers Q, λ_j, and complex numbers μ_j, ω with $\mathrm{Re}\,\mu_j \geq 0$ and $|\omega| = 1$.

(iv) *Polynomial Euler product.* There exists a positive integer m and for every prime p, there are complex numbers $\alpha_j(p), 1 \le j \le m$, such that

$$\mathcal{L}(s) = \prod_p \prod_{j=1}^{m} \left(1 - \frac{\alpha_j(p)}{p^s} \right)^{-1}.$$

(4) *Euler product.* $\mathcal{L}(s)$ satisfies

$$\mathcal{L}(s) = \prod_p \mathcal{L}_p(s), \quad \text{where} \quad \mathcal{L}_p(s) = \exp\left(\sum_{k=1}^{\infty} \frac{b(p^k)}{p^{ks}} \right)$$

with suitable coefficients $b(p^k)$ satisfying $b(p^k) \ll p^{k\theta}$ for some $\theta < \frac{1}{2}$.

(v) *Prime mean-square.* There exists a positive constant κ such that

$$\lim_{x \to \infty} \frac{1}{\pi(x)} \sum_{p \le x} |a(p)|^2 = \kappa.$$

We denote by $\tilde{\mathcal{S}}$ the class of Dirichlet series satisfying the axioms (i)–(v) and by $\mathcal{S}^\sharp$ the so-called extended Selberg class of Dirichlet series satisfying (2) and (3); the subclass of the latter class of all elements satisfying additionally the axioms (1) and (4) is the Selberg class $\mathcal{S}$.

Index

Lecture Notes in Mathematics

For information about earlier volumes
please contact your bookseller or Springer
LNM Online archive: springerlink.com

Applications. Martina Franca, Italy 2001. Editors: L. A. Caffarelli, S. Salsa (2003)

Vol. 1814: P. Bank, F. Baudoin, H. Föllmer, L.C.G. Rogers, M. Soner, N. Touzi, Paris-Princeton Lectures on Mathematical Finance 2002 (2003)

Vol. 1815: A. M. Vershik (Ed.), Asymptotic Combinatorics with Applications to Mathematical Physics. St. Petersburg, Russia 2001 (2003)

Vol. 1816: S. Albeverio, W. Schachermayer, M. Talagrand, Lectures on Probability Theory and Statistics. Ecole d'Eté de Probabilités de Saint-Flour XXX-2000. Editor: P. Bernard (2003)

Vol. 1817: E. Koelink, W. Van Assche (Eds.), Orthogonal Polynomials and Special Functions. Leuven 2002 (2003)

Vol. 1818: M. Bildhauer, Convex Variational Problems with Linear, nearly Linear and/or Anisotropic Growth Conditions (2003)

Vol. 1819: D. Masser, Yu. V. Nesterenko, H. P. Schlickewei, W. M. Schmidt, M. Waldschmidt, Diophantine Approximation. Cetraro, Italy 2000. Editors: F. Amoroso, U. Zannier (2003)

Vol. 1820: F. Hiai, H. Kosaki, Means of Hilbert Space Operators (2003)

Vol. 1821: S. Teufel, Adiabatic Perturbation Theory in Quantum Dynamics (2003)

Vol. 1822: S.-N. Chow, R. Conti, R. Johnson, J. Mallet-Paret, R. Nussbaum, Dynamical Systems. Cetraro, Italy 2000. Editors: J. W. Macki, P. Zecca (2003)

Vol. 1823: A. M. Anile, W. Allegretto, C. Ringhofer, Mathematical Problems in Semiconductor Physics. Cetraro, Italy 1998. Editor: A. M. Anile (2003)

Vol. 1824: J. A. Navarro González, J. B. Sancho de Salas, $\mathscr{C}^\infty$ – Differentiable Spaces (2003)

Vol. 1825: J. H. Bramble, A. Cohen, W. Dahmen, Multiscale Problems and Methods in Numerical Simulations, Martina Franca, Italy 2001. Editor: C. Canuto (2003)

Vol. 1826: K. Dohmen, Improved Bonferroni Inequalities via Abstract Tubes. Inequalities and Identities of Inclusion-Exclusion Type. VIII, 113 p, 2003.

Vol. 1827: K. M. Pilgrim, Combinations of Complex Dynamical Systems. IX, 118 p, 2003.

Vol. 1828: D. J. Green, Gröbner Bases and the Computation of Group Cohomology. XII, 138 p, 2003.

Vol. 1829: E. Altman, B. Gaujal, A. Hordijk, Discrete-Event Control of Stochastic Networks: Multimodularity and Regularity. XIV, 313 p, 2003.

Vol. 1830: M. I. Gil', Operator Functions and Localization of Spectra. XIV, 256 p, 2003.

Vol. 1831: A. Connes, J. Cuntz, E. Guentner, N. Higson, J. E. Kaminker, Noncommutative Geometry, Martina Franca, Italy 2002. Editors: S. Doplicher, L. Longo (2004)

Vol. 1832: J. Azéma, M. Émery, M. Ledoux, M. Yor (Eds.), Séminaire de Probabilités XXXVII (2003)

Vol. 1833: D.-Q. Jiang, M. Qian, M.-P. Qian, Mathematical Theory of Nonequilibrium Steady States. On the Frontier of Probability and Dynamical Systems. IX, 280 p, 2004.

Vol. 1834: Yo. Yomdin, G. Comte, Tame Geometry with Application in Smooth Analysis. VIII, 186 p, 2004.

Vol. 1835: O.T. Izhboldin, B. Kahn, N.A. Karpenko, A. Vishik, Geometric Methods in the Algebraic Theory of Quadratic Forms. Summer School, Lens, 2000. Editor: J.-P. Tignol (2004)

Vol. 1836: C. Năstăsescu, F. Van Oystaeyen, Methods of Graded Rings. XIII, 304 p, 2004.

Vol. 1837: S. Tavaré, O. Zeitouni, Lectures on Probability Theory and Statistics. Ecole d'Eté de Probabilités de Saint-Flour XXXI-2001. Editor: J. Picard (2004)

Vol. 1838: A.J. Ganesh, N.W. O'Connell, D.J. Wischik, Big Queues. XII, 254 p, 2004.

Vol. 1839: R. Gohm, Noncommutative Stationary Processes. VIII, 170 p, 2004.

Vol. 1840: B. Tsirelson, W. Werner, Lectures on Probability Theory and Statistics. Ecole d'Eté de Probabilités de Saint-Flour XXXII-2002. Editor: J. Picard (2004)

Vol. 1841: W. Reichel, Uniqueness Theorems for Variational Problems by the Method of Transformation Groups (2004)

Vol. 1842: T. Johnsen, A. L. Knutsen, K_3 Projective Models in Scrolls (2004)

Vol. 1843: B. Jefferies, Spectral Properties of Noncommuting Operators (2004)

Vol. 1844: K.F. Siburg, The Principle of Least Action in Geometry and Dynamics (2004)

Vol. 1845: Min Ho Lee, Mixed Automorphic Forms, Torus Bundles, and Jacobi Forms (2004)

Vol. 1846: H. Ammari, H. Kang, Reconstruction of Small Inhomogeneities from Boundary Measurements (2004)

Vol. 1847: T.R. Bielecki, T. Björk, M. Jeanblanc, M. Rutkowski, J.A. Scheinkman, W. Xiong, Paris-Princeton Lectures on Mathematical Finance 2003 (2004)

Vol. 1848: M. Abate, J. E. Fornaess, X. Huang, J. P. Rosay, A. Tumanov, Real Methods in Complex and CR Geometry, Martina Franca, Italy 2002. Editors: D. Zaitsev, G. Zampieri (2004)

Vol. 1849: Martin L. Brown, Heegner Modules and Elliptic Curves (2004)

Vol. 1850: V. D. Milman, G. Schechtman (Eds.), Geometric Aspects of Functional Analysis. Israel Seminar 2002-2003 (2004)

Vol. 1851: O. Catoni, Statistical Learning Theory and Stochastic Optimization (2004)

Vol. 1852: A.S. Kechris, B.D. Miller, Topics in Orbit Equivalence (2004)

Vol. 1853: Ch. Favre, M. Jonsson, The Valuative Tree (2004)

Vol. 1854: O. Saeki, Topology of Singular Fibers of Differential Maps (2004)

Vol. 1855: G. Da Prato, P.C. Kunstmann, I. Lasiecka, A. Lunardi, R. Schnaubelt, L. Weis, Functional Analytic Methods for Evolution Equations. Editors: M. Iannelli, R. Nagel, S. Piazzera (2004)

Vol. 1856: K. Back, T.R. Bielecki, C. Hipp, S. Peng, W, Schachermayer, Stochastic Methods in Finance, Bressanone/Brixen, Italy, 2003. Editors: M. Fritelli, W. Runggaldier (2004)

Vol. 1857: M. Émery, M. Ledoux, M. Yor (Eds.), Séminaire de Probabilités XXXVIII (2005)

Vol. 1858: A.S. Cherny, H.-J. Engelbert, Singular Stochastic Differential Equations (2005)

Vol. 1859: E. Letellier, Fourier Transforms of Invariant Functions on Finite Reductive Lie Algebras (2005)

Vol. 1860: A. Borisyuk, G.B. Ermentrout, A. Friedman, D. Terman, Tutorials in Mathematical Biosciences I. Mathematical Neurosciences (2005)

Vol. 1861: G. Benettin, J. Henrard, S. Kuksin, Hamiltonian Dynamics – Theory and Applications, Cetraro, Italy, 1999. Editor: A. Giorgilli (2005)

Vol. 1862: B. Helffer, F. Nier, Hypoelliptic Estimates and Spectral Theory for Fokker-Planck Operators and Witten Laplacians (2005)

Vol. 1863: H. Führ, Abstract Harmonic Analysis of Continuous Wavelet Transforms (2005)

Vol. 1864: K. Efstathiou, Metamorphoses of Hamiltonian Systems with Symmetries (2005)

Vol. 1865: D. Applebaum, B.V. R. Bhat, J. Kustermans, J. M. Lindsay, Quantum Independent Increment Processes I. From Classical Probability to Quantum Stochastic Calculus. Editors: M. Schürmann, U. Franz (2005)

Vol. 1866: O.E. Barndorff-Nielsen, U. Franz, R. Gohm, B. Kümmerer, S. Thorbjønsen, Quantum Independent Increment Processes II. Structure of Quantum Lévy Processes, Classical Probability, and Physics. Editors: M. Schürmann, U. Franz, (2005)

Vol. 1867: J. Sneyd (Ed.), Tutorials in Mathematical Biosciences II. Mathematical Modeling of Calcium Dynamics and Signal Transduction. (2005)

Vol. 1868: J. Jorgenson, S. Lang, $Pos_n(R)$ and Eisenstein Series. (2005)

Vol. 1869: A. Dembo, T. Funaki, Lectures on Probability Theory and Statistics. Ecole d'Eté de Probabilités de Saint-Flour XXXIII-2003. Editor: J. Picard (2005)

Vol. 1870: V.I. Gurariy, W. Lusky, Geometry of Müntz Spaces and Related Questions. (2005)

Vol. 1871: P. Constantin, G. Gallavotti, A.V. Kazhikhov, Y. Meyer, S. Ukai, Mathematical Foundation of Turbulent Viscous Flows, Martina Franca, Italy, 2003. Editors: M. Cannone, T. Miyakawa (2006)

Vol. 1872: A. Friedman (Ed.), Tutorials in Mathematical Biosciences III. Cell Cycle, Proliferation, and Cancer (2006)

Vol. 1873: R. Mansuy, M. Yor, Random Times and Enlargements of Filtrations in a Brownian Setting (2006)

Vol. 1874: M. Yor, M. Émery (Eds.), In Memoriam Paul-André Meyer - Séminaire de probabilités XXXIX (2006)

Vol. 1875: J. Pitman, Combinatorial Stochastic Processes. Ecole d'Eté de Probabilités de Saint-Flour XXXII-2002. Editor: J. Picard (2006)

Vol. 1876: H. Herrlich, Axiom of Choice (2006)

Vol. 1877: J. Steuding, Value Distributions of L-Functions (2007)

Vol. 1878: R. Cerf, The Wulff Crystal in Ising and Percolation Models, Ecole d'Eté de Probabilités de Saint-Flour XXXIV-2004. Editor: Jean Picard (2006)

Vol. 1879: G. Slade, The Lace Expansion and its Applications, Ecole d'Eté de Probabilités de Saint-Flour XXXIV-2004. Editor: Jean Picard (2006)

Vol. 1880: S. Attal, A. Joye, C.-A. Pillet, Open Quantum Systems I, The Hamiltonian Approach (2006)

Vol. 1881: S. Attal, A. Joye, C.-A. Pillet, Open Quantum Systems II, The Markovian Approach (2006)

Vol. 1882: S. Attal, A. Joye, C.-A. Pillet, Open Quantum Systems III, Recent Developments (2006)

Vol. 1883: W. Van Assche, F. Marcellàn (Eds.), Orthogonal Polynomials and Special Functions, Computation and Application (2006)

Vol. 1884: N. Hayashi, E.I. Kaikina, P.I. Naumkin, I.A. Shishmarev, Asymptotics for Dissipative Nonlinear Equations (2006)

Vol. 1885: A. Telcs, The Art of Random Walks (2006)

Vol. 1886: S. Takamura, Splitting Deformations of Degenerations of Complex Curves (2006)

Vol. 1887: K. Habermann, L. Habermann, Introduction to Symplectic Dirac Operators (2006)

Vol. 1888: J. van der Hoeven, Transseries and Real Differential Algebra (2006)

Vol. 1889: G. Osipenko, Dynamical Systems, Graphs, and Algorithms (2006)

Vol. 1890: M. Bunge, J. Funk, Singular Coverings of Toposes (2006)

Vol. 1891: J.B. Friedlander, D.R. Heath-Brown, H. Iwaniec, J. Kaczorowski, Analytic Number Theory, Cetraro, Italy, 2002. Editors: A. Perelli, C. Viola (2006)

Vol. 1892: A. Baddeley, I. Bárány, R. Schneider, W. Weil, Stochastic Geometry, Martina Franca, Italy, 2004. Editor: W. Weil (2007)

Vol. 1893: H. Hanßmann, Local and Semi-Local Bifurcations in Hamiltonian Dynamical Systems, Results and Examples (2007)

Vol. 1894: C.W. Groetsch, Stable Approximate Evaluation of Unbounded Operators (2007)

Vol. 1895: L. Molnár, Selected Preserver Problems on Algebraic Structures of Linear Operators and on Function Spaces (2007)

Vol. 1896: P. Massart, Concentration Inequalities and Model Selection, Ecole d'Eté de Probabilités de Saint-Flour XXXIII-2003. Editor: J. Picard (2007)

Vol. 1897: R. Doney, Fluctuation Theory for Lévy Processes, Ecole d'Eté de Probabilités de Saint-Flour XXXV-2005. Editor: J. Picard (2007)

Vol. 1898: H.R. Beyer, Beyond Partial Differential Equations, On linear and Quasi-Linear Abstract Hyperbolic Evolution Equations (2007)

Vol. 1899: Séminaire de Probabilités XL. Editors: C. Donati-Martin, M. Émery, A. Rouault, C. Stricker (2007)

Vol. 1900: E. Bolthausen, A. Bovier (Eds.), Spin Glasses (2007)

Vol. 1901: O. Wittenberg, Intersections de deux quadriques et pinceaux de courbes de genre 1, Intersections of Two Quadrics and Pencils of Curves of Genus 1 (2007)

Vol. 1902: A. Isaev, Lectures on the Automorphism Groups of Kobayashi-Hyperbolic Manifolds (2007)

Vol. 1903: G. Kresin, V. Maz'ya, Sharp Real-Part Theorems (2007)

Recent Reprints and New Editions

Vol. 1618: G. Pisier, Similarity Problems and Completely Bounded Maps. 1995 – 2nd exp. edition (2001)

Vol. 1629: J.D. Moore, Lectures on Seiberg-Witten Invariants. 1997 – 2nd edition (2001)

Vol. 1638: P. Vanhaecke, Integrable Systems in the realm of Algebraic Geometry. 1996 – 2nd edition (2001)

Vol. 1702: J. Ma, J. Yong, Forward-Backward Stochastic Differential Equations and their Applications. 1999 – Corr. 3rd printing (2005)

Vol. 830: J.A. Green, Polynomial Representations of GL_n, with an Appendix on Schensted Correspondence and Littelmann Paths by K. Erdmann, J.A. Green and M. Schocker 1980 – 2nd corr. and augmented edition (2007)